THE KELMSCOTT PRESS

THE KELMSCOTT PRESS ❧ A HISTORY OF WILLIAM MORRIS'S TYPOGRAPHICAL ADVENTURE ❧ BY WILLIAM S. PETERSON

CLARENDON PRESS OXFORD MCMXCI

Oxford University Press, Walton Street, Oxford OX2 6DP
Oxford New York Toronto
Delhi Bombay Calcutta Madras Karachi
Petaling Jaya Singapore Hongkong Tokyo
Nairobi Dar es Salaam Cape Town
Melbourne Auckland
and associated companies in
Berlin Ibadan

Oxford is a trade mark of Oxford University Press

Published in the United States
by University of California Press

British Library Cataloging in Publication Data
Peterson, William S. (William Samuel), 1939–
The Kelmscott Press: a history of William Morris's typographical adventure.
1. London. Hammersmith and Fulham (London Borough).
Private Presses: Kelmscott Press
I. Title
338.7'6862'0942133
ISBN 0-19-812887-8

Library of Congress Cataloging in Publication Data
Peterson, William S.
The Kelmscott Press: a history of William Morris's typographical adventure / by William S. Peterson.
p. cm.
1. Kelmscott Press. 2. Morris, William, 1834–1896–Knowledge—Book arts and sciences. 3. Private presses—England—Hammersmith (London)—History—19th century. 4. Book design—England—Hammersmith (London)—History—19th century. 6. Hammersmith (London, England)—Imprints—History—19th century. 7. London (England)—Imprints—History—19th century. I. Title.
Z232.M87P45 1989
070.5'09426'5-dc20 89-3351
ISBN 0-19-812887-8:

Printed in Great Britain at the Alden Press
Oxford London and Northampton

TO

MY MOTHER

AND

THE MEMORY OF MY FATHER

ACKNOWLEDGEMENTS

THIS book began almost accidentally in 1978 when I devoted a summer at the Library of Congress to poring over nineteenth-century English printers' trade magazines, trying to discover what they had to say about book-design. My quest was largely fruitless, because Victorian printers seemed inarticulate on that subject, but as I turned the pages and reflected upon the typography of the last century, I frequently stumbled upon the name of William Morris. At first I resisted the lure of Morris, with his enormous reputation among bibliophiles, since it appeared self-evident that there could be nothing new to say about the Kelmscott Press, for I was aware that it had been discussed by almost every English-speaking printing historian of this century. Eventually, however, I realized that there was no avoiding Morris – who seems to many of us the most significant figure in modern book typography – and I then discovered, to my great pleasure, that much fresh evidence about the Kelmscott Press, unavailable to earlier scholars, was now to be found in public and private collections. Only gradually did I grasp the larger cultural significance of Morris's 'little typographical adventure' and understand that the meaning of this episode extended far beyond the boundaries of printing history.

That moment of illumination came to me a decade ago, and since then I have had the pleasure of examining the handsome books of the Kelmscott Press in many British and American libraries, of studying a good deal of manuscript material relating to the history of the Press, and of receiving innumerable courtesies from libraries and collectors. I am keenly aware of the inadequacy of merely supplying simple, undifferentiated lists of institutions and persons who have given me assistance along the way, yet I could not record here all the nuances of generosity and friendship without writing a memoir of that decade. I feel confident, however, that many of the persons mentioned will recognize the warm memories that lie behind this litany of names.

Without several fellowships and travel grants, it would have been impossible for me to complete my research. They were supplied by the American Philosophical Society, the Division of Arts and Humanities of the University of Maryland, the John Simon Guggenheim Memorial Foundation, and the National Endowment for the Humanities.

Those who have given me permission to quote unpublished documents and to reproduce illustrations are cited elsewhere in this book, but I must register my special indebtedness to the Society of Antiquaries (London), the owner of the William Morris copyright.

John Dreyfus, Joseph R. Dunlap, and Julia Markus read drafts of the book and offered stimulating criticisms, from which I hope I benefited.

My wife Eileen, my daughter Heather, and my son Glenn have cheerfully tolerated my prolonged preoccupation with the Kelmscott Press and have given me much love and support. At an early stage in the research Heather also helped in tracking down books at the Library of Congress.

The following have assisted me in ways too numerous to describe: Seymour Adelman, Iain Bain, Nicolas Barker, Ben Bass, B. H. Baumfield, John Bell, Sandford L. and Helen Berger, John Bidwell, Carey Bliss, Ronald Briggs, John P. Chalmers, Freeman Champney, Betty Coley, Peter Cormack, Maureen B. Duffany, A. R. Dufty, Ellen S. Dunlap, Blanche T. Ebeling-Koning, Crispin and Jan Elsted, Donald Farren, Rita S. Faulders, Peter Faulkner, Colin Franklin, J. A. Gere, Norah C. Gillow, Theodore S. Hope, Jr., Shirley Strum Kenny, Mary Lago, Mark Samuels Lasner, J. Ben Lieberman, Jaquelyn Lyman-Phillips, Stephen Massey, Elizabeth Masterman, Francis O. Mattson, William J. McClung, David J. McKitterick, Linda Merians, Ruth Mortimer, James Mosley, Paul Needham, R. L. Nikirk, Alan O'Day, Frank Paluka, Annabel Patterson, Fiona Pearson, Cynthia H. Peters, Ann Phillips, Robert M. Pierson, Nicholas Poole-Wilson, D. S. Porter, Gordon N. Ray, Duncan Robinson, Florian J. Shasky, Barry Smith, M. A. Smith, Peter Stansky, Jeanne Sugiyama, Virginia Surtees, Lola L. Szladits, G. Thomas Tanselle, Lance Thirkell, F. H. Thompson, Susan O. Thompson, Decherd Turner, Peter VanWingen, Ruth Vyse, Hilary Walford, John J. Walsdorf, Kim Scott Walwyn, Lucille Wehner, Mark F. Weimer, Frances Whistler Stephen Wildman, Paul Woudhuysen, Marilyn Wurzburger, and John J. Yamulla.

W.S.P.

University of Maryland
March 1988

CONTENTS

ILLUSTRATIONS

Illustrations

Illustrations

Dimensions of the pages of Kelmscott Press books shown in the illustrations are supplied in William S. Peterson, *A Bibliography of the Kelmscott Press* (Oxford, 1984).

THE KELMSCOTT PRESS

Figure 1 Many of the Kelmscott Press books had engraved title-pages, such as this one in *News from Nowhere* (1893), which appeared opposite the first page of text. The house (drawn by C. M. Gere) is Kelmscott Manor, after which Morris's Press was named.

INTRODUCTION

I BEGAN printing books with the hope of producing some which would have a definite claim to beauty, while at the same time they should be easy to read and should not dazzle the eye, or trouble the intellect of the reader by eccentricity of form in the letters.'

With this brisk introduction William Morris began writing on 11 November 1895 a brief explanation for an American admirer of how and why he had established the most famous of all private presses, the Kelmscott Press. The essay first appeared in *Modern Art*, a Boston magazine, in 1896 and was later reprinted, under the title *A Note by William Morris on his Aims in Founding the Kelmscott Press*, as the last publication of the Press in the spring of 1898. Morris's formulation of his purposes was characteristically terse: 'it was the essence of my undertaking', he wrote, 'to produce books which it would be a pleasure to look upon as pieces of printing and arrangement of type.' The rest of the essay was devoted to a discussion of paper, type-design, spaces between words and lines, margins, and ornaments.

Morris's *Note* is an undeniably important document. As one of his few public statements about the Kelmscott Press, it has been quoted frequently since 1898 in histories of the Press; indeed, in 1924 Bernard H. Newdigate went so far as to suggest that the *Note* was one of the three books that every student of English book-production ought to read.[1] Yet one might argue that the essay is cryptic, even evasive, leaving a host of questions unanswered. How did Morris develop an interest in book-design and production? Why did he turn to fifteenth-century German and Italian books for his models? What was the relationship between the Kelmscott Press books and the best work of other nineteenth-century English printers? How exactly did Morris rediscover some of the fundamental principles of fine printing? How did he find the craftsmen to carry out those principles for him?

We need the answers to these and other questions, for there is nowadays general agreement – even among those who dislike his style of typography – that Morris is one of the most important printers of the modern era and that the establishment of the Kelmscott Press was an event of major cultural significance. Throughout northern Europe and the English-speaking world, particularly in the

United States, the impact of Morris's typographical theories and of the Kelmscott books themselves was pervasive. Historians are still arguing about the nature of that influence (whether, in particular, it was always a good or enlightening influence), but no one disputes that the forces set in motion at the Kelmscott Press eventually spread round the globe.

From a merely quantitative point of view, the achievement of the Kelmscott Press does not seem especially impressive. Between 1891 and 1898 it produced fifty-two books and a set of specimen pages of another book that was never completed. The books were designed by Morris, printed on hand-presses, and issued in limited editions averaging about 300 copies. All were ornamented with initials and borders by Morris, and some were illustrated, usually by Edward Burne-Jones. So why are these few Kelmscott books famous everywhere today, and why have they profoundly affected twentieth-century theories of book-design?

One explanation of their potent appeal lies in the fact that Morris's ideas, like his ornamented books, are richly suggestive. When he appears to be speaking about typographical matters, he is in reality examining how we ought to live. 'If I were asked to say what is at once the most important production of Art and the thing most to be longed for, I should answer, A beautiful House; and if I were further asked to name the production next in importance and the thing next to be longed for, I should answer, A beautiful Book,' he wrote in an unfinished essay. 'To enjoy good houses and good books in self-respect and decent comfort, seems to me to be the pleasurable end towards which all societies of human beings ought now to struggle.'[2] That is a most characteristic passage by Morris: not content with treating typography as 'a minor art' (to quote Stanley Morison's phrase),[3] Morris insists upon enlarging the context and seeing the beautiful book as an essential part of his attempt to redeem the late-Victorian world from dreary mechanical ugliness. It is surely this passion for moral, aesthetic, and social reform which lends resonance to Morris's typographical pronouncements and which makes the history of the Kelmscott Press a matter of some importance even to non-bibliophiles.

Morris, like many of his contemporaries, had grave misgivings about a world that was increasingly being shaped by industrialism. He looked about him at the Victorian city – smoky, polluted, crowded, and gloomy – and asked himself whether the human race was really in a better state now than it had been in the Middle Ages. In an essay published in 1890 Morris took his readers on a tour of an unidentified but typical English industrial town (apparently Peterborough) and revealed, street by street, building by building, the sheer horror of the urban landscape. With great eloquence Morris described the dreary terraced houses, the small, shabby shops, the 'ugly modern iron bridge', the depressed and underfed workers in the streets, the wilderness of dirty brick and stone.[4] This was the grimy present that Morris found about him; the past which he then immediately conjured up was a medieval world of simplicity, purity, and cleanliness. Morris's Hamlet-like injunction to 'Look here, upon this picture, and on this' reminds one of that famous pair of plates which had appeared in A. W. N. Pugin's *Contrasts* (second edition, 1841): 'Catholic town in 1440' and 'The Same Town in 1840'. In the lower plate we see a serene city of spires pointing towards heaven; in the other plate the multitude of lovely churches has been supplanted by factories, belching smokestacks, and a particularly grim Victorian workhouse.

Figure 2

Pugin's pictures, in turn, evoke Carlyle's *Past and Present* (1843), the very title of which vibrates with the fervour of Victorian moral idealism, and its implied contrast, like Pugin's, between a lost Golden Age and the bleak world of nineteenth-century England may help us to grasp the passionate intensity of Morris's rejection of the present. We may perhaps wish to protest that Pugin, Carlyle, Ruskin, Morris, and others of their outlook idealized the Middle Ages and ignored the material benefits of our mechanized civilization; yet we fail to do them justice unless we recognize that the Gothic Revival was not ultimately an attempt to resuscitate a period style of architecture or art but rather a morally powerful protest against the psychological and social disintegration produced by the modern world.

The Kelmscott Press was, in effect, the final phase of the Gothic Revival. Burne-Jones compared the Kelmscott *Chaucer* (1896) to a cathedral; by the same analogy, the smaller books can be regarded as parish churches, each faithfully rendered in quasi-medieval style. But such a sweeping rejection of contemporary life brought with it apparently insoluble problems. How, for example, could one reproduce the general appearance of a fifteenth-century book and still escape the dangers of mere imitation? By turning one's back on contemporary printing machinery, how could one improve the appearance of the ordinary book? In reverting to methods of hand-production, how could one print books that readers with modest incomes might afford?

These doubts about the Kelmscott Books – which are so redolent of Victorian revivalism and historicism – were in fact being expressed

Catholic town in 1440.

1. St Michaels on the Hill. 2. Queens Cross. 3. St Thomas's Chapel. 4. St Maries Abbey. 5. All Saints. 6. St Johns. 7. St Peters. 8. St Alkmunds. 9. St Maries. 10. St Edmunds. 11. Grey Friars. 12. St Cuthberts. 13. Guild hall. 14. Trinity. 15. St Olaves. 16. St Botolphs.

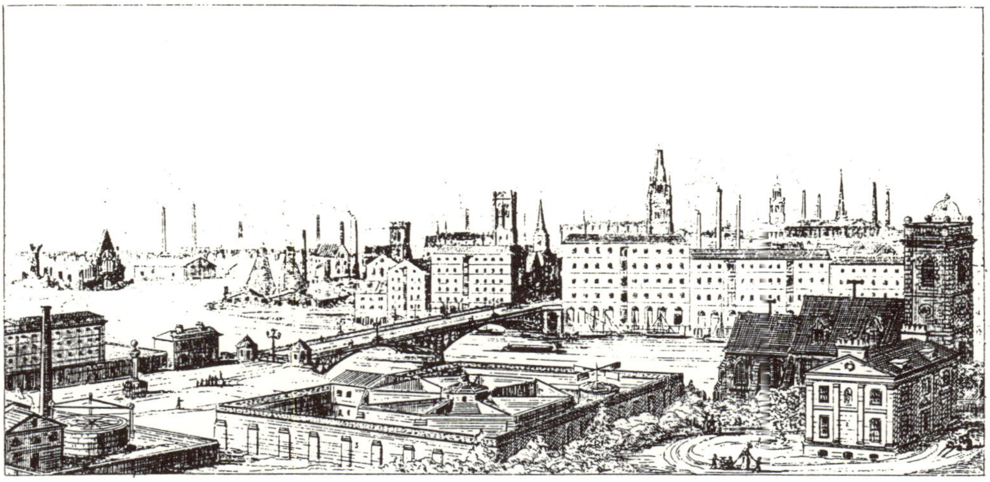

THE SAME TOWN IN 1840

1. St Michaels Tower, rebuilt in 1750. 2. New Parsonage House & Pleasure Grounds. 3. The New Jail. 4. Gas Works. 5. Lunatic Asylum. 6. Iron Works & Ruins of St Maries Abbey. 7. Mr Evans Chapel. 8. Baptist Chapel. 9. Unitarian Chapel. 10. New Church. 11. New Town Hall & Concert Room. 12. Westleyan Centenary Chapel. 13. New Christian Society. 14. Quakers Meeting. 15. Socialist Hall of Science.

Figure 2 This pair of plates in the second edition of A. W. N. Pugin's *Contrasts* (1841) is a pictorial rendering of that nostalgia for a medieval Golden Age which pervades the writings of many Victorians, particularly Carlyle, Ruskin, and Morris. The mood of Pugin's drawings is connected with Morris's 'typographical adventure', for the Kelmscott Press represents the final phase of the nineteenth-century Gothic Revival.

already during Morris's lifetime. Why, it was asked repeatedly in the press, did Morris, a committed socialist, not print books for the poor? Why did he insist upon producing anachronistic books whose glamour derived primarily from their resemblance to medieval volumes? Morris was conscious of such accusations, as we shall see, and he offered persuasive answers to some of them. The dilemmas he faced at the Kelmscott Press are of more than technical interest today, because they illustrate both the rewards and dangers of a reversion to handicraft methods. If, as Morris argued, our complex technological civilization bears within it the seeds of its own destruction, then the story of the Kelmscott Press – possibly the best-known example in the Western world of a serious, thoughtful attempt to revive a traditional craft – becomes a fascinating parable of the extent to which we may be able deliberately to turn the clock backwards.

One suspects that Morris has been written about more extensively than any other printer except Gutenberg. Even before the Kelmscott Press was officially opened, enthusiastic reports about Morris's 'typographical adventure'[5] (as he described it in a letter) began to appear in British and American periodicals. By 1893, in distant Chicago, the *Inland Printer* was prepared to proclaim that 'a new era has dawned in English printing'.[6] Reporters flocked to Kelmscott House, Morris's home in Hammersmith, for interviews with 'the Master Printer'. The Kelmscott edition of Chaucer was widely described as the most beautiful printed book since the fifteenth century. In the last volume produced at the Press, Sydney Cockerell supplied 'A Short History and Description of the Kelmscott Press' and 'An Annotated List of All the Books Printed at the Kelmscott Press in the Order in Which They Were Issued'. In 1924 H. Halliday Sparling, who had been Morris's son-in-law and, like Cockerell, a Secretary of the Press, published a book-length history entitled *The Kelmscott Press and William Morris Master-Craftsman*. Since then many other more specialized studies of the Press have appeared in print.[7]

How, then, can one justify still another book on the Kelmscott Press? The short answer is that there is really only one extended history, Sparling's, which is by now very much out of date; since 1924 a great quantity of manuscript material relating to the Press's history – including letters, diaries, manuscripts, and corrected proofs of Kelmscott books – has become available to scholars. The dispersal of Sir Sydney Cockerell's collection in 1956 and the acquisition by the Pierpont Morgan Library of the Kelmscott materials of John M.

Crawford, Jr., are only two notable instances of the emergence in recent decades of important new documentary evidence. With these fresh records in hand, and with the calmer perspective that the passing of time usually brings, it ought (one hopes) to be possible now to write a history of the Kelmscott Press that avoids both the embarrassing proselytizing tone of Sparling's book and the biliousness of certain anti-Morris polemics published in the past.

The Kelmscott Press, despite – or perhaps because of – its fame, is a subject encrusted with myths and misconceptions. It is surely worth the effort to try to set the record straight, if only because Morris's curious typographical experiment may force us to reflect on the meaning of that fascinating yet morally ambiguous technology which threatens today to envelop the world.

ONE THE WORLD OF VICTORIAN PRINTING

THE history of the printed book may seem, at first glance, to be a perfect confirmation of Morris's view that the older crafts had steadily deteriorated in quality since the Middle Ages. 'Printing', as Richard Durnford, Bishop of Chichester, put it trenchantly, 'is an art which beginning perfect, ever afterwards declined.'[1] And who, comparing the typical mass-produced volume of the nineteenth century – nastily printed in greyish ink on shoddy, acidic paper – with the Mainz Psalter or Gutenberg's Bible, could doubt that Morris and the Bishop were right?

Yet the story of the decay of printing is a good deal more complex than Morris ever acknowledged. For him, it was all quite simple: the earliest printed books were still the best because they were produced at a time when the spirit of medieval craftsmanship was strong, but, with the advent of false historicism and a weakness for pseudo-classical ornament in the sixteenth century, printing, like architecture, began to decay. Only in the second half of the eighteenth century, according to Morris, were the consequences of this typographical dry rot fully revealed in the 'sweltering hideousness of the Bodoni letter, the most illegible type that was ever cut'. Bodoni and Didot, even more relentlessly than Baskerville (whose achievements Morris certainly underestimated), carried the development of the 'modern' typeface to its logical conclusion. 'It was reserved for the founders of the later eighteenth century', wrote Morris, 'to produce letters which are *positively* ugly, and which, it may be added, are dazzling and unpleasant to the eye owing to the clumsy thickening and vulgar thinning of the lines. . . .'[2] The nineteenth century's further contributions to this melancholy tale were increased mechanization and cheaper paper, ink, and binding materials. When he reflected upon the contemporary printing scene, Morris found a small source of optimism only in the growing popularity of Caslon's old-face type and its imitations.

Because of the wide influence of Morris's typographical writings, this sweeping indictment of nineteenth-century printing became accepted dogma for several decades following his death in 1896. There came in time the inevitable reaction against his oversimplifications and his doctrinaire hatred of the modern world. Writing in an American typographical journal, the *Dolphin*, in 1938, Peter Beilenson remarked with obvious irony that 'there had developed the general opinion that between the early great classical period of Didot, Bodoni, and Bulmer, and the late arts-and-crafts period of Morris, the printing of the century was one continuous sacrifice on the altars of Machinery and Mammon'. In fact, he argued, many nineteenth-century printers had been preoccupied with the problem of quality, and 'technically book-making averaged at least the level of preceding centuries (except that papermaking and binding sought and found lower commercial levels) and . . . artistically the century offered a greater variety of good things, which on the whole unfortunately suit our present taste less than those of any preceding century'.[3] Recently a less cautious revisionist, Ruari McLean, has insisted that 'more exciting things happened in book design between 1837 and 1890 than in any other comparable period in the history of the world's printing; and most of them happened in London'.[4] (In fairness to McLean, it must be added that he is referring to the most sumptuous Victorian books; he elsewhere acknowledges that the ordinary commercially produced book of the last century was a very dismal thing indeed.)

Not surprisingly, nineteenth-century printers themselves did not share Morris's low estimate of their achievements. The trade journals of the time were filled with hymns of self-praise. 'What wondrous strides have been made in almost every department of the Art!' the *Printers' Register* (6 July 1881) declared. 'Hand-presses have become almost exterminated, pushed out by power-driven machines; type-faces . . . may now be counted by hundreds in their varieties; the uniformity of black ink, relieved on rare occasions by red initials, now gives way to mosaics of every imaginable colour, combined with gold, silver, and many-shaded bronzes; while borders, ornaments, picture-types, engravings, and lithographs are made use of to an extent never before conceived.' The naïve enthusiasm for machinery – and the implication that technological advances always produce aesthetic triumphs – in this representative pronouncement may make us reluctant to attend very seriously to the testimony of Morris's contemporaries in the printing trade; but at least it should be observed that they rejected his assessment of their craftsmanship, for Victorian printers believed, or professed to believe, that nineteenth-century bookwork was as fine as that of any preceding era.

New technology – which Morris was to reject so dramatically at the Kelmscott Press – had completely changed the character of the printing trade in his day. Before the third Earl Stanhope perfected an iron press in about 1800, all printing – from Gutenberg's day onwards – had been done on a wooden press which was so simple in structure that it was frequently built by the local carpenter or joiner. The earliest detailed pictorial representations of the 'Common Press' (as it came to be called) appeared in the early decades of the sixteenth century, and they show the same unsophisticated mechanism that was still being used nearly three hundred years later. If Morris had really been motivated solely by the antiquarian impulses that sometimes are attributed to him, he presumably would have reverted to the wooden press; there were still a few of them around in late-Victorian England, and they were occasionally seen in operation on special occasions such as the Caxton Exhibition in 1877. But Morris recognized that the older presses were actually inferior to the iron hand-press of the nineteenth century. 'One of the great drawbacks to wood-block printing in those times [the fifteenth century] was the weakness of the presses,' he wrote. 'Their only resource was to print with the paper very wet, and with very soft packing, so that the block went well into the paper; but many books . . . have suffered much from this cause.'[5]

Lord Stanhope's press, despite some enhancements, was in fact only a slightly improved version of the medieval press. More dramatic technological developments appeared in the printing trade within two decades. Friedrich Koenig invented a steam-driven cylinder press which was first used to print *The Times* on 29 November 1814, and for the rest of the century the presses that were employed by newspapers and mass-circulation magazines grew ever more elaborate and imposing. By the 1840s the machine-minders who tended these clanking monsters had come to resemble flies crawling over the skin of an enormous elephant. In the latter half of the century the hand-press (always iron by now) was used much less frequently even in book-printing; when Morris founded the Kelmscott Press in 1891 such presses were still readily available but were regarded as useful mainly for the printing of posters and the taking of proofs. *Figure 3*

The growing mechanization of printing led to a reduction in the cost of books and periodicals. Some observers at the time were overjoyed by this means of spreading literacy. Charles Knight (a printer and publisher as well as a writer), in *The Old Printer and the Modern Press* (1854), argued fervently that the new technological developments were a heaven-sent gift that would lead to the wider distribution of

Figure 3 By the middle of the last century, this sort of massive, clanking press was commonly used for the printing of newspapers and magazines; even books were produced on somewhat smaller steam-driven cylinder presses. Morris's adoption of the hand-press was a direct reaction to such technological developments.

books through book clubs and public libraries. The Society for the Diffusion of Useful Knowledge, of which Knight was an active leader, was symptomatic of the energetic, cheerful Victorian response to the ever-faster spinning cylinders of the steam press. It is also significant that a number of the edifying examples cited in Charles Babbage's *On the Economy of Machinery and Manufactures* (fourth edition, 1835) were drawn from the world of printing.

Of course there were other voices, sometimes from within the trade, warning that the growing acceleration of the printing process was leading to a debasement of traditional standards in typography and materials. T. C. Hansard, in his *Typographia* (1825), a technical manual for printers, gave an example of the problems that grew out of the new preoccupation with speed: 'The system of hurrying works through the press, lately adopted by some booksellers, by giving them among a variety of houses, is destructive of uniformity. . . . It is not unusual to discover, in volumes, so divided, one to contain more lines in a page than another, or, even if the number of lines are the same, the page differs in length; or, should the type agree in depth, it is very probable that it varies in thickness; and still more probable that it varies in the quantity of work it has done; – all of which produce a glaring want of that uniformity which constitutes the beauty of the typographic art' (p. 394).

At the same time there was a noticeable cheapening of inks and papers. Until the third decade of the century ink continued to be made from the usual ingredients – lampblack and linseed oil – according to the traditional recipes (one of which can be found in Moxon's *Mechanick Exercises on the Whole Art of Printing*, first published in 1683–4). Soap began to find its way into ink, because ink thus diluted gave a sharper impression, and soon the chemists were able to provide a whole range of artificial substitutes for the old ingredients. When Morris in the 1890s insisted upon using an ink similar to that employed by printers of earlier centuries, it was seen as a revolutionary gesture.

Developments in paper-manufacturing were even more alarming. With the supply of good white rags declining, paper mills began, in the final decades of the eighteenth century, to bleach inferior rags and to introduce gypsum and alum into the fibre. The disastrous consequences were evident long before the Kelmscott Press came on the scene. John Murray (not related to the publisher), a popular writer on scientific subjects, in 1823 contributed to the *Gentleman's Magazine* an article which was reprinted as a book the following year under the title

Observations and Experiments on the Bad Composition of Modern Paper. After describing the disintegration of particular books, Murray suggested that 'the *chief* causes of destruction consist in the employment of sulphate of lime, &c. in the pulp, and bleaching the rags previously, or the paper subsequently, with oxymuriatic acid gas. (*Chlorine.*)' (p. 8). Worse was to come. In the latter half of the century book papers were more frequently made from various rag substitutes, the poorest of which was mechanical woodpulp. As a general rule (with naturally a few exceptions), the older a book, the more likely its paper is to be in good condition today; librarians around the world are overwhelmed by the prospect of the steady disintegration of nearly all books printed since the mid-nineteenth century. To Morris, the most insidious aspect of modern book papers was that they frequently masqueraded as something better, with fake wire and chain lines (to suggest that they were handmade) and adulturants (to add weight and shininess). In certain kinds of books and periodicals, he suggested, cheap paper was acceptable, but only if it frankly acknowledged its own cheapness and made no effort to disguise itself.

Probably the ordinary reader a century ago paid no more attention to such matters than readers nowadays do; the most appalling woodpulp paper, when it is new, may seem decent enough to the unpractised eye. But changes in type-styles were sufficiently drastic to be obvious to even casual observers. Everywhere around him — in posters, advertisements, shop signs, sometimes even in books — the nineteenth-century Englishman witnessed an unprecedented efflorescence of new type-designs. Egyptians, grotesques, Gothics, Tuscans: their very names were mysterious and exotic, their shapes heavy and insistent, perfectly suited to the frenetic commercial atmosphere of the times. Book-typography, meanwhile, had moved in the opposite direction, with letterforms that grew ever more delicate and fragile. *Figure 4*

Historians of printing usually classify serifed typefaces designed primarily for bookwork into three major categories. Except to specialists, however, the first two groups — Venetian and old face — look very much alike; on the other hand, the differences between old face and modern face are obvious to all. Until the eighteenth century, despite a slow evolution in the shape of the Roman letter, typefaces used in books changed remarkably little. Caslon's Roman types, for example, which began to appear in the 1730s, are not fundamentally dissimilar from the type used by Aldus in his *Hypnerotomachia Poliphili* (1499), even though they are separated by more than two hundred

Figure 4 Victorian typefounders produced a bewildering variety of new letterforms, as shown in John Parry's painting, *A London Street Scene* (1835). These decorative, sometimes bizarre typefaces were used mainly in posters and advertisements, though occasionally they crept into books as well.

years. But in the generation after Caslon a new style of type-design was introduced throughout Europe, subsequently known as 'modern face', which was to have an immense influence on the appearance of books in Morris's century.

Figure 5 Although a Frenchman, Philippe Grandjean, is usually regarded as the progenitor of the modern face, it was left to Giambattista Bodoni, an Italian designer, and the French foundry of Firmin Didot to carry the new style to its logical conclusion. In England, at about the same time, John Baskerville produced a transitional version of the modern face that seemed better suited to conservative tastes. What all these varied type-designs had in common was vertical rather than oblique shading, unbracketed (i.e. unsupported) serifs, and greater contrast between thick and thin strokes. The effect was – as Morris always said disparagingly – dazzling to the eye. Whereas old-face types had grown directly out of the letterforms used by Renaissance scribes and Roman stone-cutters, the newly fashionable letters were geometrically precise and brilliant in their effect.[6] Old-face designs convey a mellow charm, like that of an ancient building; modern-face types suggest the smooth regularity of a well-made machine. It is not difficult to understand why Morris repudiated modern face.

Though they are not to everyone's taste, the books of Bodoni and Baskerville are extremely handsome. (Morris, needless to say, would never have admitted that.) Yet the eighteenth-century type-designers must be judged not only by their own types and books but also by the mainly unwholesome effect they had on their successors. As Daniel Berkeley Updike remarked, 'Grandjean, Baskerville, Bodoni, and the Didots had a mischievous influence on type-forms: for the *derivations* from types that their work made popular culminated in a kind of letter which was capable of greater vulgarity and degradation than was ever the case with old fonts. The ordinary English, French, or Italian book printed between 1830 and 1850 was very often a cheap and mean-looking production. Perhaps Bodoni and other great persons were not wrong in their day; but they put type-forms on the wrong track.'[7]

Figure 6 The modern-face types for books that evolved in the nineteenth century from the Didot–Bodoni models were bland, undistinguished, and boring, and they changed very little as the century went on, except to become progressively lighter. It is also important to bear in mind that the evolution of type-designs is influenced by developments in technology as well as shifts in fashion. With the introduction of the steam press, it became necessary to cast types in a harder metal than

CANONCINO

Quousque tandem abutêre, Catilina, patientiâ nostrâ? quamdiu nos etiam furor iste tuus eludet? *quem ad finem sese effrenata jactabit audacia? nihilne te noctur-*

M. TULL. CICERO

ARPINAS ORATOR.

RECANATI

Figure 5 A modern-face type, from Bodoni's *Manuale tipografico* (Parma, 1818). Notice the vertical stress, the contrast between thick and thin strokes, and the unbracketed (i.e. unsupported) serifs. Morris wrote contemptuously of the 'sweltering hideousness' of Bodoni's types.

CHAP. III.

Road from Penzance to Land's End.—Trereife.—Ancient Crosses.—Buryas Bridge.—Singular View of St. Michael's Mount.—Drift.—Monumental Stone at Tregonebris.—Boscawen-ûn Druidic Circle.—Creeg Tol.—Giant's Footmarks.—Crowz-an-wra.—Quakers' Burying Ground.—Sennen Church.—Table Mên.—Visit of the Saxon Kings.—First and Last Inn in England.—Adventure at the Land's End.—Land's End Hole.—Land's End during a Calm.—Waves at the Land's End.—The Tides.—The fabled Land of Lyonesse.—The Scilly Islands.—The Wolf Rock.—Longships Lighthouse.—Bolerium.—Dr. Johnson's Head.—The Little Gull.

AS there are so many objects of interest to be seen and described in the interior, it will be necessary to leave the coast for a time.

The Land's End, following the carriage road from Penzance, is about ten miles distant. Hundreds of visitors pass over this road in the summer time, and we also will now take this route, not hesitating to step aside occasionally to inspect some curious monument or legendary spot.

The beautifully wooded grounds of Trereife, about a mile from the town, first attract attention; the roadway is here arched with a long avenue of noble elms; near its extremity it is crossed by another avenue,—the road turning from the left leads to Newlyn,—a few yards up on the right is Trereife House, almost buried in foliage; a yew tree is trained over the front of the building, giving it the appearance of a living wall of leaves. The rooks have long been established here

ANCIENT CROSSES. 59

in their "old ancestral trees," and we always know when spring is coming by their noisy preparations for building. The walks and peeps of woodland scenery around this place are very beautiful. The *Sibthorpia Europæa*, previously noticed, is met with here again, as it is indeed in almost every rill in the district.

On the ascent of a hill, a little beyond Trereife, an old oak forms a complete archway over the road, and if you are on the seat of the carriage beside the driver, he will probably tell you that it is the only instance of the kind to be met with on the journey from London to the Land's End. A short distance from this, on the left hand side of the road, stands an ancient cross; such monuments are very

WAYSIDE CROSS.

numerous throughout the county, but more particularly in this western part. They are pleasing and picturesque objects by the wayside, reminding us of interesting periods of our history. The crosses in this district are probably the most ancient of any in England. Some of the earliest

Figure 6 A typical late-Victorian book: J. T. Blight, *A Week at the Land's End* (Truro, 1893). It is set in Old Style (a symptom of the growing disenchantment with modern-face types) and has dull wood-engravings and curious ornamented initials that owe something to the craze for 'artistic' printing during that decade.

previously, so that they could withstand greater abuse; and this in turn led to the discovery that it was now possible to make the strokes and serifs of letters much finer. Such typefaces required less ink from mechanical roller systems, which functioned well only with thinner inks. Later in the century it became increasingly common for printers to use dry, smooth papers on presses with a hard backing; whereas the ink on hairlines had once spread slightly on damp, more coarse-textured papers, it did not do so when the new technique of printing was employed. 'From an engraver's point of view, new types so printed were exquisitely sharp and clean,' Theodore L. De Vinne, the American printer, explained; 'but from a reader's point of view, the general effect of the print was relatively mean and wiry, gray and feeble. Each letter lost some of its individuality. A new style of delicate but weak presswork came in fashion.'[8]

De Vinne himself bore some responsibility for the developments that he lamented, for this style of printing came to England largely through the influence of American periodicals such as *Harper's Magazine* and *Century Magazine* (the latter printed by De Vinne), which had pioneered the technique of hard-packing in order to improve the reproduction of illustrations. However, the progressively lighter typefaces, though they may have had their origin in technical decisions, exactly suited the taste of late-Victorian readers and printers, and there was much boasting in the trade journals of the time that the growing delicacy of types was a sign of increased 'refinement': 'in the present day we see the effect of an artistic taste, and appreciation for delicacy of outline and freedom of ornamentation developed in the excessively light types and floral embellishments now so popular with the best printers,' according to a contributor to the *Printers' Register* (6 April 1885). 'Even the heaviest styles were materially toned down to meet the current refinement of taste.' By contrast, older typefaces were condemned as heavy, clumsy, and crude. It is in the context of this widespread preference for delicacy and lightness that Morris's almost oppressively dark types must be seen: discarding what seemed to him the attenuation and weakness of nineteenth-century typography, Morris reacted (over-reacted, some would say) by designing faces that were solid, black, and aggressively self-assertive.

But Morris was to argue on more than one occasion that typefaces alone did not determine the beauty and harmony of the printed page: even with mediocre type, if certain other principles of design were observed, it was possible to produce a satisfactory page. Morris had profoundly held views on margins, for example, and nothing infuriated him more than the popular practice of placing the text virtually in the centre of a page. That Morris was fighting a firmly entrenched enemy is illustrated by a series of articles written by De Vinne, the most skilful and best-informed printer in America, on 'Book Margins' which appeared in the *Printers' Circular* (Philadelphia) between May and September 1871. De Vinne, who certainly ought to have known better, offered the unconvincing argument that 'the text or printed page is the picture, and the margin is the frame; all the proprieties require that the picture, when framed, shall be in the centre of the frame' (June 1871). In fact, he was merely endorsing what printers on both sides of the Atlantic already did. Similarly, there was a broad consensus – which was not to be seriously questioned until Morris and Emery Walker attacked it in the 1890s – that generous spaces between words and lines of type improved legibility. The combination of lighter typefaces and ample spaces produced a page in which the text appeared to be, at best, greyish.

Yet such a damning summary of Victorian printing practices fails to take account that printers of the last century were sometimes, when not subjected to the usual commercial pressures, capable of producing very impressive work. All the printers' manuals of the day discussed what was variously known as 'fine press work', 'fine work', or (later in the century) simply 'fine printing'. As the terms suggest, the printer who did this sort of thing earned his additional wages (double or more the usual rate) by taking greater care in the presswork, especially the make-ready; there appears to have been no distinction made between 'ordinary' and 'fine' composition.

The greatest achievements of printing during the last century undoubtedly are to be found in the striking technical skill with which illustrations were produced. In the eighteenth century woodcuts had disappeared almost completely from books, which were illustrated instead with copperplate engravings, often quite elegant in themselves but bearing no relationship to the typography. The engravings and the text pages were printed by entirely separate processes and were joined together only when the book was bound. Morris regarded this as a pernicious development, because it destroyed that harmony which seemed to him the supreme attribute of a well-printed book. The cautionary examples which Morris, Walker, and others in their circle frequently cited were Samuel Rogers's *Italy* (1830) and *Poems* (1834), both handsomely illustrated with steel engravings designed by Turner. Albert L. Cotton summarized the Morrisian view when he wrote in the *Contemporary Review* (August 1898): 'Beautiful as pictures, they bear no apparent relation to the volumes which they illustrate; their charm is independent and extraneous; the artist clearly did not concern himself to harmonise them with the text or with each other; taken apart entirely from the books, they would lose nothing of their force. They are steel engravings, pure and simple, which might just as well have been issued separately in a portfolio.'

With less justice, Morris brought the same accusation against Thomas Bewick,[9] who had revolutionized English book-illustration at the beginning of the century by perfecting the wood-engraving, which, unlike the woodcut, is done on the end grain of boxwood with graver's tools and is therefore capable of rendering very subtle effects. Morris did not especially admire subtlety in art; he preferred the bold, simple outlines of medieval woodcuts to the intricate delicacy of Bewick's vignettes of English rural life. Probably Morris also harboured an irrational prejudice against Bewick because the brilliant technique he had created degenerated very quickly into the facile, characterless style of wood-engraving that was to be found in so many English books and magazines for the remainder of the century. Within a generation Bewick's successors were producing, on an unprecedented scale, illustrations that were both technically dazzling and aesthetically empty. The 'facsimile' engraving reigned triumphantly in Victorian England. Its supremacy was challenged finally in the 1880s by a new mechanico-chemical process then known as zincography, which had been invented by Eberhard, a Bavarian. This technology bypassed the human engraver entirely, since the artist's design was transferred photographically on to a zinc plate, which was then engraved by immersion in a chemical bath. An illustration that previously had been engraved by hand over several days could now be prepared for the press in a few hours. By the time the Kelmscott Press came into existence, wood-engraving had disappeared almost overnight, and process engravings were to be seen everywhere, even in the most expensively produced books. Morris held zincography in contempt; it is one of the minor ironies of the Kelmscott Press that his chief adviser was Emery Walker, by profession a process engraver, who probably knew more about the latest developments in techniques of illustration than anyone else in England. In the end Walker was able to wring one small concession to modern technology from Morris: the borders and initials used at the Kelmscott Press were all printed from electrotypes, another new means of duplicating engravings.

Chromolithography was used less frequently in books, but it was capable of producing really splendid effects, as in some of the folios by Henry Noel Humphreys. Morris, curiously enough, seems to have been intrigued by the process: in 1894 he and Walker spent a day at the Bodleian Library, Oxford, to investigate the possibility of reproducing a beautifully ornamented fourteenth-century manuscript known as the Douce Apocalypse, no doubt under the Kelmscott Press imprint. Nothing came of their plans, but it is certainly significant that Morris regarded such an undertaking seriously. (Morris probably had chromolithography in mind on this occasion, though he may have intended to call into service some form of photographic reproduction, such as the newly invented halftone process.) In a letter written earlier, in 1882, Morris suggested that an Indian carpet recently presented to the Boston Museum of Fine Arts might be reproduced by the same means: 'some of the chromo-lithographers are so good nowadays that a reproduction of a good & careful drawing of it would be of great service to those who are interested in such matters.'[10]

Despite these occasional flickers of interest in the latest methods of facsimile illustration, the one subject that most deeply preoccupied Morris was type-design, and it was there, he felt, that the contemporary book displayed its gravest weakness. Morris was not alone in his dissatisfaction with the modern-face type. There had been several efforts earlier in the century, both in England and on the Continent, to revive old-face designs,[11] but the one attempt which had the most far-reaching consequences for English printers, including Morris, was the reintroduction by the Chiswick Press of Caslon's old-face Roman in the 1840s. Notwithstanding an earlier debate among historians as to which books first appeared in Caslon type, recent scholarship has confirmed that it was *The Diary of Lady Willoughby*, a novel published by Longmans in the spring of 1844.[12] At about the same time, however, the Chiswick Press was also setting up two other books entirely in Caslon: George Herbert's *The Temple*, which was issued later in 1844, and Juvenal's *Satires*, which appeared the following year. Two points should be made about this revival of Caslon. First, unlike modern printers in search of historic designs, Charles Whittingham the Younger, the proprietor of the Chiswick Press, was not compelled to have the Caslon type recut; he simply went to the firm founded by

Figure 7

Figure 8

1

Some Paſſages from the Diary of Lady *Willoughby*.

1635.

1635. May 12, *Tueſday.*

ARoſe at my uſual houre, ſix of the clock, for the firſt time ſince the Birth of my little *Sonne;* opened the Caſement, and look'd forth upon the Park; a drove of Deer paſſ'd bye, leaving the traces of their Footſteps in the dewy Graſs. The Birds ſang, and the Air was ſweet with the Scent of the Wood-binde and the freſh Birch Leaves. Took down my *Bible;* found the Mark at the 103d *Pſalm;* read the ſame, and return'd Thanks to *Almighty God* that he had brought me ſafely through my late Peril and

B Extremity,

Figure 7 A page of *The Diary of Lady Willoughby* (1844), printed at the Chiswick Press. It represents the first revival of Caslon Old Face type in the nineteenth century; Morris was later to use Caslon in his commercially produced books.

Figure 8 William Pickering and Charles Whittingham often met in the latter's summer house at Chiswick to plan the books that they jointly produced. Their books were among the finest printed in nineteenth-century England, and Morris himself learned much about the art of book-design from the Chiswick Press.

Caslon and discovered that the original matrices were still in storage. Second, though the Caslon type was generally admired as attractive when it appeared in *The Diary of Lady Willoughby*, there was no rush by other printers to use it, for Caslon appeared too self-consciously archaic and odd to Victorian eyes. Charles T. Jacobi, who in Morris's day was manager of the Chiswick Press, said of it: 'the excellent design of these different founts [i.e. the several sizes of Caslon] was not generally appreciated until about the year 1888, when the Arts and Crafts Exhibition Society first came into existence, and Mr. Emery Walker was perhaps more responsible than anyone else for making known this series of faces. Since that date these types . . . have gradually been adopted largely for the printing of the better class of bookwork, especially that of a more antiquarian nature. . . .'[13]

Caslon type does not look particularly remarkable to us today, because its frequent use in the early twentieth century has made it seem ordinary and very 'English'. (In time Caslon became so closely identified with Englishness that it was adopted by at least some printers for patriotic motives.) Bruce Rogers, the distinguished American book-designer, complained in 1917 that Caslon was, if anything, excessively familiar – 'a much overrated type', he said; its best quality was 'a certain amount of irregularity, restful to the eye after the smoothness and monotony of our more modern letters'.[14] England seemed to him at the time a 'Caslon-ridden country'. It was precisely the occasional clumsiness and crudeness of Caslon, so offensive to Rogers, which recommended it to Morris, and in the last decade of his life he adopted it enthusiastically for his own commercially printed books. Morris in turn influenced Bernard Shaw to use it in his books from the late 1890s onwards; this had the effect of bringing Caslon into even more general circulation.

Because Caslon seemed appropriate only for 'period' typography to Victorian readers, the Edinburgh typefoundry of Messrs Miller and Richard offered their customers in 1860 a design by A. C. Phemister called at first 'Modernized Old Style' (later simply 'Old Style'), which attempted to capture the flavour of Caslon without sacrificing many of the characteristics of modern-face type. It also had the merit of being designed to work well with modern inks and papers. Other type-founders soon sold their own versions. Old Style was a deliberate compromise, and a good deal of scorn was poured on it through the years. One of the earliest to register a protest was Emery Walker, who, though he admitted that it was more suitable for smooth-surfaced papers than Caslon, described it as 'far inferior in beauty to types of the early printers'; similarly, Morris remarked that 'the design of the letters of this modern "old style" leaves a good deal to be desired, and the whole effect is a little too grey, owing to the thinness of the letters'.[15] Bruce Rogers found it at best inoffensive, at worst monotonous.[16] Whatever its limitations, Old Style proved to be an immensely popular typeface – by some estimates, indeed, the most widely used book face in England during the final decades of the century. (Modern face, however, continued to be the preferred type in periodicals, school-books, and scientific works.) A writer in the *Printers' Register* (6 November 1884) offered the opinion that 'probably no variety of type at present in use has so firm a hold upon the cultivated taste of the printer or of the public'. It was eventually supplanted in the second decade of the twentieth century by the revivals of historic types that began to emanate from the Monotype Corporation.

The resurrection of Caslon by the Chiswick Press in mid-century, followed by the introduction of Old Style by Miller and Richard, brought about gradual but important changes in the appearance of English books. We must remind ourselves that the rejection of modern-face types had begun long before 1891 and that the Kelmscott Press was, in a sense, merely the most visible manifestation of the antiquarian winds which had been stirring the bibliographical waters in England for a number of decades. The two figures who – apart, of course, from Morris – played the most decisive roles in sponsoring the revival of older types and ornaments were the publisher William Pickering and Charles Whittingham the Younger, the owner of the Chiswick Press. Stanley Morison, in fact, has suggested that Morris's celebrated achievements as a printer were firmly rooted in the less spectacular work of Pickering, Pugin, and certain publishers associated with the Oxford Movement. In a letter to Geoffrey Keynes, written in 1924, he remarked that 'Pickering was the first to design books in an eclectico-antiquarian spirit: I mean that he was the first to go back to XVIth century Venice & Lyons for inspiration. If I am right Pickering is the tribal chief of the Morris–Ricketts–Rogers school.'[17] It is an astute observation: period typography – that is to say, typography which deliberately imitates the style of older books – has become so commonplace in the twentieth century that we assume it has always been with us, but the evidence suggests very strongly that it was invented by Pickering and several contemporaries of Catholic or Anglo-Catholic sympathies.

As early as 1830 Pickering acquired a set of wood-engraved initials designed by Thomas Stothard, but it was a decade later when the ornamentation of the Chiswick Press–Pickering books began in earnest. Inspired perhaps by the publication of Henry Shaw's *Alphabets, Numerals and Devices of the Middle Ages* (1845),[18] Whittingham commissioned more than a dozen sets of initials from Mary Byfield and launched a series of sumptuous reprints – including several versions of the English Prayer Book – adorned with blackletter titles and attractive borders, initials, and tailpieces. (Gladstone testified in 1890 that 'the pages of their Prayer Book are a perfect picture, reminding one of those early printed books which are still marvellous examples of printing'.[19]) Though Whittingham turned occasionally to other printers, his finest books were produced by the Chiswick Press, which had the deserved reputation of being the most distinguished 'old-style' printer in England during the latter half of the nineteenth century. Even Morris had to serve what was, in effect, an apprenticeship at the Chiswick Press before he could set up a press of his own.

But there were other old-style printers, less frequently remembered today perhaps than the Chiswick Press but famous in their own time. The firm of George Falkner and Sons in Manchester, for example, were celebrated for the large collection of woodcuts, decorated initials, and head- and tailpieces with which they ornamented pseudo-antique books. Falkner productions were invariably described by reviewers or the firm itself as 'quaint' – i.e. self-consciously antiquarian in flavour. (It is significant that Morris never uses the word 'quaint' when writing about his own variety of old-style printing.) The Falkners had set up a wooden hand-press in 'the quaint Caxtonian Printing Offices in Old Manchester', and their printer's mark, designed by Walter Crane, showed a pressman at work on a hand-press. Printers like Falkner and Sons issued a good deal of ephemera as well as books: their 'Old Style Calendar for 1882', for instance, contained 'many delightful and entertaining particulars, very commodious indeed and profytable, worth the labour of your reading. (A reprint from many early Kalendars.)' Their book titles included *The Shakespeare Show Book* and *The Old English Fayre Volume*; many of the books were stained both internally and externally to suggest great age.[20] (This sort of fakery was, needless to say, the typographical equivalent of the fraudulent restoration of ancient buildings which Morris condemned near the end of his life.)

Figure 9

AT Y^E^ REFECTORIE

THERE willen ben moche goode Sustenaunce and Cunnynge Drynkes ser-en for alle sike as ben wearied wythe y. press or wythe moche Pourchasynge, wythe Cakes for Yu-e-tyde, ornymented for ye Chyldrene, and manie Boxes of goode Thynges. Y^e^ Hostesses ben Maistresse SOLE and Maistresse JESSIE ARMITAGE, assisted bye Maistresse PRESTON, Maistresse GWENDOLINE PRESTON, Maistresse RICHARDSON, Maistresse MEAKINS.

AT Y^E^ BOWER

HERE ben founde ye loveliest Flouers of ye Moneth of Decembre, and maye ben pourchased of ye Smilynge and Innocent Maydens of whom there schal ben a goodlie Store, and who willen receive wythe Pleasure and eke Gladnesse sike Monies as ye Charitie maye move eche soule.

Maistresse FERGUSON, Maistresse TRESIDOR, Maistresse PALMER.

Maistresse SKINNER willen sell Poultrie, Gayme, and Goode Fare.

Ye Signe of ye Sherwoode Oak. Maistresse ARMITAGE (Russet.)	Assisted bye Maistresse RICHARDSON. Maistresse MÜLLER. Maistresse C. RICHARDSON. Y^e^ Maistresses ARMITAGE. Maistresse JESSIE FORSTER.	WILLEN SELL Such Artycles as schal ben riche raire, beautyfulle, and chepe, that willen delighte all Tastes; speciallye Wares manufactured also with moche Cunninge by ye poore Blinde Folkes of greate London Towne!
Ye Signe of ye — Sylver Sheelde. — Maistresse SCOTT and Maistresse SYMONS. (Dark Blue.)	Assisted bye Maistresse PHILLIPS. Maistresse LIDDERDALE. Maistresse WILLIAMS. Maistresse WHEELDON. Maistresse DANGER.	WILLEN SELL Fine Arte Wares of Original Designs and Execution; speciallie a riche Quilte of Silken Stuff, made verrie cosie with softe Eider-downe.
Ye Signe of ye Anchor. Maistresse HENSLOW. (Heliotrope.)	Assisted bye Maistresse HOBSON.	WILLEN SELL Countrie Butter, Eggs, Cheese of Creame, Ham, and righte goode Junket, from true recipe; and alle that makes Mens Handes stronge and Hearts gladde and joiesomme.

Figure 9 A specimen of the work of George Falkner and Sons, Manchester, well-known provincial 'old-style' printers. Despite their professed admiration for pseudo-antique 'impryntynge', the Falkners owned the largest steam-presses in England.

Not all old-style printing was done on such a low level of sophistication, but even very respectable firms had only a muddled notion of what was 'old-style' or 'antique' or even (in some cases) 'medieval'. The Gresham Steam Press, operated by the Unwin brothers, were among the first printers to follow the Chiswick Press's example in adopting Caslon type, yet in an 1862 type-specimen book they supplied this extraordinary note: 'Type of the old style of face is now frequently used, more especially for the first class of books and ornamental works. The series in use at the Gresham Steam Press are from the original matrices which were cut at the beginning of the last century, and thus possess all the peculiarities of the mediaeval letter.'[21]

It is easy to mock late-Victorian printers self-deluded by 'old-style' pretensions, with their verbose colophons in imitation Middle English, their confused appeals to antiquity, and their sham paper and leather (stained, of course). Nor is it difficult to point out the inconsistency of their alleged commitment to the ideal of handicraft: the Falkners might display a hand-press in their device, but they were also notorious for owning the largest steam-driven presses in England. Nevertheless, old-style printers – and the publishers who patronized them – occasionally produced books which, while not possessing the massive authority of the Kelmscott volumes, still have considerable charm. The publisher Joseph Cundall, who was virtually forgotten until recently, issued a series of children's books, mainly in Caslon and often beautifully illustrated by Birket Foster and others, which are extremely handsome by any standard.[22] Did Morris ever see them? We do not know, because he is silent about the Victorian books he considered worthy of admiration, but certainly Cundall, like Pickering, belongs to a very small group of nineteenth-century publisher–designers whose work may have influenced Morris.

Figure 10

Likewise it would be fascinating to discover whether Morris was aware of the books published and printed by Andrew Tuer, whose Leadenhall Press ('Ye Leadenhalle Presse' was the *quaint* version) in London issued many lively, amusing exercises in pastiche. The founder of the *Paper and Printing Trades Journal* as well as a wholesale stationer, Tuer avoided merely imitative period typography by a playful approach to book-design that at times reminds one of the zestful spirit of Francis Meynell in our own century. (There is in fact a direct link: Meynell confessed to a special interest in Tuer because his father had been Tuer's literary adviser.[23]) In works like *Old London Street Cries* (1885) and *A History of the Hornbooks* (1896) Tuer turned to the usual

Figure 10 A specimen sheet of 'Ye Leadenhalle Presse' which appeared in the *Printers' International Specimen Exchange* (1880). The proprietor, Andrew Tuer, was one of the most energetic and inventive 'old-style' printers in London at the close of the last century.

antiquarian subject-matter of which old-style printers were fond, but his books possess a vivacity which the Falkners and their tribe could never have imparted. This special quality of Tuer's publications can be seen most vividly in the famous booklets in chapbook style published during the 1880s, illustrated with peculiar (but effective) woodcuts by Joseph Crawhall and set in type that appears to have been deliberately battered.[24] For many English readers in the final decades of the last century, Tuer's name was synonymous with old-style printing, which had by then become extremely fashionable.

Even in Tuer's own journal, writers occasionally expressed misgivings about the way in which 'old style' seemed to lead so often to an unhappy display of meretricious virtuosity in typography and spelling. Commenting on *Ye Printers' Blotter: A Votarie of ye Wm. Caxton & ye Bn. Franklin*, privately printed 'atween times' by Edwin C. Bell of Titusville, Pennsylvania, a contributor to the *Paper and Printing Trades Annual* (June 1880) observed that it 'has evidently been projected, and its leading ideas copied, from Mr. S. L. Nussey's *Ye Quaynt*, published in this country, and it is only to be regretted that the imitation is not typographically more complete – for Mr. Bell, although evidently smitten with the idea of following the old-style, lacks the material, or more probably the necessary knowledge to carry it out, as his production is full of the grossest inconsistencies, literals, and bad imitations of early English spelling. . . . He is fond of the long 'ſ,' but in every case makes an 'f' do duty, and worse than that he uses it for finals, as for instance in the word 'news' in his opening address, which he spells 'newf' – two mistakes in a word of four letters.' It was no doubt easier to criticize the amateurish blunders of a printer across the ocean, but in fact the English landscape was also littered with long 's's (frequently misplaced) and ubiquitous 'ye's. Quaintness had become a typographical disease. In *The Art of Illustration* (1894), Henry Blackburn complained bitterly about the fraudulence of imitative modern books in which process engravings masqueraded as woodcuts, machine-made papers with deckle edges pretended to be the product of hand-labour, and artificial binding materials were disguised as vellum. Such books, he said, cracked in half with a loud bang when they were opened the first time.

Old-style printing, then, was a mixed blessing. It originated with a commendable attempt to resist the tyranny of modern-face types by reviving the use of Caslon, a superior eighteenth-century face, but it became very quickly associated with deception and cheap materials. When Ruskin subjected one of the most famous nineteenth-century English paintings, Holman Hunt's *The Awakening Conscience* (exhibited at the Royal Academy in 1854), to a characteristically close analysis, he seized upon one detail – the modern book in a papier-mâché binding which lies upon a table – as a particularly telling symptom of Victorian moral failure.[25] Ruskin was much interested in book-production: he tried his hand at designing bindings, and for the 1889 edition of *Modern Painters* he insisted upon a newly cast fount of type and painstakingly oversaw the printing of the plates.[26] Morris must have been aware of Ruskin's prolonged effort to control the printing of his own books.

Ruskin also played a minor role in launching what the Victorians called 'artistic' printing, the only serious rival to 'old-style' printing in the 1880s.[27] Thomas Hailing, a Cheltenham printer who had achieved some celebrity by establishing a small trade journal called *Hailing's Circular* in 1877, proposed another periodical in which leaves printed by various printers would be bound together and circulated for their mutual edification.[28] The first number (1880) of the *Printers' International Specimen Exchange*, under the editorship of Andrew Tuer (who had a foot in both the artistic and old-style camps), contained a letter from Ruskin to Tuer: 'It seems to me . . . that a lovely field of design is open in the treatment of decorative type – not in the mere big initials in which one cannot find the letters – but in a delicate and variably fantastic ornamentation of capitals, and filling of blank spaces or musically-divided periods of sentences and breadths of margin. Paper that won't break or won't mildew would be literally a "godsend" to me. I scarcely care to design an engraving to go on to modern paper. I have the most entire sympathy with your objects, but believe that people *will* have bad paper nowadays, bad printing nowadays, and bad painting nowadays – and nothing else.'

Figure 11

It is a curious letter of endorsement – its bleak pessimism is certainly at odds with the upbeat tone of Tuer's introduction – and if Ruskin thought that Victorian printing was about to be regenerated by the *Printers' International Specimen Exchange*, he was seriously mistaken. The 'artistic' style which dominated the *Specimen Exchange* from the beginning had been imported from the United States and was based

Figure 11 An example of 'artistic' printing in the *Printers' International Specimen Exchange* (1880), edited by Andrew Tuer. More sophisticated Victorian printers like De Vinne and Jacobi decried the obsession with bent rules and weird decorative types.

upon a hideous (or so it now seems to us) combination of bizarre ornamental types and bent rules. The *Printers' Register* (7 July 1884), reviewing the fifth volume of the *Specimen Exchange*, argued that it represented a higher level of taste and workmanship than was to be found among any of the great printers of the past. But this was pure bluster (of the sort very common in trade journals of that decade), and the most sensitive printers recognized that the artistic style was a temporary aberration which could only be deplored. Jacobi remarked that 'the excessive rule-work, twisted and twirled in all manner of fantastic shapes, the adoption of grotesque types, together with the employment of diverse colours (not always harmonious) is work really overstepping the bounds of proper typography, and it appears to me but an emulation of lithography or chromolithography'.[29] De Vinne, in a lecture delivered in Philadelphia, offered this sarcastic observation: 'What advances have we made in rule-twisting! What unknown possibilities in typography have been developed by our new race of compositors! . . . How it does delight us to employ a typographical gymnast who tortures brass rules and spends hours and days in experiments with borders, fancy job types, tint grounds, and flourishes!'[30]

Fortunately the artistic style had its strongest impact on job-printing; it crept into books only occasionally, and then mainly on bindings and title-pages. Many book-printing houses such as Richard Clay and William Clowes were affected remarkably little by either the old-style or the artistic modes yet managed to produce typography that was at least neat, decent, and unobtrusive. Scottish printers in particular had a reputation for making books with these old-fashioned virtues. When Morris began to proclaim his typographical views in the 1890s, it is understandable that some of these firms must have felt they were being overshadowed by a showy upstart. In an interview published in the *Printers' Register – Supplement* (6 February 1894), Walter Blaikie, head of T. and A. Constable, Edinburgh, made a number of barbed comments that were clearly directed at Morris. 'Many machine papers are quite as durable as hand-made,' he said; deckle edges had become an end in themselves (here Morris would have agreed with him), 'just as in some of Morris's books, he strives to perpetuate the embossing of papers by the type, which was one of the defects of the hand-press'. Caslon type, he added, was used chiefly 'for what I call *quasi*-facsimile reprints'. And then the final veiled allusion to Morris: 'you must always remember that the first object of a book is to be read, not to become an article of bric-a-brac.' When a journalist from the periodical *Bookselling* interviewed Morris the following year, he found equal distrust on the other side. 'The Constables are good printers,' Morris observed, 'but those square capitals with the thick faces, which they use in the headlines of their books, are simply hideous. . . . In my opinion the best printers in Scotland are the Clarks. They have done some really beautiful work and have done it cheaply.'[31]

Among nineteenth-century English private printers, the only one whose work anticipated Morris's in any sense was the Revd C. H. O. Daniel, Fellow of Worcester College, Oxford. The term 'private', incidentally, is extremely troublesome: Emery Walker was even prepared to argue that the Kelmscott Press was not, strictly speaking, a private press, since Morris was willing to print books for others.[32] When John Martin, the librarian of Woburn Abbey, compiled a *Bibliographical Catalogue of Privately Printed Books* (second edition, 1854), he explained that 'by privately-printed books, the Editor means to designate such only as were not intended by the writers for sale, and the circulation of which has been confined entirely to their friends and connections, or to those who took an interest in the matter contained in them' (p. vii). The Daniel Press would fit such a narrow definition, but certainly not the Kelmscott Press – or most other modern private presses, for that matter.

Between 1845 and 1863 Daniel produced a large number of ephemeral pieces of printing, at first on a toy press and later on a small Albion press. In 1875, after an interval of neglect, he moved the press into his college rooms at Oxford and began printing again in earnest, and during this second phase he created a series of pamphlets and books which are admired and collected today. Falconer Madan, who has written the fullest account of Daniel's work as a printer, made a large claim for his press: 'Dates appear to show that the Daniel Press was the first attempt to raise the standard of English Victorian printing – a work which was carried on with higher artistic and professional ideals and better opportunities on the establishment of the Kelmscott Press in 1891. The use of the Fell type in 1877 and the production of the *Garland of Rachel* in 1881 may fairly be regarded as the first genuine signs of the Revival of Printing in this country.'[33]

But is this true? Stanley Morison has shown that Daniel was not after all the first printer to revive the use of the splendid Fell types (owned by the Oxford University Press) in the nineteenth century, and Jacobi, among others, pointed out that Daniel's presswork was not as good as his typography.[34] A close examination of *The Garland of Rachel*, which Madan describes as 'the most celebrated and valuable product of the Daniel Press' (p. 86), reinforces such misgivings. The little collection of eighteen poems, written by Daniel's friends to celebrate the first birthday of his daughter Rachel, is attractively set in the Fell types with ornaments by Alfred Parson, RA, and Mrs Daniel; yet the initials are weak, the inking is uneven, the register is erratic, and the paper has deteriorated badly. It is a charming production – indeed, intensely 'private' – but scarcely will stand up to a comparison with even the smallest of the Kelmscott books. This is not to disparage Daniel's printing: it is likely to appeal to precisely those collectors who are offended by Morris's penchant for the monumental book. But to see the Kelmscott Press as the direct descendant of the Daniel Press (as Madan suggests) is misleading.

Was Morris familiar with any of the Daniel books? He could easily have known about them, but as usual he seems never to have mentioned Daniel or his press. The only tantalizing clue we have is an entry in the diary of Sydney Cockerell, Morris's secretary, on 28 September 1898 (after Morris's death): 'Went to Bodleian & then to see Revd C. H. Daniel & his printing press (which has been idle for nearly a year) at Worcester House. Had a nice talk about printing &c.'

Just two years after Daniel began printing in Oxford, a major exhibition was held which was to have a more direct effect on the history of the Kelmscott Press. To observe the four-hundredth anniversary of printing in England, a Caxton Exhibition opened in South Kensington on 30 June 1877 with unprecedented attention from both the press and the public. Squeezed in between the large array of early printed books (mainly but not exclusively Caxton's) and modern printing equipment was a small exhibit which proved to be the hit of the show: while curious visitors watched him, a white-haired pressman in knee-breeches operated a wooden press approximately two hundred years old. (At a Health Exhibit in 1884 Tuer's 'Ye Leadenhalle Presse' managed to put their hands on the same press for a reconstruction of a medieval printing office.) Morison asserts that the 1877 exhibition 'had a decisive effect upon bibliographical and typographical studies'.[35] Whether this was the case or not, it certainly had the effect of elevating Caxton to the status of a national hero. William Blades's revised biography appeared in the same year; in 1882 a Caxton Memorial Window was unveiled in St Margaret's Church, Westminster. In his last typographical experiment before founding the

Kelmscott Press, Morris used a modern imitation of one of Caxton's types, and five of the Kelmscott titles were Caxton reprints. Though Caxton's books could not have appealed to him aesthetically, Morris obviously shared his countrymen's enthusiasm for England's first printer.[36]

A growing interest in old books, facsimiles, and improved typography was evident on every side during the last two decades of the century. A writer in the *Paper and Printing Trades Annual* (December 1879) declared that 'fifty years ago a facsimile or an imitation of Caxton type could only be seen in the private library of the lover of books, but now the taste for Caxtonian and old-style works is so general that pamphlets and tradesmen's circulars are printed in type so nearly resembling the original that only the initiated can discern the difference'. One well-known London publisher, Elliot Stock, exploited this popular antiquarian taste by issuing a series of type-facsimiles of older English books, including Walton's *Compleat Angler*, *Pilgrim's Progress*, *Paradise Lost*, and Herbert's *The Temple*. They were attractively printed at the Chiswick Press and were well received by readers and reviewers. During the 1880s Kegan Paul published a similar series, known as the 'Parchment Library'.

The cult of Caxton led also to a minor revival of Gothic types, at least for display purposes. Calbe Stower's *The Printer's Grammar* (1808) had been able to declare emphatically that blackletter 'is now abolished in England, except in a very few instances' (p. 41), and throughout most of the nineteenth century its use was restricted to legal and ecclesiastical books, usually on title-pages. In the 1880s it began to appear again (though still for display only) in books and journals. The extent to which the Caxton exhibition had filtered down to the popular level is amusingly illustrated by this advertisement in the *Antiquary* magazine, 1880–1: 'Cash's Woven Initial Letters and Monograms for sewing on household linen, socks, and underclothing. The letters are manufactured in three sizes of Old English Type, in Turkey Red, and orders can also be executed in Black, both of which are warranted perfectly fast. . . . Sold by Drapers and Hosiers everywhere.' By the time Morris turned his attention to designing the Troy type, the Victorian Gothic Revival had already touched the Gothic typeface.

Another indication of the new interest in typographical experimentation was A. H. Mackmurdo's periodical the *Century Guild Hobby Horse*, which first appeared in 1884 but did not resume publication until two years later.[37] Though its style was more art nouveau than Morrisian, the *Hobby Horse* was a distinguished journal that reminds us once again (if we need reminding) that many of the ideas usually attributed to Morris were in fact simply in the air at the end of the last century. With an engraved cover designed by Selwyn Image (whose Greek type Morris was to use in the following decade) and ornaments by Image, Herbert Horne, and others, the *Hobby Horse* – printed, not surprisingly, at the Chiswick Press – anticipates many of Morris's notions about the well-designed page. And, since Morris knew Mackmurdo and others associated with it, there can be no doubt that he was aware of the existence of the *Hobby Horse*. It is most unlikely that he was impressed by the *Hobby Horse*, though Mackmurdo in later years was to boast that his periodical was the true precursor of the Kelmscott Press. One can perhaps find a hint of Morris's probable attitude in a letter (undated) which he wrote to Charles Ricketts and Charles Shannon about their *Dial*, another journal, founded in 1889, with similar typography and ornaments: 'Thank you for the *Dial* which duly arrived. I confess that I looked at the art portion of it with somewhat mixed feelings, as the talent and the aberration of the talent seemed to me to be in about equal proportions.'[8]

The world of typographical scholarship – with which Morris had numerous connections – was also showing signs of renewed life in the final years of the nineteenth century. William Blades and Talbot Baines Reed were laying the foundations for the history of English printing; at the major British libraries figures like Robert Proctor and Henry Bradshaw were pioneering in the study of incunabula. During the 1890s Kegan Paul published a bibliographical series under the general editorship of A. W. Pollard: the titles included *The Great Book Collectors* by Charles and Mary Elton, *Book Plates* by W. J. Hardy, *Books in Manuscript* by Falconer Madan, *Early Printed Books* by E. Gordon Duff, *Early Illustrated Books* by Pollard, and *Book Bindings* by Herbert P. Horne. Along similar lines, Elliot Stock inaugurated a series called 'The Book-Lover's Library' under the editorship of H. B. Wheatley as well as several bibliographical journals (the *Antiquary*, 1880–1915; the *Bibliographer*, 1881–4; *Book-Lore*, 1885–7; the *Bookworm*, 1888–92). In the 1890s the Bibliographical Society was formed in London, with the primary purpose of recording all early English books.

Morris himself read a paper entitled 'The Ideal Book' in June 1893 to the Bibliographical Society, and on the same evening Jacobi of the

Chiswick Press also spoke on 'The Printing of Modern Books'. Another organization with which Morris was actively associated, the Art-Workers' Guild, devoted a substantial number of its meetings from 1886 onwards to the book arts; the topics discussed or demonstrated at the Guild's meetings included 'Bookbinding', 'Calligraphy, Typography, and the Book Beautiful', 'The Manufacture of Paper' (this by Joseph Batchelor, Morris's papermaker, who brought along a washtub for his demonstration), 'Book Illustration' and 'Reformation of Printing Type'.

Shortly before his foundry cast Morris's Golden type in 1890, Talbot Baines Reed delivered an erudite lecture on 'Old and New Fashions in Typography' to the Society of Arts (the group to which Morris was later to speak about 'The Woodcuts of Gothic Books'). Reed's talk offers an excellent illustration of the enlightened climate of typographical opinion in which Morris's own theories were formed. Surveying the development of letterforms through the centuries, Reed praised the superiority of Venetian and old-face types – especially those of Nicholas Jenson and Caslon – and argued that 'the modern fashion [i.e. modern-face type] lacks all the graciousness and dignity of the old. Its straight hard serifs, its stiff interiors, its harsh contrasts of thick and thin, more than nullify the passing advantage of exact lining and delicate finish.'[39] This assertion has a familiar ring, because it is precisely the doctrine that Morris was to promulgate within a year or two after Reed's lecture. Morris may not have been directly indebted to Reed, because there were others from whom he could have absorbed the same ideas (most notably Emery Walker, who was present that evening and participated in the discussion); but the lecture demonstrates that Morris's typographical theories did not evolve in isolation.

A slightly earlier lecture which eerily anticipates many of Morris's complaints about modern books was delivered by Henry Stevens, an American rare-book dealer long resident in London, to the Library Association at Cambridge in October 1884.[40] Published under the title *Who Spoils our New English Books Asked and Answered by Henry Stevens of Vermont* (1884) and dedicated 'to the memory of two old friends Charles Whittingham and William Pickering', Stevens's lecture was a powerful attack upon declining standards in Victorian printing. 'The manufacture of a beautiful and durable book costs little if anything more, it is believed, than it does to manufacture a clumsy and unsightly one,' he declared in words that Morris was to echo (almost verbatim) in the following decade.[41] Like Morris, Stevens insisted that margins of classic proportions, which he specified, were essential to a beautiful printed page, and that good design was meaningless without good materials, especially ink and paper. He even made a stirring assault on the indiscriminate use of the binder's knife, which Morris later described as one of the chief enemies of civilization.

As always, we find ourselves asking the obvious question: was Morris familiar with this lecture? And, as in other instances considered earlier, there is no firm answer, though he undoubtedly had an opportunity to read it, for the booklet was on display at the 1888 Arts and Crafts Exhibition where Morris must have seen it. The issue, however, is not whether Morris borrowed particular ideas or attitudes from some of his contemporaries; what we need to keep in mind while examining the history of the Kelmscott Press is that Morris's reaction against the enfeebled Victorian book reflects a surprisingly widespread response to the typography of the day. Morris was not alone in protesting against the modern debasement of printing, nor was he the only one to suggest sensible remedies. What sets Morris apart from Jacobi, Reed, Stevens, and other lesser voices is that he was able, during the 1890s, to fuse a brilliant synthesis of their individual criticisms, to bring the craft of printing back in touch with literature and art, and to move these typographical debates out of the realm of theory by producing books of unsurpassed beauty.

TWO ❦ WILLIAM MORRIS & THE ARTS OF THE BOOK

THOUGH William Morris displayed a lively interest in the arts of the book at various times throughout his life, the first volume printed at the Kelmscott Press was completed only five and a half years before his death. Seen in isolation, the Press appears to be an inexplicable phenomenon of Morris's final decade, an undertaking brought to fruition only by the desperate energy of a dying man. Yet Morris was able to achieve so much as a printer in so little time not only because of the obsessive intensity he brought to every task but also because of the profound understanding of the physical form of books that he had acquired over a lifetime. In fact, he made several noteworthy (but mostly abortive) efforts to produce lavishly ornamented and illustrated books before 1891. The Kelmscott Press, which to the casual observer perhaps seems to have sprung fully matured from the brow of Morris, had actually undergone a long period of gestation.[1]

The Press alone would have been enough to make the reputation of any other man or woman, but it was merely the last phase of a long, extraordinary career in which Morris achieved fame as a poet, revived a whole series of traditional crafts, established himself as the foremost English designer of the nineteenth century, and founded and managed a firm (Morris, Marshall, Faulkner, and Company, later reorganized as Morris and Company) which was to have a marked effect upon the furnishings of the middle-class Victorian home. In his later years he became, moreover, a pioneer in the architectural preservation movement and one of the most influential leaders of English socialism. Merely listing Morris's activities, however, cannot convey adequately the vitality of either the man or his ideas: born in an age of growing specialization, he was a brilliant generalist who did superlatively well whatever he turned his hand to. In the latter half of the twentieth century we have increasingly come to recognize Morris as one of the seminal figures in the history of modern architecture, art, and design.

Though not disposed to introspection, Morris has left us in his work itself – the tapestries, stained-glass windows, wallpapers, poems, books – an unmistakable record of the growth of his inner life. 'I can't enter into politico-social subjects with any interest,' he wrote to his friend Cormell Price in July 1856, 'for on the whole I see that things are in a muddle, and I have no power or vocation to set them right in ever so little a degree. My work is the embodiment of dreams in one form or another.'[2] In the 1880s he did after all try to translate these dreams into political realities through socialist agitation, but he remained convinced, like Dickens's Stephen Blackpool, that nineteenth-century life was 'aw a muddle', perhaps a hopeless muddle, and ultimately only his compelling personal vision of a simpler, purer world – located especially in the Middle Ages and, in moments of optimism, in the future – sustained him. Near the end of his life, reflecting on the comparative failure of the Society for the Protection of Ancient Buildings (of which he had been primary founder), he lamented, 'Now that I am grown old and see that nothing is to be done, I half wish that I had not been born with a sense of romance and beauty in this accursed age.'[3] The echo of Hamlet seems to be deliberate in this letter: finding himself living in a time that was out of joint, Morris, silently cursing, bent all his prodigious energies to setting it right. The Kelmscott Press was his last great attempt to restore beauty and sanity to an ugly, mad world.

It is scarcely necessary to trace intellectual 'sources' of Morris's powerful, visionary attachment to the medieval past, for apparently from earliest childhood he was drawn to Gothic churches and cathedrals. The most intense experience of these years seems to have been a rapturous glimpse of Canterbury Cathedral at the age of eight. At Marlborough he devoured books in the school library on archaeology and ecclesiastical architecture, and invented stories 'about knights and fairies'; at Oxford he was intoxicated by 'a vision of grey-roofed houses and a long winding street, and the sound of many bells'.[4] Burne-Jones, whose moods often mirrored Morris's, has left a revealing account of a walk across Oxford's Port Meadow from the ruins at Godstowe on a winter day when he (Burne-Jones, that is) forgot altogether the modern world and lost himself in a rapturous dream of 'the old days, the abbey, and long processions of the faithful, banners of the cross, copes and crosiers, gay knights and ladies by the river bank, hawking-parties and all the pageantry of the golden age – it made me feel so wild and mad I had to throw stones into the water to break the dream. I never remember having such an unutterable ecstasy, it was quite painful with intensity, as if my forehead would burst.'[5]

It is tempting to dismiss this sort of experience as adolescent fantasizing, but in Morris's case his passion for the Gothic world was connected with the growth of a comprehensive set of ideas which attempted to explain the moral, artistic, and social decline of Victorian England. They were 'Ruskinian' ideas, of course, yet one suspects that Morris was drawn so irresistibly to Ruskin's writings because he was already, by temperament and experience, predisposed to find the distant glamour of the Middle Ages more absorbing than the industrialized, urbanized world of the nineteenth century. In 1851–3 the three volumes of Ruskin's *The Stones of Venice* appeared, and they struck Morris and Burne-Jones at Oxford with the force of revelation. *Figure 12* Morris was to reprint its central chapter, 'The Nature of Gothic', as the fourth book of the Kelmscott Press in 1892: 'one of the very few necessary and inevitable utterances of the century,' he called it in his preface. 'To some of us when we first read it,' Morris wrote, 'now many years ago, it seemed to point out a new road on which the world should travel. And in spite of all the disappointments of forty years, and although some of us, John Ruskin amongst others, have since learned what the equipment for that journey must be, and how many things must be changed before we are equipped, yet we can still see no other way out of the folly and degradation of Civilization.'[6]

What exactly did Morris learn from Ruskin? Above all, that 'art is the expression of man's pleasure in labour' (as Morris put it in the *Nature of Gothic* preface); and that the only hope for the future lay in the handicrafts, for 'the longing of simple people will take up the chain where it fell from the hands of the craft-guilds of the fifteenth century, and . . . the academical art which has developed from that misreading of history which we call the Renaissance, will prove a barren stem'.[7] The modern artist (or craftsman, as Morris and Ruskin might say, since they did not recognize any distinction between the two categories) was compelled to look back to the Middle Ages for models and inspiration, though mere imitation of medieval art was to be discouraged. Writing to the illustrator C. M. Gere in 1893, Morris observed: 'One piece of advice I will now give you, which you must deal with according to your own mind, as it is I am aware a little dangerous. This it is; that you should now try to steep yourself, so to say, in mediaeval design; look [at] illuminations in 13th & 14th century books[,] at wood cuts and so on, and make sketches from them. . . . But (there is always a but, you know) all this will be of no use to you unless you feel yourself drawn in that direction and are really enthusiastic about the old work. When I was a young-bear, I think I really succeeded in ignoring

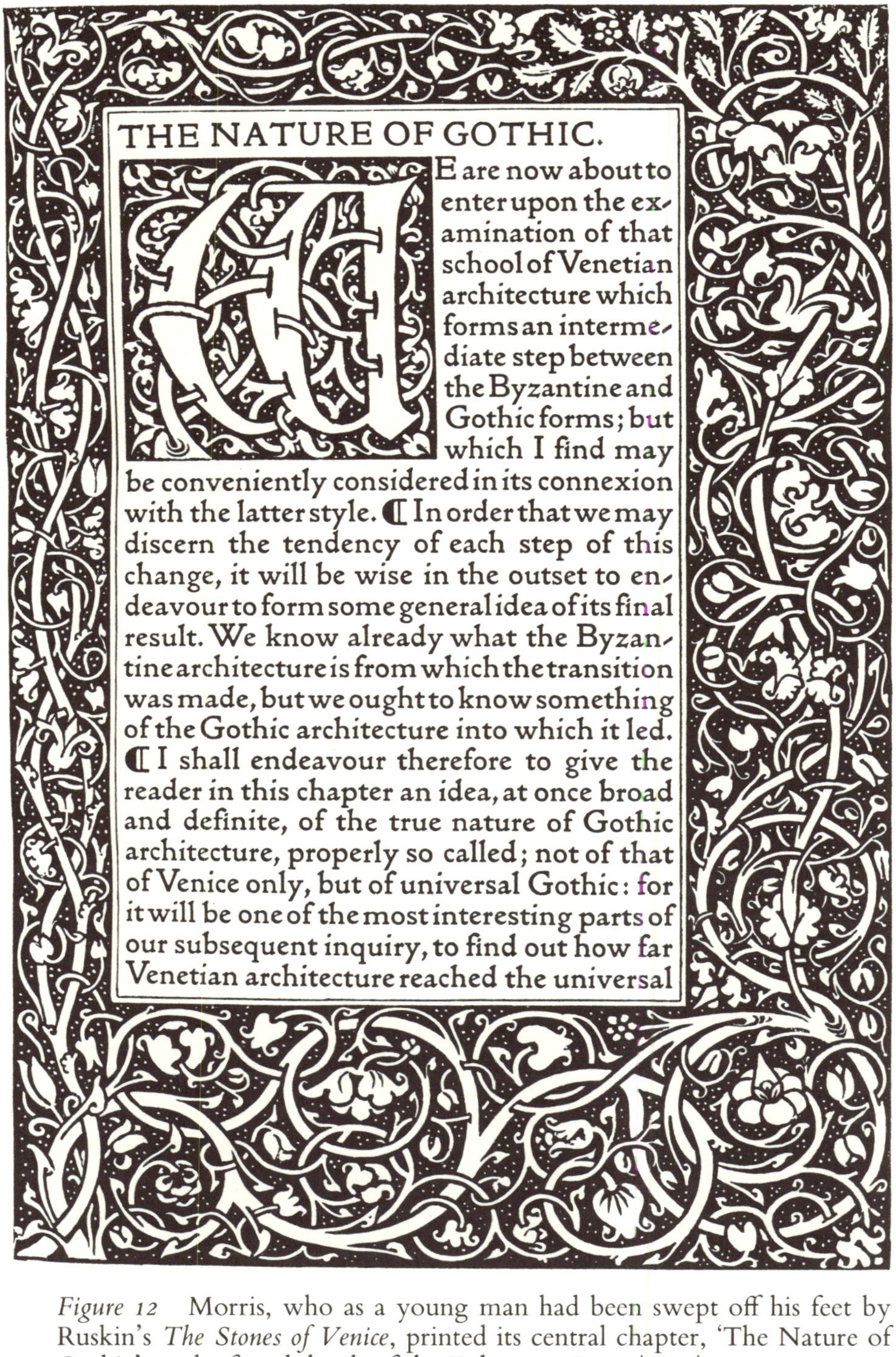

THE NATURE OF GOTHIC.

WE are now about to enter upon the examination of that school of Venetian architecture which forms an intermediate step between the Byzantine and Gothic forms; but which I find may be conveniently considered in its connexion with the latter style. ¶ In order that we may discern the tendency of each step of this change, it will be wise in the outset to endeavour to form some general idea of its final result. We know already what the Byzantine architecture is from which the transition was made, but we ought to know something of the Gothic architecture into which it led. ¶ I shall endeavour therefore to give the reader in this chapter an idea, at once broad and definite, of the true nature of Gothic architecture, properly so called; not of that of Venice only, but of universal Gothic: for it will be one of the most interesting parts of our subsequent inquiry, to find out how far Venetian architecture reached the universal

Figure 12 Morris, who as a young man had been swept off his feet by Ruskin's *The Stones of Venice*, printed its central chapter, 'The Nature of Gothic', as the fourth book of the Kelmscott Press (1892).

modern life altogether. And it was of great service to me.' Yet in his next letter to Gere, Morris admitted that this was 'doubtful advice'.[8] Nineteenth-century England was, after all, littered with cautionary examples of derivative, historicist art and architecture, and Morris was fully conscious of the dangers in turning to the past. But he was convinced that only by rediscovering the older traditions could the Victorian artist reunite labour and pleasure; only thus could he rebuild the decayed social foundations of art.

At Exeter College, Oxford, in the 1850s Morris, Burne-Jones, and several like-minded friends discovered Ruskin, Rossetti, and the Pre-Raphaelites, toyed with the notion of founding an Anglo-Catholic monastic community of craftsmen, visited country churches and rubbed brasses, and read Tennyson, Chaucer, Keats, Carlyle, and Charlotte Mary Yonge's *The Heir of Redclyffe*. After taking his degree Morris worked for a short time for G. E. Street, the architect, but within a few years he was married to Jane Burden and well launched on his own distinctive career as a renewer of traditional crafts. Architecture was always for Morris the supreme art, and throughout his life he never wavered in his conviction that all other arts were subservient to it. Yet Morris came to realize that minor crafts such as weaving, wall-decorating, and printing, because they were less ambitious and involved simpler technologies, could be more easily restored to their former glories than architecture itself could. The book in particular, whether written or printed, appealed to him because it was a small, self-contained object; Morris doubted 'whether it were possible to have buildings generally good in the present day', but he noted that 'there are not the same difficulties with a book as with many other articles of commerce'.[9]

In the last year of his life Morris told Wilfrid Scawen Blunt, 'I remember as a boy going into Canterbury Cathedral and thinking that the gates of heaven had been opened to me. Also when I first saw an illuminated manuscript, these first pleasures which I discovered for myself were stronger than anything else in life.'[10] (This close identification of the Gothic cathedral and the Gothic book in Morris's mind is significant: in the 1890s he was to speak of the arrangement of the Kelmscott Press *Chaucer* as purely architectural.) In the Bodleian Library at Oxford, Morris had his first opportunity to examine a large number of illuminated manuscripts, and he was to recall frequently in later years the pleasure he found in them, especially the 'Douce Apocalypse' and a fifteenth-century 'Romance of Alexander'.[11] *Figure 13*

Figure 13 Morris, while an undergraduate at Oxford, admired the 'Douce Apocalypse' in the Bodleian Library, and in November 1894 he and Walker came to the Bodleian in order to examine it again; Morris considered the possibility of publishing a facsimile of the thirteenth-century manuscript.

Morris's passion for the decorated book never diminished: May Morris remarked that her father and Burne-Jones were for many years such frequent visitors to the manuscript room of the British Museum that 'no young man in London, outside the staff of the department, was so familiar with its treasures as my father was'.[12] Indeed, Morris was prepared to argue that 'if all the other art of the early Middle Ages had disappeared, they might still claim to be considered a great period of art on the strength of their ornamented books'.[13] In 1894 Morris contributed an article entitled 'Some Notes on the Illuminated Books of the Middle Ages' to the *Magazine of Art*, and at about the same time he began an essay, never completed, 'Some Thoughts on the Ornamented Manuscripts of the Middle Ages'. In both essays Morris displayed an impressive technical knowledge of medieval manuscripts; H. M. Hyndman claimed that Morris possessed such remarkable expertise that, during a visit to the Bodleian in about 1882, he was able to identify a whole group of manuscripts in rapid succession for the benefit of one of the librarians.[14] What impresses the reader of Morris's essays today, however, is not his scholarly grasp of a specialized subject (though that is considerable) but his fervently held conviction that such manuscripts represent the craftsmanship of the Middle Ages at its very highest level.

There is no evidence that Morris began to collect manuscripts seriously after he left Oxford, though in October 1858 he wrote from France that he was buying 'old manuscripts and armour and ironwork and enamel'.[15] The earliest extant catalogue of his personal library, dating from about 1876, lists only six manuscripts, most of which he later sold.[16] Nor did Morris systematically acquire early printed books in his youth, yet there are ample indications that he was drawn to them just as inevitably as he was to illuminated manuscripts. As a boy, for example, he found in his parents' library a copy of Gerarde's *Herbal* (1597), which was to influence the floral patterns of his wallpapers and tapestries and which he was to consult again in 1875 when he studied the recipes of vegetable dyes. Greater pleasures lay ahead: during their undergraduate years Morris and Burne-Jones were introduced to the delights of fifteenth- and sixteenth-century wood-engravings. 'Two luminaries burst upon us at that time,' Burne-Jones remarked. 'One was Chaucer and the other was Dürer.'[17] Morris later felt vaguely embarrassed about his admiration for Dürer, who worked a century too late to be regarded as a medieval artist, but his ingenious solution to this problem was to assert that 'though his [Dürer's] method was infected by the Renaissance, his matchless imagination and intellect made him thoroughly Gothic in spirit'.[18] At about the time that Morris was founding the Kelmscott Press, he was able to put together within a few years an outstanding collection of incunabula, with particular emphasis on French and German illustrated books. He lectured and wrote enthusiastically on the subject in the 1890s, and it is characteristic of how Morris's mind worked that his lecture on 'The Woodcuts of Gothic Books', delivered to the Society of Arts in January 1892,[19] concluded with a discussion of the prospects for improving the modern illustrated book; for Morris never approached the history of printing or book-illustration from a merely antiquarian point of view. How, he wanted to know, could we learn from the printers and illustrators of previous centuries, and how could we match their achievements in our own day? As early as 1855 we hear of Morris trying his hand at 'drawing on wood and cutting the designs himself', inspired by 'The Maids of Elfen-Mere', an engraving designed by Rossetti which appeared in William Allingham's *Day and Night Songs*.[20]

It is clear what Morris and Burne-Jones found offensive in routine wood-engravings of the nineteenth century. Burne-Jones summed up their indictment of contemporary work in a letter written to his sister-in-law in 1862: 'As to scribbly work, it enrages one beyond endurance. Nearly all book and periodical illustration is full of it – drawings, you know the kind, that have wild work in all the corners, stupid, senseless rot that takes an artist half a minute to sketch and an engraver half a week to engrave, for scribble is fearful labour to render. *Figure 14* My dear, look at most things in "Once a Week" – the wasted time of poor engravers in rendering all that scrawl, if rightly used, might fill England with beautiful work.'[21] The solution, Morris and his friend came to believe, lay in stern simplicity: austere outlines in bold, vigorous strokes enhanced by only a minimum of shading, which was never to be rendered with cross-hatching. Such a formula, as Morris recognized, pointed back directly to the fifteenth century, and it is noteworthy that all the wood-engravings produced by Morris or done under his direction – from the first tentative experiments of 1855 to the great *Chaucer* illustrations of 1896 – are, in their primitive strength, more closely related to the medieval woodcut than to the post-Bewick engraving.[22]

Morris and Burne-Jones were occasionally capable of responding sympathetically to modern wood-engravings. The publication of

Figure 14 Though Morris praised the wood-engravings of Fred Walker (such as this one published in the *Cornhill Magazine* in 1863), he felt that their delicate lines, cross-hatching, and extensive use of shading made them unsuitable as book illustrations.

Edward Moxon's illustrated edition of Tennyson in 1857, for example, caused a great stir in Morris's circle, with Rossetti's contributions evoking special admiration. Rossetti himself complained bitterly that his designs had been ruined by the engravers, the Dalziel Brothers, whom Morris later called 'as big humbugs as any within the narrow seas'.[22] Burne-Jones, however, remarked more shrewdly upon the weakness of the book when, in 1896, he compared it with the Kelmscott *Chaucer*: 'As a book . . . it's nothing. There was no command over the type and printing such as Mr. Morris has, and there were so many hands engaged on the pictures as to make it impossible as a book.'[24]

This was to be in later years the great central tenet of Morris's creed – the necessity of harmony in all aspects of a book, particularly between text and illustration – but it was one which he was capable of formulating only after much unhappy experience. In fact, his interest in the book as a physical object flickered on and off for several decades. Morris seems to have been consulted about such matters from time to time by Rossetti, who drew a highly successful title-page for his sister's *Goblin-Market* (1862) and designed attractive bindings for his own and others' books. Burne-Jones continued to contribute occasional illustrations to books and periodicals. One day in the early 1870s Morris is reported to have picked up a volume printed by Aldus Manutius in about 1490 and to have remarked as he turned over the pages, 'Ah! I wish I could get my books printed like that!'[25] Surrounded by friends with a professional interest in book-production, Morris, as both a writer and an artist, felt much drawn to printing. On the whole, though, he resisted the temptation: one senses that, when the Kelmscott Press was established in 1891, Morris was able to throw himself into it with such extraordinary pent-up energy because it was something he had longed to do for decades.

In the meantime there were several forays into the world of book-design which deserve to be described in more detail, since in them we catch glimpses of Morris working his way, tentatively and circuitously, towards the goal, as yet unseen, of the Kelmscott Press. These episodes include the unsuccessful attempt to produce an illustrated edition of *The Earthly Paradise* in the 1860s; the abandoned ornaments and illustrations for *Love is Enough* in 1871; Morris's calligraphic experiments, mostly dating from the 1870s; Morris's editorship of the

Commonweal in the 1880s; and the last three books designed by Morris – *The House of the Wolfings* (1889), *The Roots of the Mountains* (1890), and *The Story of Gunnlaug the Worm-Tongue* (1891) – before the opening of the Kelmscott Press.

Morris's first volume of verse, *The Defence of Guenevere*, was published by Bell and Daldy in 1858 and printed at the Chiswick Press. The typography was neat and inoffensive – precisely that sort of bland mid-nineteenth-century printed page that was later to make Morris apoplectic – ornamented only by the usual Chiswick Press wood-engraved headpieces and decorated initials cut by Mary Byfield. When in the following decade Morris met F. S. Ellis, the antiquarian bookseller and publisher who was to play such a key role in the Kelmscott Press, he decided to allow Ellis to publish his books; and thus Morris's early connection with the Chiswick Press was broken, not to be renewed until the 1880s.

Morris's great literary undertaking of the 1860s was a massive collection of poetic tales grouped together under the title *The Earthly Paradise*. (*The Life and Death of Jason*, published in 1867, was originally conceived as a part of it.) Morris decided, with typical bravado, to illustrate it with hundreds of wood-engravings designed by Burne-Jones. 'At dinner William Morris,' William Allingham recorded in his diary on 1 August 1866. 'The Big Story Book, woodcut of Olympus by N. Jones. M. and friends intend to engrave the wood-blocks themselves – and M. will publish the book at this warehouse.' A few weeks later Allingham noted that Burne-Jones 'founds his style . . . on old Woodcuts, especially those in *Hypnerotomachia*, of which he has a fine copy'.[26] At first Morris turned to the services of one or more professional wood-engravers;[27] he approached also George Wardle, who engraved two blocks, 'but after that I got no more,' Wardle recalled, 'and wondered for a while why, as I thought the second was certainly better than the first. The reason was, Mr. Morris became possessed by the idea of cutting the blocks himself: and he took them all in hand and carried them through, not without some lively scenes in Queen Square. He cut with great ardour and with much knowledge, but the work did not always go to his mind. It was necessarily slow and he was constitutionally quick. . . .'[28] Wardle's testimony is confirmed by an amusing Burne-Jones caricature of Morris at work on a wood-engraving for *The Earthly Paradise*: he is shown bent over the block with a look of ferocious intensity, while engraving tools lie scattered about the table and floor in splendid disarray. George Campfield, the *Figure 15*

Figure 15 A caricature by Edward Burne-Jones of Morris attempting to engrave a woodblock for *The Earthly Paradise*.

foreman of Morris and Company, and Lucy Faulkner also lent a hand (Fairfax Murray claimed that one block was engraved by Elizabeth Burden, Jane Morris's sister[29]), but it soon became obvious to Morris that the project could not be completed quickly. When a premature announcement of Morris's plans appeared in the *Athenaeum*, he wrote to that journal, 'The time of publication . . . of this illustrated edition must, from the magnitude of the work, be very remote'.[30]

Figure 16 Fortunately ten trial pages of the book have survived (now in the Morgan Library, though they had dropped out of sight for many years), and these give us an accurate notion of the kind of volume Morris had in mind. Eight of the pages are set in Caslon type, with a long 's' and italic side-notes (of which Morris proved to be very fond during the Kelmscott Press years); two of the pages are set in a type called Basle Roman which was available only at the Chiswick Press. (Morris was to turn to Basle Roman again in *the Roots of the Mountains* and *The House of the Wolfings*.) In both groups of pages the titles are in Caslon, since Basle Roman had been cut in only one size, and Mary Byfield's Renaissance-vintage initials make another appearance. The format in each instance is double-column folio: the leaves set in Caslon are $14\frac{1}{8} \times 9\frac{1}{2}$ inches, and those set in Basle Roman are $14\frac{3}{8} \times 9\frac{5}{8}$ inches. Since the first edition (an unillustrated octavo) of *The Earthly Paradise* contains 1,220 pages of text, this projected illustrated version would no doubt have been a stupendously large book. It seems fair to assume that

Morris abandoned the book primarily because its production (especially the engraving of the blocks) would have proved interminable. Furthermore, the trial pages reveal another serious problem, the incompatibility between the typefaces and the illustrations (not to mention Miss Byfield's initials). The heavy, bold wood-engravings overpower the text, even when it is set in the slightly darker Basle Roman.

What this unsuccessful experiment must have demonstrated to Morris was that the revival of fine printing could not be brought about by the improvement of illustrations alone. At the conclusion of a lecture delivered in 1892, Morris summarized the lesson that he himself had learned, at least partially, in the 1860s: 'An illustrated book, where the illustrations are more than mere illustrations of the printed text, should be a harmonious work of art. The type, the spacing of the type, the position of the pages of print on the paper, should be considered from the artistic point of view. The illustrations should not have a mere accidental connection with the other ornaments and the type, but an essential and artistic connection. . . . This is the only possible way in which you can get beautiful books.'[31]

Morris apparently began cutting illustrations, of which he completed about thirty-seven or thirty-eight, for *The Earthly Paradise* in November 1865; the project was definitely put aside with the publication, in April 1868, of the first volume of the poem by F. S. Ellis.[32] Its only illustration was a small wood-engraving of three women playing on musical instruments, and scholars are uncertain whether the engraving was designed by Burne-Jones or Morris. Yet the great illustrated edition of *The Earthly Paradise*, though it never materialized in exactly the form envisaged by Morris, was to have a curious after-history. With the help of Walter Crane, Burne-Jones, who often borrowed from his own earlier work, painted in 1872, a 'Cupid and Psyche' frieze, based on some of the *Earthly Paradise* designs, at No. 1 Palace Green, London, the Earl of Carlisle's town house.[33] In 1897 there were plans afoot to publish the 'Cupid and Psyche' section of the poem, set in Troy type, with Burne-Jones's illustrations; a contract was drawn up with Longmans, Green, and seven trial pages were printed, but the undertaking was abandoned with Burne-Jones's death the following year. In 1935 the Gregynog Press published an edition of Robert Bridges's poem *Eros and Psyche* illustrated by newly cut engravings based on Burne-Jones's designs. Despite the assistance of Robert Catterson-Smith, who had helped to 'translate' Burne-Jones's

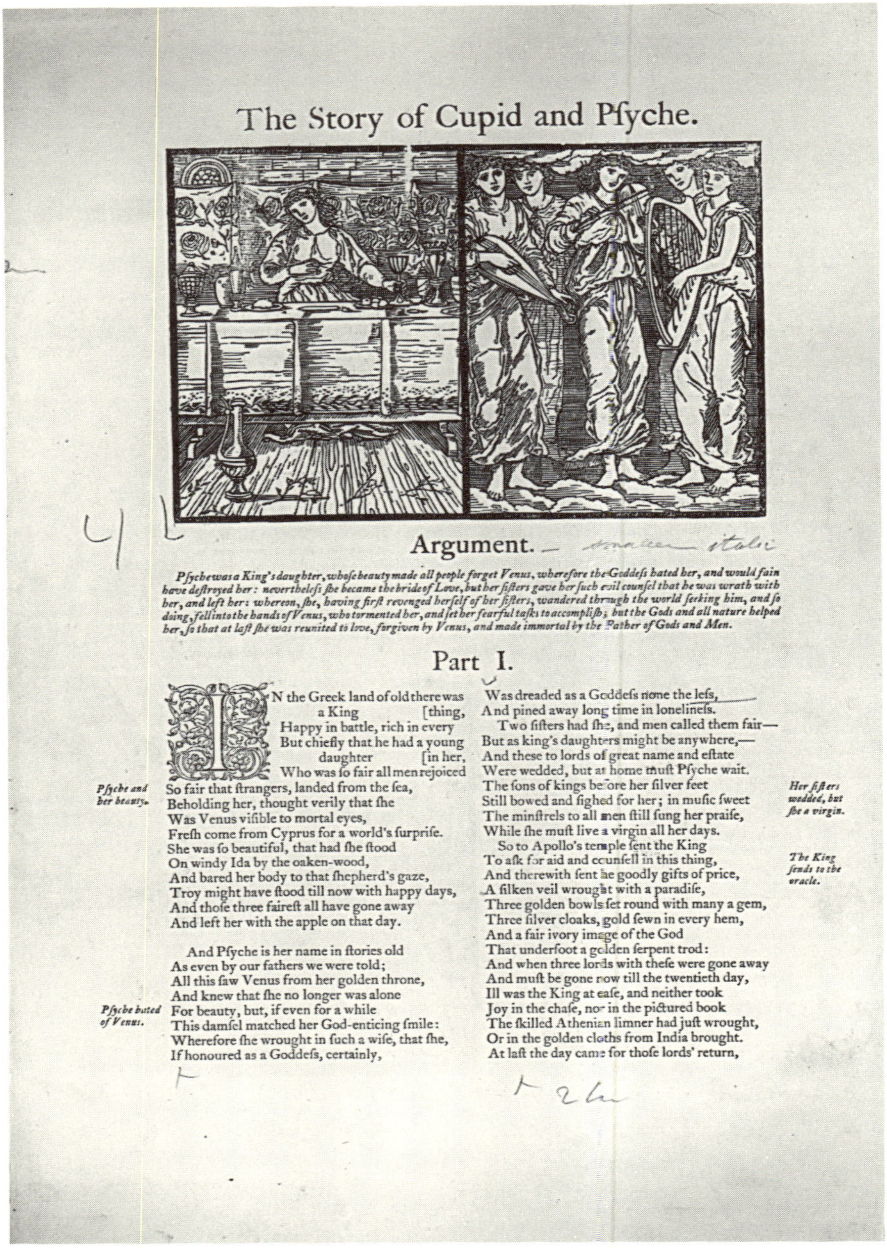

The Story of Cupid and Pſyche.

Argument.

Pſyche was a King's daughter, whoſe beauty made all people forget Venus, wherefore the Goddeſs hated her, and would fain have deſtroyed her: nevertheleſs ſhe became the bride of Love, but her ſiſters gave her ſuch evil counſel that he was wrath with her, and left her: whereon, ſhe, having firſt revenged herſelf of her ſiſters, wandered through the world ſeeking him, and ſo doing, fell into the hands of Venus, who tormented her, and ſet her fearful taſks to accompliſh; but the Gods and all nature helped her, ſo that at laſt ſhe was reunited to love, forgiven by Venus, and made immortal by the Father of Gods and Men.

Part I.

IN the Greek land of old there was a King
Happy in battle, rich in every thing,
But chiefly that he had a young daughter
Who was ſo fair all men rejoiced in her,
Pſyche and her beauty. So fair that ſtrangers, landed from the ſea,
Beholding her, thought verily that ſhe
Was Venus viſible to mortal eyes,
Freſh come from Cyprus for a world's ſurpriſe.
She was ſo beautiful, that had ſhe ſtood
On windy Ida by the oaken-wood,
And bared her body to that ſhepherd's gaze,
Troy might have ſtood till now with happy days,
And thoſe three faireſt all have gone away
And left her with the apple on that day.

And Pſyche is her name in ſtories old
As even by our fathers we were told;
All this ſaw Venus from her golden throne,
And knew that ſhe no longer was alone
Pſyche hated of Venus. For beauty, but, if even for a while
This damſel matched her God-enticing ſmile:
Wherefore ſhe wrought in ſuch a wiſe, that ſhe,
If honoured as a Goddeſs, certainly,
Was dreaded as a Goddeſs none the leſs,
And pined away long time in lonelineſs.
Two ſiſters had ſhe, and men called them fair—
But as king's daughters might be anywhere,—
And theſe to lords of great name and eſtate
Were wedded, but at home muſt Pſyche wait.
The ſons of kings before her ſilver feet *Her ſiſters wedded, but ſhe a virgin.*
Still bowed and ſighed for her; in muſic ſweet
The minſtrels to all men ſtill ſung her praiſe,
While ſhe muſt live a virgin all her days.
So to Apollo's temple ſent the King
To aſk for aid and counſell in this thing, *The King ſends to the oracle.*
And therewith ſent he goodly gifts of price,
A ſilken veil wrought with a paradiſe,
Three golden bowls ſet round with many a gem,
Three ſilver cloaks, gold ſewn in every hem,
And a fair ivory image of the God
That underfoot a golden ſerpent trod:
And when three lords with theſe were gone away
And muſt be gone now till the twentieth day,
Ill was the King at eaſe, and neither took
Joy in the chaſe, nor in the pictured book
The ſkilled Athenian limner had juſt wrought,
Or in the golden cloths from India brought.
At laſt the day came for thoſe lords' return,

Figure 16 One of ten surviving trial pages for an illustrated edition of *The Earthly Paradise* that Morris worked on during the 1860s, though the book was never printed in this form. The typeface is Caslon, and the wood-engravings are designed by Burne-Jones.

drawings for the Kelmscott *Chaucer*, the engravings struck many readers as feeble and emasculated.

Finally, in 1974 a portion of *The Earthly Paradise* was published in a style that Morris himself would probably have admired. Using the original wood-engravings (owned by the Society of Antiquaries) and Morris's Troy type, Will and Sebastian Carter, at the Rampant Lions Press in Cambridge, produced an illustrated edition of *The Story of Cupid and Psyche* which represents the kind of book Morris wanted to print in 1868 but was unable to achieve because of the typographical limitations of the day.[34]

'[Morris] makes a poem these days,' wrote Burne-Jones in 1871, '– in dismal Queen Square in black old filthy London in dull end of October he makes a pretty poem that is to be wondrously happy; and it has four sets of lovers in it, and THEY ARE ALL HAPPY and it ends well, and will come out some time next summer and I shall make little ornaments to it. . . .'[35] What Morris had in mind for *Love is Enough* (*Figure 17*) was more ambitious than the phrase 'little ornaments' might suggest, though he certainly was now thinking on a smaller scale than in the case of *The Earthly Paradise*. In addition to wood-engraved ornaments and illustrations, Morris intended that the pages should be decorated with much hand-illumination. (One can see, very imperfectly, what Morris had in mind in Harry Buxton Forman's copy – now in the Berg Collection of the New York Public Library – which has an illuminated title executed by Coventry Patmore's daughter in 1889. Was Morris ever shown this copy?) Morris designed a series of ornamental initial 'L's (which are much more delicate than the initials that were to appear in the Kelmscott books) and several borders closer in style to his calligraphic work than to the Kelmscott Press borders; Burne-Jones produced one or more borders and at least one illustration. The ornaments were all engraved by Morris. Trial pages, probably struck off in 1872, betray the same weakness that beset the *Earthly Paradise* trials, perhaps even more acutely in this instance: the pages are set in a conventional modern-face type that seems weak and enervated compared with Morris's vigorous ornaments.[36] Like *The Earthly Paradise*, the book was published without decoration in 1873, yet Morris had a second opportunity to realize the right typographical form for it when he issued *Love is Enough* at the Kelmscott Press. He died, however, before it came off the press in 1898, and the Kelmscott

Love is Enough.

THE MUSIC.

LOVE is enough: though the world
be a-waning,
And the woods have no voice but
the voice of complaining;
Though the sky be too dark for
dim eyes to discover
The gold-cups and daisies fair
blooming thereunder,
Though the hill be held shadows,
and the sea a dark wonder,
And this day draw a veil over all deeds passed over,
Yet their hands shall not tremble, their feet shall not
falter;
The void shall not weary, the fear shall not alter
These lips and these eyes of the loved and the lover.

THE EMPEROR.

The spears flashed by me, and the spears swept round,
And in war's hopeless tangle was I bound;
But straw and stubble were the cold points found,
For still thy hands led down the weary way.

THE EMPRESS.

Through hall and street they led me as a queen,
They looked to see me proud and cold of mien;
I heeded not though all my tears were seen,
For still I dreamed of thee throughout the day.

THE EMPEROR.

Wild over bow and bulwark swept the sea
Unto the iron coast upon our lee;
Like painted cloth its fury was to me,
For still thy hands led down the weary way.

Figure 17 Trial page of an ornamented edition of Morris's *Love is Enough* that he planned but failed to complete in about 1870. The initial and ornament on the left are designed by Morris; the crnament on the right is by Burne-Jones.

Figure 18 The handsome binding of *Love is Enough* (1872), designed by Morris, contrasts with the plain typography of the text.

edition, splendid though it is in three colours, would surely have been even more fully ornamented had Morris been able to see it through to completion. One of Burne-Jones's two illustrations was a redrawing of a design he had first done in 1872. Rather more unexpectedly, one of the ornamented initials for the unfinished *Love is Enough* crops up in a trial page for the Kelmscott Press *Syr Perecyvelle of Gales* (1895);[37] the initial of course does not appear in the published book, and none of the *Love is Enough* ornaments was ever used by Morris in print.

Figure 18 The undistinguished typography of the 1873 *Love is Enough* is in marked contrast to its handsome binding designed by Morris. The title appears on the green cloth front cover in generously spaced gilt Roman capitals (not dissimilar to the lettering used subsequently on many of the Kelmscott Press engraved title-pages) within a striking willow- and myrtle-pattern frieze.[38] Morris from time to time professed indifference to binding as a 'minor' art (thus greatly distressing Cobden-Sanderson on one occasion), but this was rather a pose. He was also responsible for another publisher's binding for a new edition of *The Earthly Paradise* (1890), and about 1870 he designed four tools for gold-tooled leather bindings. These were used on a sketch-book owned by Charles Fairfax Murray and on a manuscript written and illuminated by Morris for Georgiana Burne-Jones.[39] The ornaments on the binding of the Saga Library likewise emanated from Morris.

Notwithstanding his intense (but sporadic) interest in printing, Morris was drawn even more strongly to the arts of calligraphy and illumination. When the first sheets of *The Golden Legend* (1892) came from the Kelmscott Press in May 1891, he was moved to remark: 'Pleased as I am with my printing, when I saw my two men at work on the press yesterday with their sticky printers ink, I couldn't help lamenting the simplicity of the scribe and his desk, and his black ink and blue and red ink, and I almost felt ashamed of my press after all.'[40] Morris often, to the surprise and amusement of his listeners, spoke disparagingly of printing, which he called a 'trifling' invention and of which he said he disapproved.[41] Much as he might admire the illustrated books of the fifteenth-century printers, Morris knew that they represented medieval art in a state of decline; to discover the Gothic world in its wholeness and freshness, one had to return to the illuminated manuscript, preferably of the thirteenth century. Not surprisingly, Morris during the 1870s, frustrated in his attempts to produce attractively printed books, resorted with increasing frequency to the writing and ornamenting of manuscripts.

A long, relaxed letter to Charles Fairfax Murray, written in March 1874, conveys the sense of delight Morris found in this new activity: 'I wrote a book (on paper confound it) of about 250 p.p. translations of unpublished Icelandic stories with pretty letters to each chapter, which looked well on the whole. . . . Now I am at work at an Odes of Horace which will make about 100 pp of vellum octavo of the big sheets: the odes are short so there is nearly an ornamental letter to every page, which makes it a heavyish piece of work; however I have written about half & done 20 letters. To say the truth I have a mind to try and sell a book if I could find a customer: I work much neater now, & have got I think more style in the ornament, & have taken rather to the Italian work of about 1450 for a type. . . . I am very keen on the thing just now, & really enjoy my Sundays very much at it, and as you may imagine am very like to neglect my other work for it. . . . I am rather thinking by the way of trying a Cupid & Psyche for the market, painting in some of the master's pictures with it which he [Burne-Jones] has given me leave to do: for belike the oddity of a poet illuminating his own poem might make it saleable. . . .'[42] (Morris did not systematically turn his work into a commercial enterprise; many of his manuscripts instead became delightful gifts to friends, particularly Georgiana Burne-Jones).

A casual glance at Morris's foray into writing and ornamenting manuscripts may suggest that he was merely following the Victorian fashion for 'illumination' that had first been stimulated by the publication of A. W. N. Pugin's *Glossary of Ecclesiastical Ornament and Costume* (1844). In reality, Morris, with great independence of mind, charted his own course: the similarities between his manuscripts and the illuminated Biblical texts and uplifting proverbs done by middle-class Victorian ladies for display on the walls of their parlours are not very conspicuous. Morris of course was aware of the popular revival of interest in medieval illumination that swept across England in the 1860s. He owned, for example, a copy of *The Art of Illuminating as Practised in Europe from the Earliest Times, Illustrated by Borders, Initial Letters, and Alphabets* (1860), with chromolithographs by W. R. Tymms and 'an Essay and Instructions' by M. D. Wyatt; and he was no doubt conscious of the other books on the subject flooding the market from writers such as Henry Noel Humphreys (*The Art of Illumination and Missal Painting*, 1849; *The Illuminated Books of the Middle Ages*, 1844–9), Freeman Gage Delamotte (*A Book of Ornamental Alphabets*, 1862; *Medieval Alphabets and Initials for Illuminators*, 1861; *A*

Primer of the Art of Illumination, 1860), and Henry Shaw (*A Handbook of the Art of Illumination as Practised during the Middle Ages*, 1866). (Morris also had in his library the 1870 edition of Shaw's book.) In 1861 there was even a London periodical entitled *The Amateur Illuminator's Magazine, and Journal of Miniature Painting.*

Morris's three earliest surviving manuscripts, dating from 1856, are written in a weak Gothic script reminiscent of the style of the parlour illuminators, but in the third – a fragment of 'The Iron Man', a translation of Grimm's 'Der Eisenhans' – there is a fine historiated initial. Though these early efforts by Morris seem undistinguished in comparison with his later manuscripts, they made a strong impression on his contemporaries. 'In all illumination and work of that kind, he is quite unrivalled by anything modern that I know,' Rossetti wrote in December 1856; Ruskin declared at about the same time that Morris's 'gift for illumination is I believe as great as any thirteenth century draughtsmans'.[43] Nearly all of Morris's more successful manuscripts, however, were done in the period between 1869 and 1875, and they display greater sophistication in both calligraphy and ornamentation. One explanation for this development, apart of course from Morris's increased maturity as an artist, is that he had in the meantime acquired four of the most important Renaissance writing manuals, all published in 1525 or 1526 and bound together in a single volume: *La operina di Ludovico Vicentino, da imparare di scrivere lettera cancellarescha* by Arrighi; *Il modo de temperare le penne*, also by Arrighi; *Lo presente libro insegna* by Giovantonio Tagliente; and *Thesauro de scrittori*, a compilation of pages from various manuals. The volume as a whole, Alfred Fairbank has observed, provides 'a remarkable collection of sixteenth-century writing and lettering manuals' and 'marks the beginning of the italic handwriting reform now engaging world-wide attention'.[44] There is in fact a direct line of descent from Morris to Edward Johnston, who initiated the modern calligraphic revival, for Johnston's career was effectively launched when Sydney Cockerell in 1898 described to him Morris's rediscovery of the older hands and showed him Morris's manuscripts.[45]

Fairbank lists eighteen manuscripts by Morris (not including trial pages, fragments, and a few more manuscripts that have turned up since he published his census). Naturally they vary in style and achievement, and it is difficult to examine all of them today because they are scattered about in many libraries. (The largest group is in the Bodleian.) Still, it is possible to offer a few broad generalizations about

THE TWO SIDES OF THE RIVER

THE YOUTHS

O winter, O white winter, wert thou gone
No more within the wilds were I alone
Leaping with bent bow over stock and stone!

No more alone my love the lamp should burn,
Watching the weary spindle twist and turn,
Or o'er the web hold back her tears and yearn:
O winter, O white winter, wert thou gone!

THE MAIDENS

Sweet thoughts fly swiftlier than the drifting snow,
And with the twisting thread sweet longings grow,
And o'er the web sweet pictures come and go
For no white winter are we long alone.

Figure 19 One of Morris's more attractive calligraphic experiments from the 1870s is *A Book of Verse*, a collection of his poems intended as a birthday gift for Georgiana Burne-Jones. The illustrations are by Edward Burne-Jones, most of the ornaments by Morris.

Morris's manuscripts of 1869–75. All are in either a cursive which owes something to the writing manuals or a more formal Roman. This is surprising, given Morris's known predilection for Gothic hands and types; he appears never to have explained to others what reasons lay behind such an unexpected choice. It must also be said that Morris's writing, though accomplished for its period, may strike us today as slightly primitive. Early in our century Graily Hewitt, when shown by May Morris quills cut by her father, concluded that they were not well cut, and it is obvious that on some occasions he used crow quills, which do not provide a broad nib, rather than goose quills.[46] The ornamentation, which was supplied by both Morris and others, is often delightful and fresh: Morris was a master of decoration, as the Kelmscott Press books were also to demonstrate. Yet, as in the Kelmscott volumes, one feels that at times the ornamentation is so out of control that it threatens to crush the text. Morris wrote on both paper and vellum; the latter was sent to him from Rome by Fairfax Murray, who posted him such a large supply that in 1891 there was still enough left to print six copies of *The Story of the Glittering Plain*, the first Kelmscott Press book.

That Morris's skill as calligrapher grew rapidly during the 1870s can be seen by looking more closely at two manuscripts from the beginning and end of the 1869–75 period. *Figure 19* *A Book of Verse* (1870; in the Victoria and Albert Museum), an anthology of some of Morris's shorter poems (many of which were to be published in the Kelmscott Press *Poems by the Way* (1891)) intended as a birthday gift for Georgiana Burne-Jones, is written in a decent but unremarkable cursive; what lends a special charm to the manuscript are the clusters of graceful leaves and flowers by Morris that crawl up and down the margins and penetrate the blank lines of the text. Morris's colophon reflects the degree to which many of his manuscripts were collaborative productions: 'As to those who have had a hand in making this book, Edward Burne Jones painted the picture [of Morris] on Page 1: the other pictures were all painted by Charles F Murray, but the minstrel figures on the title-page, and the figures of Spring[,] Summer and Autumn on page 40, he did from my drawings. As to the pattern-work, George Wardle drew in all the ornament on the first ten pages, and I coloured it; he also did all the coloured letters both big and little; the rest of the ornament I did, together with all the writing.'

Figure 20 In 1874 Morris began translating the *Aeneid*, which led him in turn to produce an ornamented manuscript of the Latin text (1875;

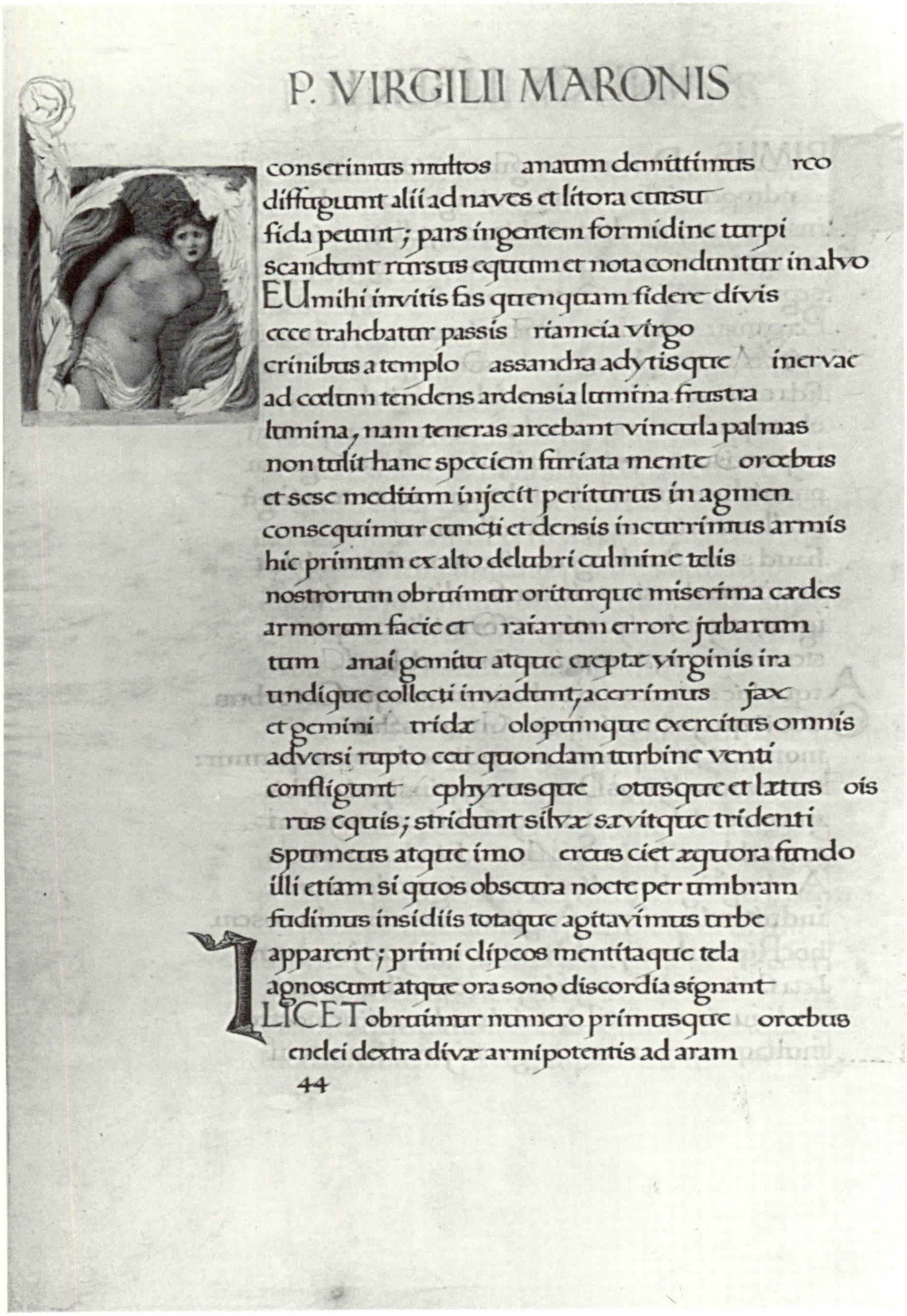
P. VIRGILII MARONIS

conserimus multos anaum demittimus rco
diffugiunt alii ad naves et litora cursu
fida petunt; pars ingentem formidine turpi
scandunt rursus equum et nota conduntur in alvo
EU mihi invitis fas quemquam fidere divis
ecce trahebatur passis riameia virgo
crinibus a templo assandra adytisque inervae
ad coelum tendens ardensia lumina frustra
lumina, nam teneras arcebant vincula palmas
non tulit hanc speciem furiata mente oroebus
et sese medium injecit periturus in agmen
consequimur cuncti et densis incurrimus armis
hic primum ex alto delubri culmine telis
nostrorum obruimur oriturque miserima cædes
armorum facie et raiarum errore jubarum
tum anai gemitu atque ereptæ virginis ira
undique collecti invadunt, acerrimus jax
et gemini tridæ olopumque exercitus omnis
adversi rupto ceu quondam turbine venti
confligunt ephyrusque otusque et lætus ois
rus equis; stridunt silvæ sævitque tridenti
spumeus atque imo ereus ciet æquora fundo
illi etiam si quos obscura nocte per umbram
fudimus insidiis totaque agitavimus urbe
apparent; primi clipeos mentitaque tela
agnoscunt atque ora sono discordia signant
LICET obruimur numero primusque oroebus
enelei dextra divæ armipotentis ad aram
44

Figure 20 'It is to be wonderful and put an end to printing,' Burne-Jones wrote of the *Aeneid* (1875), ornamented by himself and written by Morris.

formerly in the Doheny Memorial Library, St John's Seminary, Camarillo, Calif.).[47] Like many of his manuscripts, the *Aeneid* was left unfinished by Morris, who was an exceptionally busy man and was also notoriously impatient when confronted with time-consuming tasks. But before he abandoned the *Aeneid* Morris had completed 177 vellum pages, approximately half the text, in a fine, broad Roman hand; he also painted some decorated initials that are as accomplished as any he did before the founding of the Kelmscott Press. Burne-Jones began designing a group of illustrations (now in the Fitzwilliam Museum, Cambridge) which directly prefigure the Kelmscott *Chaucer* engravings. 'Every Sunday morning', he commented in a letter, 'you may think of Morris and me together – he reads a book to me and I make drawings for a big Virgil he is writing – it is to be wonderful and put an end to printing.'[48] Eventually Morris sold the manuscript to Fairfax Murray, who later persuaded Graily Hewitt to complete it.

It is difficult to assess how much Morris's experiments in calligraphy and illumination may have affected his subsequent work at the Kelmscott Press. That the experience gave him valuable knowledge of Roman letterforms is undeniable, and anyone who is familiar with Morris's handling of borders and initials in the Kelmscott books will occasionally experience a shock of recognition as he turns over the leaves of the manuscripts. The main lesson to be drawn from Morris's calligraphic achievements, however, is that, paradoxically, the Kelmscott Press (which has drawn such superlatives from bibliophiles and printing historians) was undertaken by Morris almost grudgingly. His heart always lay with the manuscript rather than the printed book. Among the proofs and trial pages from the Press that are treasured today in library special-collections rooms, there are numerous little scraps of paper that betray Morris's secret hankering to abandon the printing press in favour of the pen: repeatedly he experimented with hand-coloured initials, and when these ideas come to nothing, he printed a few copies of some of the Kelmscott books with blank spaces for initials (just as the earliest printers had done) to be supplied by other calligraphers. Morris was not joking, merely exaggerating slightly, when he suggested that printing was a trivial invention. It was redeemed in his eyes only because it had – just barely – a medieval origin, and because, as he said, 'a book, printed or written, has a tendency to be a beautiful object'.[49]

Between 1885 and 1890 Morris was much absorbed again in printing, this time as editor of the *Commonweal*, the journal of the Socialist League. The *Commonweal* is a typographically unimpressive periodical, set in modern face and enlivened only by a wood-engraved, ornamented title designed and cut by George Campfield and two double-column headings designed by Walter Crane. Morris himself claimed that his work on the *Commonweal* contributed to his technical education, for he told a reporter in 1893 that 'in the course of my life I had obtained a good deal of knowledge of type. Particularly I was much among type when I was editor of the *Commonweal*.'[50]

Figure 21

Morris's connection with the *Commonweal* during this decade reflects his growing preoccupation with Marxism, which seemed to Burne-Jones, Wilfred Scawen Blunt, and some of his other friends an unfortunate aberration – 'a parenthesis' in his life as a poet and artist, as Burne-Jones described it.[51] Yet such a view fails to take into account how deeply intertwined Morris's social and aesthetic theories were. In an autobiographical sketch written late in life, Morris reflected that, 'in spite of all the success I have had, I have not failed to be conscious that the art I have been helping to produce would fall with the death of a few of us who really care about it, that a reform in art which is founded on individualism must perish with the individuals who have set it going. Both my historical studies and my practical conflict with the philistinism of modern society have *forced* on me the conviction that art cannot have a real life and growth under the present system of commercialism and profit-mongering.'[52] The central paradox of Morris's career, of which he was always painfully aware, was that, in attempting to restore the egalitarian basis of art, he unexpectedly found himself producing expensive objects for the wealthy. The opulent Kelmscott Press books were merely the final example of this dilemma.

There was, however, one final step before Morris decided to establish his own Press. Turning again to the Chiswick Press, which had produced his earliest books, he designed and closely supervised the printing of three volumes in which we can see him moving, now with greater assurance, towards the typographical principles of the Kelmscott Press. The Chiswick Press in 1889 was under the direction of Charles T. Jacobi, a man of sensible but cautious views on the printing of books.[53] Though to Stanley Morison's generation he seemed a boring old fogy,[54] at the end of the nineteenth century he was one of the most knowledgeable persons in the London printing world.

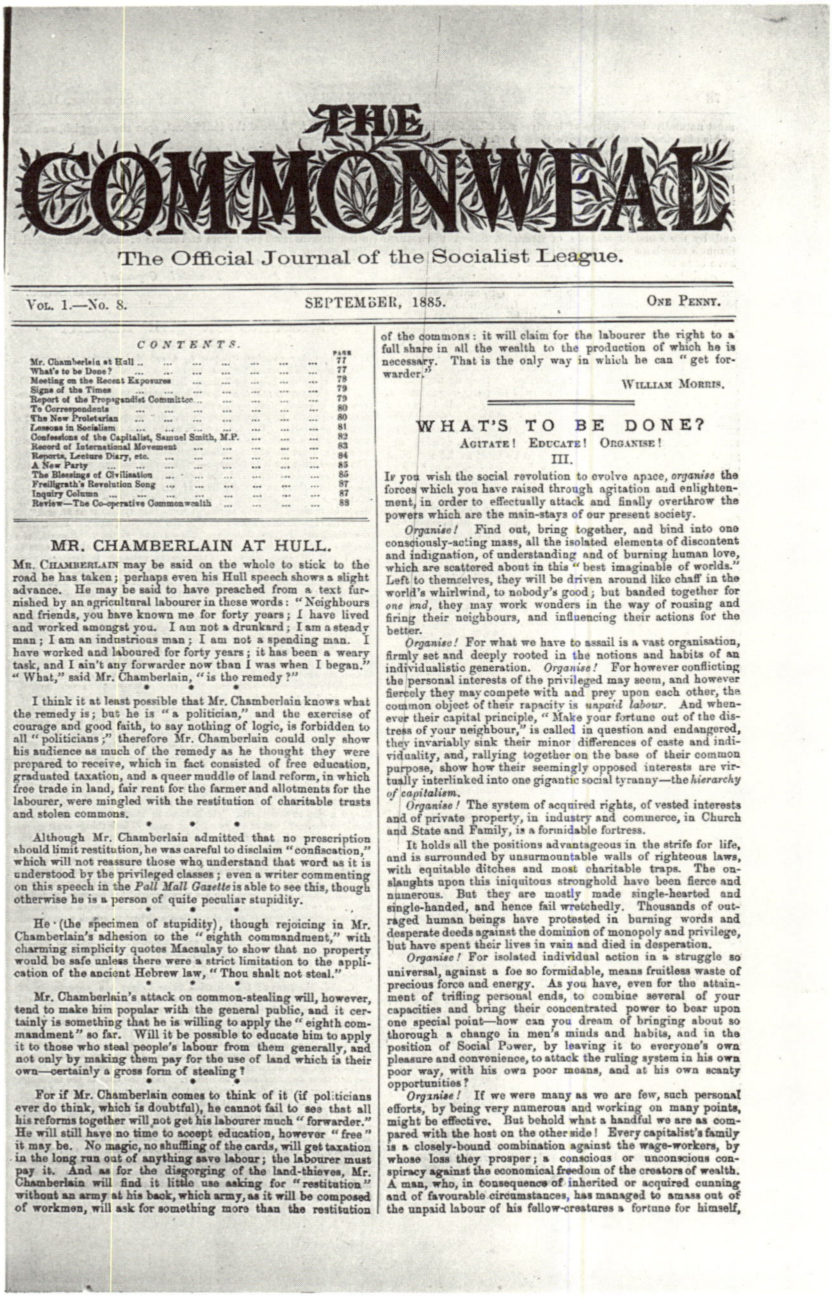

THE COMMONWEAL

The Official Journal of the Socialist League.

Vol. 1.—No. 8. SEPTEMBER, 1885. One Penny.

CONTENTS.

MR. CHAMBERLAIN AT HULL.

Mr. Chamberlain may be said on the whole to stick to the road he has taken; perhaps even his Hull speech shows a slight advance. He may be said to have preached from a text furnished by an agricultural labourer in these words: "Neighbours and friends, you have known me for forty years; I have lived and worked amongst you. I am not a drunkard; I am a steady man; I am an industrious man; I am not a spending man. I have worked and laboured for forty years; it has been a weary task, and I ain't any forwarder now than I was when I began." "What," said Mr. Chamberlain, "is the remedy?"

• • •

I think it at least possible that Mr. Chamberlain knows what the remedy is; but he is "a politician," and the exercise of courage and good faith, to say nothing of logic, is forbidden to all "politicians;" therefore Mr. Chamberlain could only show his audience as much of the remedy as he thought they were prepared to receive, which in fact consisted of free education, graduated taxation, and a queer muddle of land reform, in which free trade in land, fair rent for the farmer and allotments for the labourer, were mingled with the restitution of charitable trusts and stolen commons.

• • •

Although Mr. Chamberlain admitted that no prescription should limit restitution, he was careful to disclaim "confiscation," which will not reassure those who understand that word as it is understood by the privileged classes; even a writer commenting on this speech in the *Pall Mall Gazette* is able to see this, though otherwise he is a person of quite peculiar stupidity.

• • •

He (the specimen of stupidity), though rejoicing in Mr. Chamberlain's adhesion to the "eighth commandment," with charming simplicity quotes Macaulay to show that no property would be safe unless there were a strict limitation to the application of the ancient Hebrew law, "Thou shalt not steal."

• • •

Mr. Chamberlain's attack on common-stealing will, however, tend to make him popular with the general public, and it certainly is something that he is willing to apply the "eighth commandment" so far. Will it be possible to educate him to apply it to those who steal people's labour from them generally, and not only by making them pay for the use of land which is their own—certainly a gross form of stealing?

• • •

For if Mr. Chamberlain comes to think of it (if politicians ever do think, which is doubtful), he cannot fail to see that all his reforms together will not get his labourer much "forwarder." He will still have no time to accept education, however "free" it may be. No magic, no shuffling of the cards, will get taxation in the long run out of anything save labour; the labourer must pay it. And as for the disgorging of the land-thieves, Mr. Chamberlain will find it little use asking for "restitution" without an army at his back, which army, as it will be composed of workmen, will ask for something more than the restitution of the commons: it will claim for the labourer the right to a full share in all the wealth to the production of which he is necessary. That is the only way in which he can "get forwarder."

William Morris.

WHAT'S TO BE DONE?

Agitate! Educate! Organise!

III.

If you wish the social revolution to evolve apace, *organise* the forces which you have raised through agitation and enlightenment, in order to effectually attack and finally overthrow the powers which are the main-stays of our present society.

Organise! Find out, bring together, and bind into one consciously-acting mass, all the isolated elements of discontent and indignation, of understanding and of burning human love, which are scattered about in this "best imaginable of worlds." Left to themselves, they will be driven around like chaff in the world's whirlwind, to nobody's good; but banded together for *one end*, they may work wonders in the way of rousing and firing their neighbours, and influencing their actions for the better.

Organise! For what we have to assail is a vast organisation, firmly set and deeply rooted in the notions and habits of an individualistic generation. *Organise!* For however conflicting the personal interests of the privileged may seem, and however fiercely they may compete with and prey upon each other, the common object of their rapacity is *unpaid labour*. And whenever their capital principle, "Make your fortune out of the distress of your neighbour," is called in question and endangered, they invariably sink their minor differences of caste and individuality, and, rallying together on the base of their common purpose, show how their seemingly opposed interests are virtually interlinked into one gigantic social tyranny—the *hierarchy of capitalism*.

Organise! The system of acquired rights, of vested interests and of private property, in industry and commerce, in Church and State and Family, is a formidable fortress.

It holds all the positions advantageous in the strife for life, and is surrounded by unsurmountable walls of righteous laws, with equitable ditches and most charitable traps. The onslaughts upon this iniquitous stronghold have been fierce and numerous. But they are mostly made single-hearted and single-handed, and hence fail wretchedly. Thousands of outraged human beings have protested in burning words and desperate deeds against the dominion of monopoly and privilege, but have spent their lives in vain and died in desperation.

Organise! For isolated individual action in a struggle so universal, against a foe so formidable, means fruitless waste of precious force and energy. As you have, even for the attainment of trifling personal ends, to combine several of your capacities and bring their concentrated power to bear upon one special point—how can you dream of bringing about so thorough a change in men's minds and habits, and in the position of Social Power, by leaving it to everyone's own pleasure and convenience, to attack the ruling system in his own poor way, with his own poor means, and at his own scanty opportunities?

Orgznise! If we were many as we are few, such personal efforts, by being very numerous and working on many points, might be effective. But behold what a handful we are as compared with the host on the other side! Every capitalist's family is a closely-bound combination against the wage-workers, by whose loss they prosper; a conscious or unconscious conspiracy against the economical freedom of the creators of wealth. A man, who, in consequence of inherited or acquired cunning and of favourable circumstances, has managed to amass out of the unpaid labour of his fellow-creatures a fortune for himself,

Figure 21 During the decade before he established the Kelmscott Press, Morris devoted much of his time to editing the *Commonweal*, a socialist journal. Except for the wood-engraved title (designed and cut by George Campfield), it was typographically undistinguished.

'My own recollection of [Morris] goes as far back as 1867, when the Chiswick Press printed the first edition of his "Life and Death of Jason",' Jacobi wrote later. 'It was not until 1888 that I was brought into more personal contact with him, when I superintended the printing of his "A Tale of the House of Wolfings," in 1889; "The Roots of the Mountains," in 1890; and also the "Gunnlaug Saga," in 1891. These were all printed in the Chiswick Press special types whilst he was considering and formulating his plans for starting his own press at Hammersmith. . . .'[55]

Of four trial pages of *The House of the Wolfings*,[56] one is set in Caslon Old Face and another in Miller and Richard's Old Style type; but for both this book and *The Roots of the Mountains* Morris adopted instead Basle Roman (with the titles in Caslon), which he had used earlier in some of the *Earthly Paradise* trial pages. Basle Roman, cast for Charles Whittingham by William Howard's foundry, is a darkly coloured, relatively narrow old-face type based on a fount used by Johann Froben in Switzerland in the early sixteenth century. Printing historians have usually assumed that it was first employed in the Revd W. Calvert's *Wife's Manual* (1854), but recent research has shown that it appears in the preliminaries of *Chronicles Selected from the Originals of Cartaphilus, the Wandering Jew*, edited by David Hoffman (1853–4).[57] Basle Roman is not likely to impress most readers today as an attractive face. Why, then, did Morris use it? We have already observed the strength of his atavistic impulses: if Caslon was superior to Victorian modern-face types, why should he not attempt to penetrate still farther into the past? When Morris began to design his own typefaces, he was not content until he had travelled all the way back to the fountainhead of typography, the fifteenth century. For the moment, he had to settle for the only commercially available replica of an earlier type (except for some facsimiles of Caxton's types, one of which he was to use in *Gunnlaug*), even if its model was too late for Morris's taste.

Figure 22

With the assistance of Jacobi and Emery Walker, Morris took pains to make *The House of the Wolfings* 'a pretty piece of typography for modern times', as he phrased it in a letter to his daughter Jenny.[58] Harry Buxton Forman, who happened to be at the Chiswick Press when Morris revised the title-page, was startled to see him add a second 'in' to the phrase 'written in prose and in verse' so that the line would be longer. To Forman's objection, Morris replies, 'Ha! now what would you say if I told you that the verses on the title-page were written just to fill up the great white lower half? Well, that was what happened!'[59]

A TALE OF THE HOUSE OF THE WOLFINGS AND ALL THE KINDREDS OF THE MARK WRITTEN IN PROSE AND IN VERSE BY WILLIAM MORRIS.

WHILES IN THE EARLY WINTER EVE
WE PASS AMID THE GATHERING NIGHT
SOME HOMESTEAD THAT WE HAD TO LEAVE
YEARS PAST ; AND SEE ITS CANDLES BRIGHT
SHINE IN THE ROOM BESIDE THE DOOR
WHERE WE WERE MERRY YEARS AGONE
BUT NOW MUST NEVER ENTER MORE,
AS STILL THE DARK ROAD DRIVES US ON.
E'EN SO THE WORLD OF MEN MAY TURN
AT EVEN OF SOME HURRIED DAY
AND SEE THE ANCIENT GLIMMER BURN
ACROSS THE WASTE THAT HATH NO WAY ;
THEN WITH THAT FAINT LIGHT IN ITS EYES
A WHILE I BID IT LINGER NEAR
AND NURSE IN WAVERING MEMORIES
THE BITTER-SWEET OF DAYS THAT WERE.

LONDON 1889: REEVES AND TURNER 196 STRAND.

Figure 22 Morris called *The House of the Wolfings*, which he carefully designed, 'a pretty piece of typography for modern times', and wrote the short poem especially for its visual effect on the title-page.

It is a revealing anecdote, for, even at a time when the Kelmscott Press had not yet taken shape, Morris was already displaying his horror of white space and his determination to fill the printed page, if not with ornaments, then at least with text. What is noteworthy about *The House of the Wolfings* is that its design is less self-consciously archaic than one might expect. The long 's' of Basle Roman was not used, and an 'e' with a conspicuous oblique cross-bar was abandoned by Morris (no doubt reluctantly) in *The Roots of the Mountains*. The binding, paper (machine-made laid), and treatment of the prelims are entirely conventional. There was also a large-paper edition, which Morris immediately regretted because it played havoc with the proportions of the margins and set the text adrift in a vast sea of white paper.

Morris's apprenticeship was rapidly drawing to a close. The moment *The House of the Wolfings* was off the press, he saw clearly that his attempt to darken the tone of the page had been subverted by the loose word-spacing which characterized nearly all Victorian typography. 'I am very glad that you like the new book,' he wrote to F. S. Ellis. 'I quite agree with you about the type; they have managed to knock the guts out of it somehow. Also I am beginning to learn something about the art of type-setting; and I now see what a lot of difference there is between the work of the conceited numskulls of to-day and that of the 15th and 16th century printers merely in the arrangement of the words, I mean the spacing out: it makes all the difference in the beauty of a page of print. If ever I print another book I shall enter into the conflict on this side also.'[60]

The design of *The Roots of the Mountains*, though broadly similar to that of its predecessor, shows one or two more daring touches. (Morris, alas, won only a partial victory in his battle with the Chiswick Press compositors for tighter word-spacing.) The most significant change is that the proportions of the margins are closer to what Morris regarded as desirable; this was achieved in part by discarding headlines in favour of side-notes. Emery Walker, who was beginning to influence Morris's typographical views, remarked in a lecture delivered in 1890 that 'many of [the] early books had no headlines. The natural way of setting type was in rectangles, and he thought it looked best when composed in that way. A headline often destroyed the effect of unity in the page, and sometimes it was quite unnecessary from the reader's point of view, therefore, in such cases, it might be desirable to leave it out.'[61] It was a lesson that Morris learned well: he never used a headline in a book again.

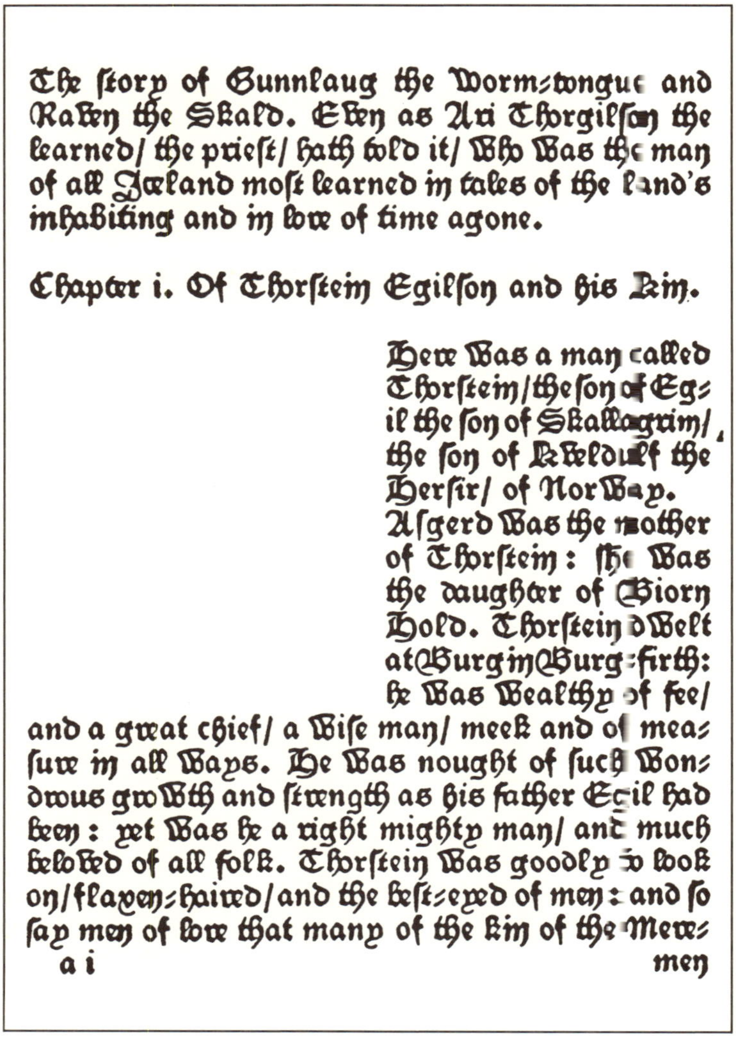

The story of Gunnlaug the Worm-tongue and Raven the Skald. Even as Ari Thorgilson the learned/ the priest/ hath told it/ who was the man of all Iceland most learned in tales of the land's inhabiting and in lore of time agone.

Chapter i. Of Thorstein Egilson and his Kin.

Here was a man called Thorstein/ the son of Egil the son of Skallagrim/ the son of Kveldulf the Hersir/ of Norway. Asgerd was the mother of Thorstein: she was the daughter of Biorn Hold. Thorstein dwelt at Burg in Burg-firth: he was wealthy of fee/ and a great chief/ a wise man/ meek and of measure in all ways. He was nought of such wondrous growth and strength as his father Egil had been: yet was he a right mighty man/ and much beloved of all folk. Thorstein was goodly to look on/ flaxen-haired/ and the best-eyed of men: and so say men of lore that many of the kin of the Mere-

a i men

Figure 23 Morris's last typographical experiment before the founding of the Kelmscott Press was *The Story of Gunnlaug the Worm-Tongue*, an exercise in antiquarianism privately printed for him at the Chiswick Press in a replica of one of Caxton's types.

Instead of a large-paper edition, Morris arranged for 250 copies of *The Roots of the Mountains* to be printed on Whatman paper, which he described as 'the only good paper at present'.[62] (Some of the remaining stock was to be used by Morris in the earliest advertising circulars of the Kelmscott Press.) Such copies were bound in a specially calendared chintz manufactured by Morris and Company ('It looks so nice and such fun,' he said[63]) – an experiment that must have ultimately disappointed Morris, since he did not repeat it. Nevertheless, *The Roots of the Mountains* represented a measurable improvement upon its predecessor, and Morris was overjoyed: 'I am so pleased with my book – typography, binding, and must I say it, literary matter – that I am any day to be seen huggling it up, and am become a spectacle to Gods and men because of it.'[64]

Morris's last endeavour in bookmaking before the Kelmscott Press, *The Story of Gunnlaug the Worm-Tongue* (dated 1891 in its colophon but apparently completed in November 1890), was more radical in nature; significantly, it was never offered for sale, and most of the sheets remained unbound until after Morris's death.[65] It is a deliberate attempt by Morris to approximate the appearance of an incunable, for there are catch-words but no page numbers, the spaces for initials are left blank, and, most extraordinary of all, the book is set in a precise replica of Caxton's fourth type which had been cast by William Howard for the Chiswick Press in about 1850. Ironically, the type was created in order to replace missing pages in Caxton books – the sort of 'restoration' which aroused Morris's fury when it was inflicted upon medieval churches.[66] Though intriguing, *Gunnlaug* was a typographical cul-de-sac, too frankly derivative and antiquarian in style. Morris, when he was true to his own theories of art and craftsmanship, recognized that we must pay homage to the great achievements of the past not by imitating them but by closely studying their essential qualities so that we ourselves can fashion beautiful objects. It is this principle that was to make the Kelmscott Press something more than a quaint anachronism. *Figure 23*

By the time the odd little *Gunnlaug* was printed late in 1890, Morris was at last ready to launch that most ambitious of all his 'typographical experiments'.

Figure 24 The Arts and Crafts Exhibition held at the New Gallery, Regent Street, in 1890. The Arts and Crafts Society promulgated Morris's handicraft ideals, and at the 1890 exhibition the Chiswick Press case displayed the Whatman-paper edition of *The Roots of the Mountains*, one of Morris's pre-Kelmscott attempts to find a suitable typographical form for his books.

THREE ✤ FOUNDING THE PRESS

HOW much might be done by the establishment of an exhibition, anywhere, in which the Right doing, instead of the Clever doing, of all that men know how to do, should be the test of acceptance!' Ruskin exclaimed in a letter to Morris in 1878.[1] His proposal (or perhaps it was only a passing whim?) is rooted in the assumption that all the arts – even the so-called minor or applied arts – have a common basis in the human creative faculty, and that the excessive honour paid to painting in the nineteenth century was unwholesome. In 1884 the Art-Workers' Guild was formed in London in order to allow the practitioners of various arts to come together for discussion: the Masters of the Guild wore a badge bearing the motto 'Art is Unity'. Morris joined the Guild in 1888 and was the Master in 1892.

When the Arts and Crafts Exhibition Society was created a few *Figure 24*
years later, in 1887, it drew both its goals and membership largely from the Art-Workers' Guild. But, whereas the Guild operated quietly behind closed doors, the Society mounted a series of widely publicized exhibitions and energetically promoted the arts-and-crafts ideal. In the preface to the catalogue of the first exhibition (which was held at the New Gallery, Regent Street, in October 1888), Walter Crane, the president, declared: 'The true root and basis of all Art lies in the handicrafts. If there is no room or chance of recognition for really artistic power and feeling in design and craftsmanship – if Art is not recognized in the humblest object and material, and felt to be as valuable in its own way as the more highly rewarded pictorial skill – the arts cannot be in sound condition. . . .'[2]

The Arts and Crafts Society was founded on ideals that had their origin in Morris's exemplary career (and, at a second remove, Ruskin's writings), yet Morris was less enthusiastic about the Society than one might suppose. By this time in his life he had come to the unhappy conclusion that individual craftsmen – or even groups of craftsmen – were powerless to change the fundamental structure of a corrupt capitalist society; only political action, culminating in revolution, could bring about the root-and-branch transformation that was required.[3] Despite these misgivings, he was active in the Arts and Crafts Society. At the first exhibition he lectured, while seated at a loom, on 'Tapestry and Carpet Weaving', and the products of Morris and Company were amply represented in the display cases. The *Century Guild Hobby Horse* and many books, including sixteen produced by the Chiswick Press, were shown at the exhibition; but significantly, not one of Morris's books could be seen at the New Gallery. This was only a temporary absence, of course. *The House of the Wolfings* made an appearance in the Chiswick Press case at the 1889 exhibition, and in 1890 six sheets of *The Roots of the Mountains*, illuminated by E. G. Reuter, and the Whatman-paper edition of the book itself were exhibited. Nevertheless, that first exhibition of the Arts and Crafts Society must have served as yet another uncomfortable reminder to Morris that his own books were still inadequately printed.

Morris's lecture on weaving at the 1888 Arts and Crafts Exhibition was one of a series of six talks sponsored by the Society. Of these, two dealt with the book arts: on 15 November Emery Walker spoke on 'Letterpress Printing and Illustration', and on 22 November T. J. Cobden-Sanderson lectured on 'Bookbinding'. It was this public exposition of the principles of good typography by Walker which, without warning, suddenly aroused Morris's dormant enthusiasm for printing and set off a chain of events that led directly to the founding of the Kelmscott Press. The evening of 15 November was, therefore, an unexpectedly momentous occasion: many historians have argued that the modern revival of fine printing can be dated precisely from Walker's lecture. When Sydney Cockerell, who was hardly likely to underestimate Morris's contribution, wrote an obituary of Walker for *The Times* (24 July 1933), he summed up the argument for Walker's importance in this bold generalization: 'It is scarcely too much to say that his influence, direct or indirect, can be discerned in nearly every well-designed page of type that now appears, and to him more than to any other man this century's great improvement in ordinary book production has been due.'[4]

Figure 25

Walker, the son of a coach-builder, early displayed bookish interests and, while still an adolescent, stumbled upon a bibliographical treatise which introduced him to the history of printing; from booksellers' penny and twopenny boxes he was soon collecting examples of the work of the most eminent printers, past and present. When his father became blind in 1864, however, Walker was forced to leave St Mark's School, Chelsea, and to find employment. In 1873 he joined the firm of Alfred Dawson, the inventor of a process-engraving technique, and there Walker laid the foundation of his extraordinary knowledge of

Figure 25 Emery Walker with his friend and fellow-socialist, Bernard Shaw. Walker was Morris's unofficial partner at the Kelmscott Press and to a large degree responsible for the Press's creation.

printing and methods of graphic reproduction. Walker and Walter Boutall left Dawson in 1886 and set up their own engraving business, which was to perform so many indispensable tasks for the Kelmscott Press.

By a happy coincidence, Walker was Morris's neighbour (the latter's daughters, seeing Walker frequently hurrying past Kelmscott House, nicknamed him 'the brown velveteen artist' because of his coat) and found himself drawn into the activities of the Socialist Democratic Federation. By 1883 or 1884 Walker and Morris were close confederates in the socialist cause. The two men quickly discovered the bibliophilic interests they had in common, and the relationship blossomed. Walker's description of the friendship, though far too self-effacing, is worth quoting: 'I have lived since I was a child in Hammersmith, and since 1878 a few minutes' walk from Mr. Morris's house. Mr. Morris possessed in his fine library a considerable collection of early printed books. . . . Mr Morris finding I had some knowledge of the practical side of printing and was thus able to explain certain details of the methods of the early printers, a subject which was beginning to occupy his mind, he invited me to visit him, which, as we had other interests in common, I was glad to do. From this time until his death in 1896 I was a frequent visitor and learned much more in bulk from him of the artistic side of printing than I could teach him of the simple elements.'[5]

Such a modest estimate of Walker's contribution, however, does not tell the whole story. As Sydney Cockerell testified, the Kelmscott Press simply could not have existed without the technical expertise of Emery Walker. It was he who was able to tell Morris where to find good paper and ink, skilled printers, presses – anything, in fact, that Morris needed: 'the universal Samaritan, Walker, comes, from his nature, to be laid-on, like the water, only we don't pay the rates for him,' Philip Webb observed whimsically in a letter to May Morris.[6] And it was not only Morris who was served in this fashion by Walker; until his death in 1933 he was connected with many of the most influential experiments in fine printing in England and Germany, acting usually in his characteristic role of the quiet, unassuming yet infinitely knowledgeable adviser.

It is surprisingly difficult to assess how many of the typographical views associated with Morris and the Kelmscott Press came originally from Walker. Both argued for the necessity of close word-spacing and little or no leading; both had emphatic convictions about margins;

both believed in the superiority of old-face types. What seems likely is that, during those long talks in Morris's study, they worked their way together towards a radical – and at the same time conservative – understanding of typography based on the practices of the earliest printers. Each had his distinctive gifts to offer to the partnership: Morris possessed a strong, intuitive grasp of the principles of design, and Walker had a craftsman's first-hand knowledge of how books were actually printed. Yet it must not be forgotten that there were differences between them as well. Walker collected primarily sixteenth-century books, and the volumes he had a hand in designing at the Doves Press show a decidedly Renaissance influence. Morris, of course, looked back always to the manuscripts and printed books of the Middle Ages. Their attitudes towards technology also diverged. Walker, as a professional process-engraver, was familiar with, and was willing to use, the latest methods of reproduction (though, like Morris, he favoured wood-engravings for illustrations in books printed by letterpress), whereas Morris distrusted nearly all forms of Victorian technology, and it was only with difficulty that Walker was able to persuade Morris that the ornaments of the Kelmscott Press should be electrotyped.

Despite their occasional disagreements, Morris knew that he had found in Walker the gifted artisan who could help him print books at last as he had wanted to print them since the *Earthly Paradise* scheme of the 1860s. One is still left wondering, though, why Walker's Arts and Crafts Society lecture in November 1888 had such a profound effect upon Morris, for they had, after all, long been discussing together its key propositions. Unfortunately the lecture itself was never published, but it was reported in the press, and Walker's notes have survived (see Appendix B below); what these sources make clear is that the most striking element of the lecture was a series of approximately two dozen lantern-slides, several of which were of books and manuscripts that Morris himself owned. In a letter to his daughter Jenny, written two days later, Morris offered this account of the lecture: 'Thursday I was at the Arts & Crafts at Walker's Lecture on printing; he was very nervous and ought to have written down his words; but of course he knew his subject thoroughly well: There were magic-lantern slides of pages of books, and some telling contrasts between the good and the bad.'[7] (Walker was in fact extremely shy and had agreed to give the lecture only at the urging of Cobden-Sanderson.) Walker's own recollections of the occasion also laid emphasis on the slides: 'To demonstrate my point, which was that the main factor in a well printed book was the type, I selected many examples of earlier and later printing and had lantern slides made from them. This method of bringing home a point to the audience was then comparatively novel and the lecture, in spite of my bad delivery, was generally considered a success.'[8]

Walker began by demonstrating the close connection between types and handwriting, comparing several early printed books and manuscripts of the same period. When he showed a page of one of the Renaissance writing manuals owned by Morris, Arrighi's *Il modo de temperare le penne* (1523), his listeners paid tribute to its beauty by bursting into applause, and Walker put beside it a passage about handwriting from Ruskin's *Fors Clavigera* (1875) set in three Victorian typefaces. Walker's method throughout the lecture was the comparative technique that had been perfected by Pugin, Carlyle, Ruskin, and Morris: he attempted to open the audience's eyes to the horrors of Victorian typography by juxtaposing it against the most beautiful and dignified books of the past. The lecture ended anticlimactically with slides of Bodoni's type, Miller and Richard's Old Style, and a page of the *Century Guild Hobby Horse*, which Walker used to demonstrate the sin of excessive leading. A final page from the *Hypnerotomachia* emphasized again the poignant contrast between fifteenth-century and nineteenth-century typography. (When Walker delivered essentially the same lecture again on later occasions, he was able to conclude on a happier note with a slide of *The House of the Wolfings*.)[9]

The publisher Charles Kegan Paul had publicly suggested in the *Fortnightly Review* (April 1883) that 'there could scarcely be a better work for the artistic future of books than that which might be done by some master of decorative art, like Mr. William Morris, and some great firm of type-founders in conjunction, would they design some new types for our choicer printed books'. Morris had more than once remarked casually to Walker that he would like to 'have a shot' at type-designing, but that evening of 15 November 1888, as Morris made his way back to Hammersmith with those huge pictures of early types still vivid in his memory, he turned to Walker and said, 'Let's make a new fount of type.'[10] For a person of Morris's obsessive energy, such a proposal led at once to its corollary – the founding of his own press – and within a few weeks he asked Walker to join him in the venture. The exact form of Walker's polite demurral is not known, since there are several recorded versions of it: Cockerell noted in his diary (after a conversation with Walker) on 10 December that 'his engagements compel him to decline the tempting offer'; Cockerell later asserted that it was because Walker had no capital to offer; Walker himself said that he refused because he had 'some sense of proportion'.[11] According to May Morris's amusing version of the conversation, her father said, 'Will you go into partnership with me? . . . but mind I shall want to do everything my own way.'[12] Walker's refusal was a purely technical one, for, as Cockerell said, 'he was all the while a virtual partner, and no important step was taken without his advice and approval'.[13]

Figure 26

Morris had long owned a small but attractive collection of incunabula (part of which he sold in 1880), but now, in search of models, he began to buy books from all the principal presses in Germany, Italy, France, and the Low Countries during the fifteenth century. Within a few years he put together an extraordinarily fine library: Paul Needham has declared that, 'if we exclude men such as William Beckford and Robert Curzon, who are known primarily as collectors rather than authors, it may be claimed that William Morris possessed a library of higher quality than any other major English literary figure'.[14] An inventory made by F. S. Ellis and his son a week after Morris's death appraised his manuscripts at about £12,000 and the printed books (excluding modern works) at about £7,300.[15] Ironically, Morris never thought of himself as a collector. These books and manuscripts were merely his working library, and in an ideal world, Morris said, they would be made available to everyone – with 'a public library at each street corner'. 'I should not then have to buy all these old books, but they would be common property,' he told a journalist in 1891, 'and I could go and look at them whenever I wanted them, as would everybody else. Now I have to go to the British Museum, which is an excellent institution, but it is not enough. I want these books close at hand, and frequently, and therefore I must buy them.'[16] Morris's protests are not persuasive: despite the evident guilt about his acquisitive instincts, there can be no doubt that he found deep, almost sensual pleasure in the ownership of beautiful books.

It should also be noted in passing that Morris's library contained a substantial collection of books about printing, typography, and manuscripts.[17] The Kelmscott Press is too often treated as a mysterious phenomenon, completely unrelated to the achievements of other Victorian printers or to Victorian typographical scholarship. This misconception may be seen as an indirect tribute to Morris's genius and

Figure 26 A note by Sydney Cockerell reads: 'This photograph of Morris's workroom on the ground floor of Kelmscott House, Hammersmith, was made in his lifetime. There is another which was made after his death, with a different table, which took the place of the plain deal trestle table towards the end of his life.' From this room Morris directed the activities of the Kelmscott Press.

originality, but in fact the modern books in his library demonstrate that he was thoroughly familiar with the literature of printing.

'I am really thinking of turning printer myself in a small way,' Morris announced to F. S. Ellis in November 1889; 'the first step to that would be getting a new fount cut. Walker and I both think Jenson's the best model, taking all things into consideration. What do you think again? Did you ever have his Pliny? I have a vivid recollection of the vellum copy at the Bodleian.'[18]

From the late 1880s until the end of his life, Morris's restless mind was increasingly preoccupied with the possibility of creating handsome books of his own. Before the Kelmscott Press was actually in operation, he conducted his final experiments with commercial methods of production in *The House of the Wolfings*, *The Roots of the Mountains*, and *The Story of Gunnlaug the Worm-Tongue* at the Chiswick Press. In the autumn of 1889, after a conference sponsored by the National Association for the Advancement of Art and Its Application to Industry (at which Morris spoke on dyeing and Walker on printing), a small socialist group visited Glasgow in order to deliver some further lectures, and at Glasgow's Station Hotel, according to Walter Crane, Morris talked to his friends with warm enthusiasm about his plans for the Press.[19]

By this time Morris had begun designing the first of his three typefaces, which was to be known as the Golden type, and early in 1890 Edward Prince started cutting the punches. The technique by which Morris created his types is worth examining in some detail, if only because it illustrates his considerable dependence upon both Walker's expertise and the Victorian technology which Morris professed to despise. This is how Walker remembered the aftermath of his lecture in 1888: 'On the way home after the lecture Morris renewed the subject we had often talked about before, of starting a small printing office from which he might issue limited editions, or as many as he thought he could sell, of his own works, and a day or two afterwards he decided upon making the venture. The first business was to get a type cut to his own design. He supplemented his collection of Early Printed books by buying examples of every fine type – gothic or roman – obtainable. These we, i.e., my firm, photographed upon a large scale to enable him better to distinguish the actual shape of the letters apart from the excess of ink by which in early printed books the shape is often obscured. . . . After a prolonged study of these types lasting throughout the year 1889 Mr. Morris began to design the models for his own letters.'[20]

Given Morris's well-known disdain for mechanical devices, why was he willing to make such heavy use of Walker and Boutall's camera in designing types? The simplest answer, no doubt, is that all of us are guilty of deep inconsistencies in our behaviour and thinking. Had the question been put to Morris, however, he probably would have been able to offer a plausible-sounding explanation. He was not (he might have said) opposed to machinery *per se*: he rejected it only when it threatened to enslave human beings rather than lift their burdens. The camera was merely a tool – quiet, unobtrusive (not, after all, operated by steam power), unthreatening – that simplified the task of the designer. It may also be significant that the Pre-Raphaelite painters, with whom Morris was closely associated, were fascinated by photography and were among the first artists to employ it for making studies: some of Rossetti's prints have survived, including several striking photographs of Jane Morris. For more than a decade Morris had used the camera to enlarge the small sketches that Burne-Jones produced for stained-glass windows to be manufactured by Morris and Company. And of course Morris was undoubtedly much influenced by Walker, who placed the full resources of his process-engraving firm at his friend's disposal. Still, it is important to recognize that the Kelmscott Press, the quintessential example of an arts-and-crafts longing for the pre-industrial age, was paradoxically built upon a foundation of photography, one of the most sophisticated forms of technology in late-Victorian England.

The photographs made for Morris by Walker's firm are now in the St Bride Printing Library, London, and, as Walker indicated, they include pages of a wide range of fifteenth-century books, each enlarged to a scale of five times the size of the original. 'By instinct rather than by conscious thinking it over,' Morris wrote, 'I began by getting myself a fount of Roman type. And here what I wanted was letter pure in form; severe, without needless excrescences; solid, without the thickening and thinning of the line, which is the essential fault of the ordinary modern type, and which makes it difficult to read; and not compressed laterally, as all later type has grown to be owing to commercial exigencies.'[21] The chief models upon which Morris based his Roman type were a fount used by Nicolaus Jenson in Pliny's *Historia naturalis* (Venice, 1476) and a similar, but slightly heavier, face

Figure 27

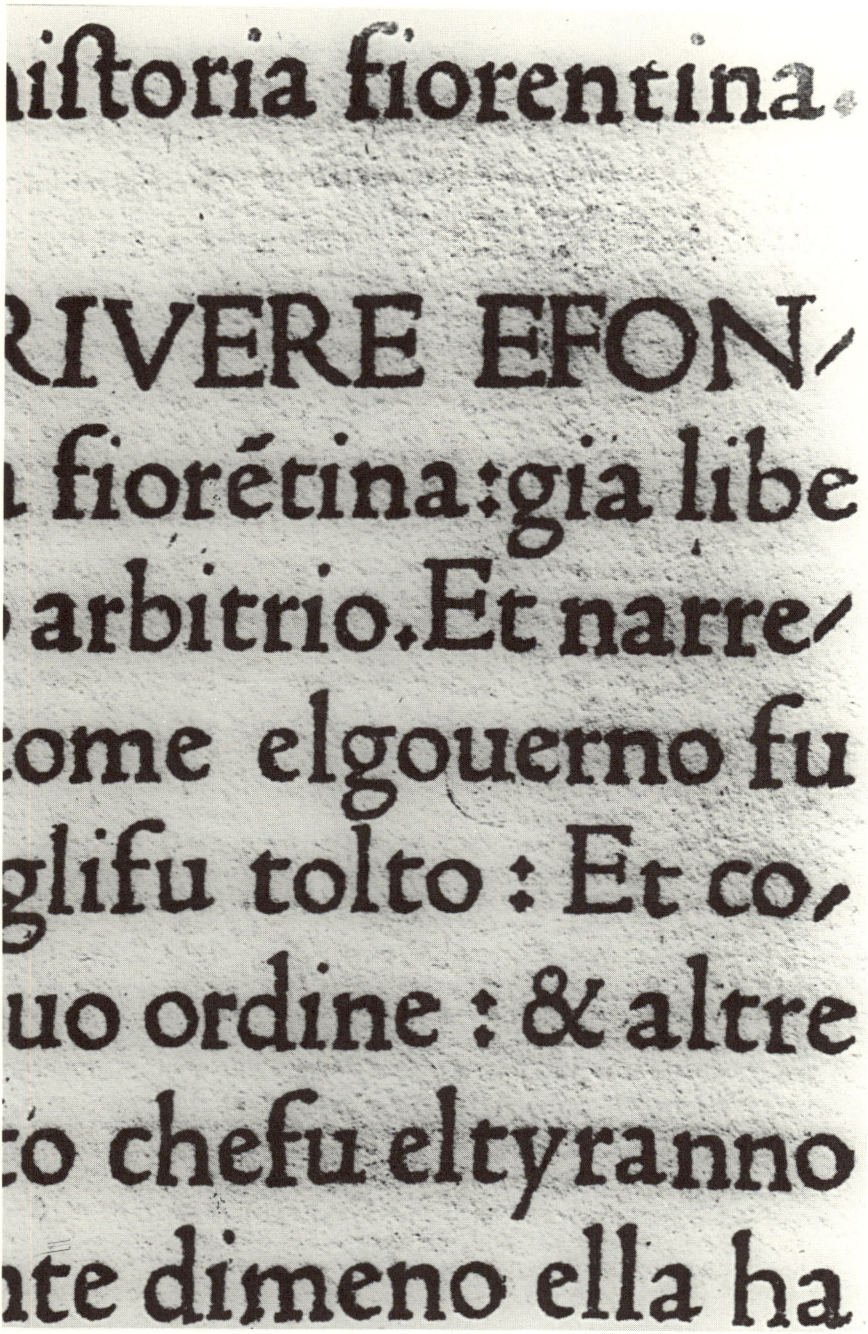
iſtoria fiorentina.

RIVERE EFON/
a fiorẽtina:gia libe
o arbitrio.Et narre/
come elgouerno fu
glifu tolto : Et co/
uo ordine : & altre
to chefu eltyranno
nte dimeno ella ha

Figure 27 Emery Walker made this photograph of a page of *Historiae Florentini populi* (Venice, 1476) for Morris to use as a model in designing his Golden type.

employed by Jacobus Rubeus in Leonardus Brunus Aretinus's *Historiae Florentini populi*, also published in Venice in the same year. As Morris's critics have often pointed out, his Golden type (which he first called 'the regenerate type or Jenson–Morris'[22]) differs from Jenson's Roman in several respects, most notably in its darker colour. Morris admitted to Walker that designing the Golden type was the most troublesome task he had ever undertaken; admirable in appearance though it is on the pages of many of the Kelmscott books, one senses that Morris was not wholly at ease in producing a typeface associated historically with the Mediterranean world rather than northern Europe. Even Emery Walker acknowledged that the Golden type was not as elegant as the Venetian faces after which it was patterned.[23]

Morris emphasized in an interview in 1895 that the typefaces of the earliest printers could scarcely be improved upon, and 'my own types', he added, 'hardly differ from theirs in essentials. They are not the same, of course; but they are not altogether original. It would be impossible to obtain absolute originality where the sphere of action, so to speak, of the designer is so very limited.'[24] Morris was also to argue, with justice, that he had succeeded in emulating without directly imitating

Figure 28

Jenson's fount. And here Walker's camera came into play once more. Having been supplied with photographic enlargements of the types, Morris traced them repeatedly until he felt he understood the essence of the letterforms. The next step was to make a group of free-hand drawings, still to the same large scale. Since it is hazardous to design type (or anything else, for that matter) on a scale too far removed from its ultimate size, Walker then photographed the best of the drawings and produced prints of them to the scale of the original type. In at least one instance he also made a line block of a drawing in the smaller size so that Morris could see how his design looked when actually impressed on paper. At each stage Morris continued to modify, touch up, and revise the drawings. At last he felt ready to turn them over to his punch-engraver.

Those familiar with modern type-designing will recognize that this procedure is still used, with minor variations, although of course nowadays the final stage involves the manufacture of a film or electronic matrix rather than a metal punch. When the Monotype Corporation launched its great series of historic type revivals in the 1920s under the direction of Stanley Morison, it followed closely the steps laid down by Morris and Walker in the previous century. What is slightly startling about all this is that there is no record of any

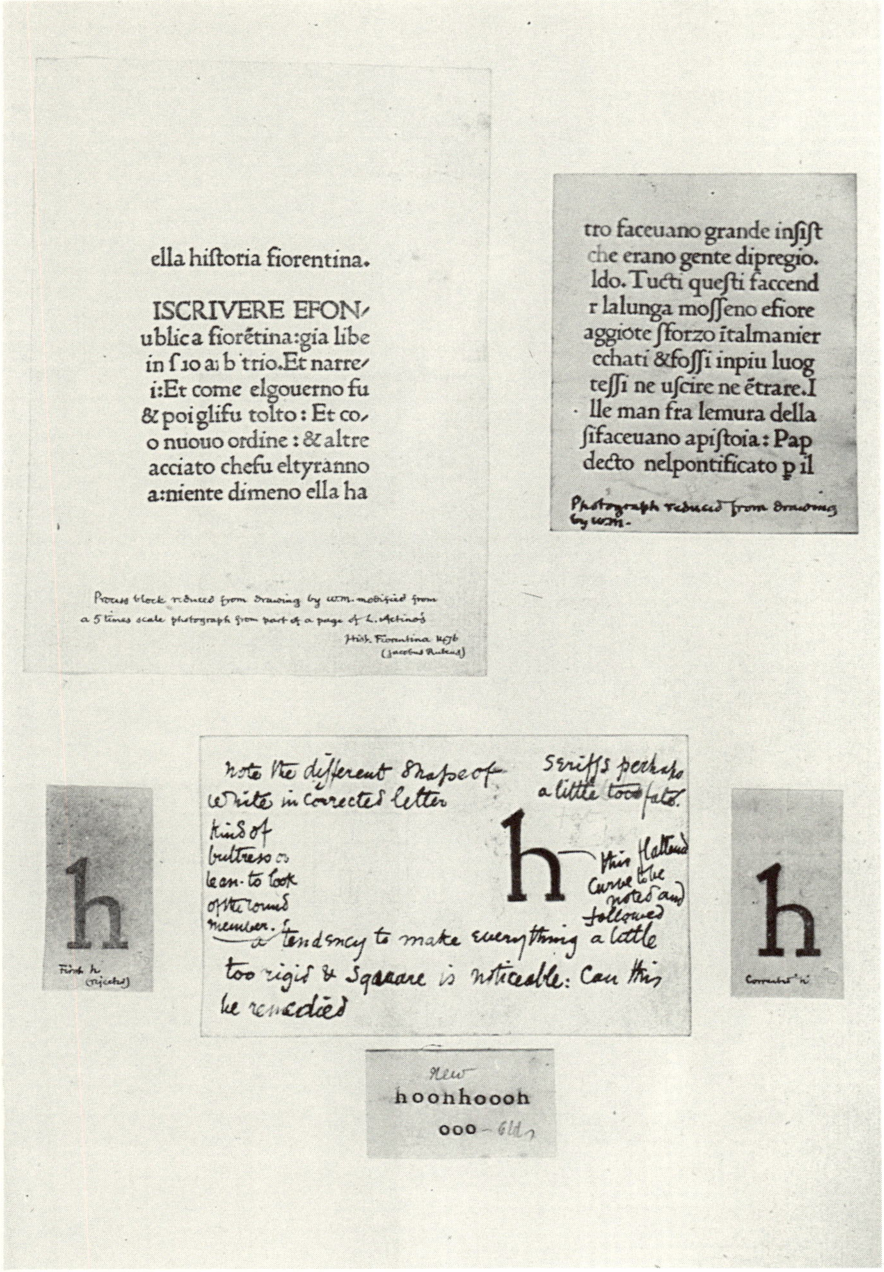

Figure 28 This group of trial designs for the Golden type reveals that Morris first traced the letterforms from photographs of incunabula, then gradually modified them before they were again photographically reduced by Walker. In the upper left is Morris's tracing or drawing of the page shown in Fig. 27.

Figure 29 Edward Prince cut the punches for Kelmscott Press types as well as the proprietary typefaces of most of the other English private presses at the end of the last century.

type-designer before Morris using the camera as an essential tool: Morris and Walker seem to have stumbled upon a technique that is now considered standard in many fields of design, and Morris managed to do this despite some unusually strong prejudices against technological shortcuts. Altogether it is a very odd achievement for a man who sighed after the Middle Ages.

Figure 29

Edward Prince, to whom Morris now turned for the cutting of his punches, was a sober, old-fashioned craftsman destined to perform a similar task for many of the most celebrated private presses of the following generation, including the Doves, Ashendene, and Vale Presses.[25] Indeed, at the time of his death in 1923 he was the last living independent punch-cutter in England, for by then the process of engraving punches had become thoroughly mechanized. Frederic Goudy and others have suggested that Prince's contribution to the creation of the Kelmscott types has been underrated, yet it must be said that Prince also engraved types that were superlatively ugly (such as C. R. Ashbee's Endeavour face) with the same skill that he lavished on Morris's founts. Moreover we have the testimony of Herbert Horne, whose Montallegro type he engraved, that he was 'an excellent workman, but very British; and the difficulty of making him perceive the finer qualities of form is enormous'.[26]

Once the punches were engraved, it was possible to make 'smoke proofs', which Morris tended to carry about with him in a matchbox so that he could inspect them from time to time. Even at this stage a few letters – the lower-case 'h', for example – were rejected and engraved a second time. For the casting of the type, Morris sought the services of the respected typefoundry of Sir Charles Reed and Sons, established in the mid-eighteenth century, and now under the management of Morris's friend Talbot Baines Reed, the author of a scholarly history of English typefoundries. The actual casting, however, was done by machine; Sparling records Morris's wry comment that 'from all I hear, there wasn't much fun in it [hand-casting] for the poor devils who jogged and bumped the moulds about'.[27] The Golden type was completed by the winter of 1890–1.

Figure 30

It was cut in the generous size then known as 'English' – i.e. 14 point. (The point system for measuring type-sizes had not yet been adopted in Morris's day.) Morris was to insist during the 1890s that most books were printed in excessively small type: 'Long Primer' (10 point), he suggested on one occasion, was the minimum size acceptable, or perhaps, he wrote elsewhere, it was 'Small Pica' (11 point).[28]

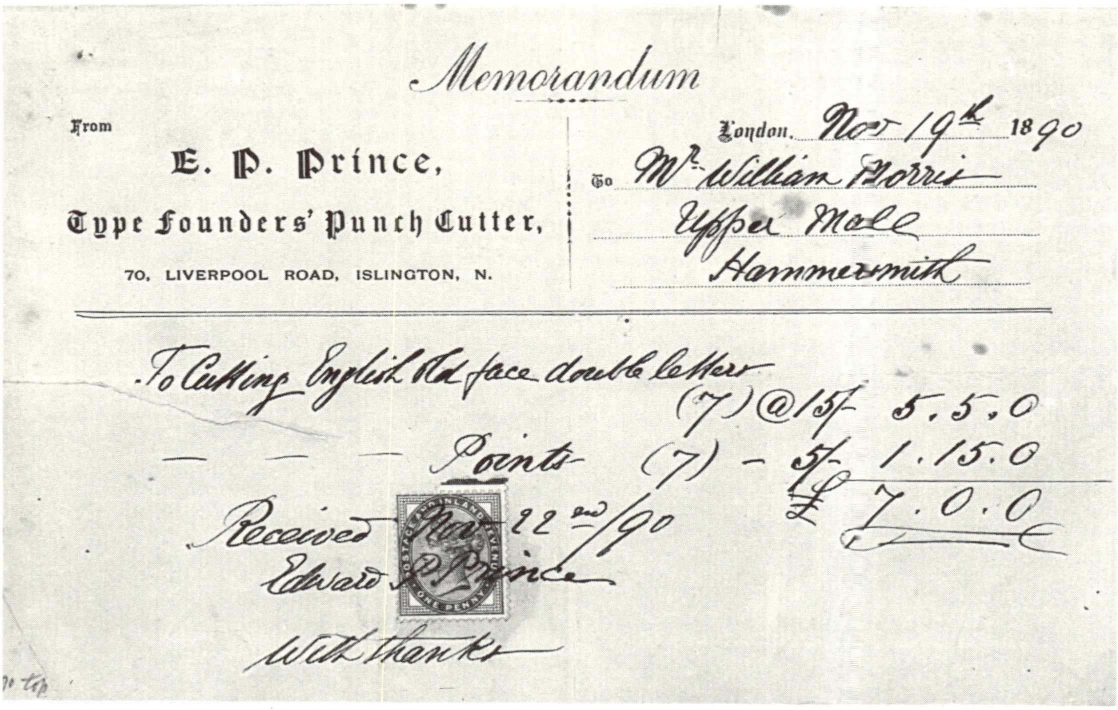

Memorandum

From

E. P. Prince,

Type Founders' Punch Cutter,

70, LIVERPOOL ROAD, ISLINGTON, N.

London. Nov 19th 1890

To Mr William Morris

Upper Mall

Hammersmith

To Cutting English Old face double letters	(7) @ 15/-	5.5.0
— — — Points	(7) – 5/-	1.15.0
		£ 7.0.0

Received Nov 22nd/90

Edward P. Prince

With thanks

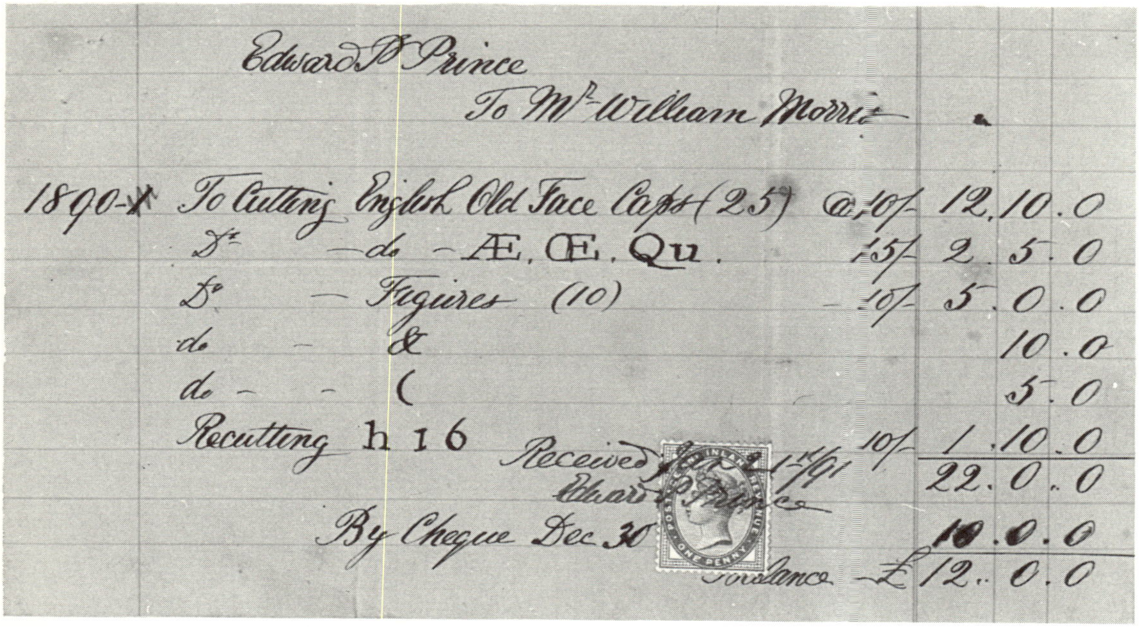

Edward P. Prince

To Mr William Morris

1890	To Cutting English Old Face Caps (25)	@ 10/-	12	10.0
	Do — do — Æ, Œ, Qu.	15/-	2	5.0
	Do — Figures (10)	10/-	5	0.0
	do — &			10.0
	do — — (			5.0
	Recutting h 1 6	10/-	1	10.0
	Received Jan 21st/91 Edward P. Prince		22	0.0
	By Cheque Dec 30		10	0.0
	Balance		£ 12	0.0

Figure 30 Two of Edward Prince's bills for cutting the Golden type. They disclose that three characters ('h', '1', and '6') had to be recut.

It is not accidental that Morris's favourite among his own types was also the largest, the Troy, a 'Great Primer' (18 point). The reduced version of the Troy type, the Chaucer – a 'Pica' (12 point) – seemed too small to Morris, and only with the greatest reluctance did he accept it as an economic necessity in the enormous Kelmscott edition of Chaucer. For a number of years Morris toyed with the notion of producing a larger version of the Golden type as well, though nothing came of the wish. He did, however, have his binder, J. and J. Leighton, employ several sizes of an enlarged Golden type (with dies produced from line blocks rather than new punches) on the spines of many of the Kelmscott books. Cockerell also records the significant fact that the first character of the Golden type was cut by Prince in 1890 in Great Primer, but Morris then decided to adopt instead the English size.[29]

Morris was not alone among the Victorians in his admiration for the types of Jenson. Henry Noel Humphreys, in *A History of the Art of Printing* (1868), a work which Morris owned, had praised Jenson's first four books for being printed in 'Roman characters . . . more perfect in form than those of any previous printer' (p. 112). In the United States, Theodore Low De Vinne, while acknowledging in 1886 that Jenson's letters might appear 'rude and coarse' to modern eyes, was still able to commend the 'simplicity and beauty of his design'.[30] Talbot Baines Reed, when lecturing to the Society of Arts in April 1890, showed a slide of one of Jenson's books and discussed his types in complimentary terms.[31] Horatio F. Brown's *The Venetian Printing Press 1469–1800* (1891), published just as the Kelmscott Press was being launched, offered a detailed and mostly admiring account of Jenson's work as a printer. Emery Walker, it goes without saying, held Jenson's typography in high esteem and turned to it again for inspiration when he subsequently designed the Doves Press type. Morris himself remarked that, 'when it is so obvious that this is the best and clearest Roman type yet struck, it seems a pity that we should make our starting point for a possible new departure at any period worse than the best'.[32] (This is undoubtedly a reference to Miller and Richard's Old Style, which was a pallid imitation of Caslon.)

The Golden type was not, in its final form, a mere facsimile of Jenson's fount: Morris insisted that he 'did not copy it servilely', and Walker admitted that Morris took only 'the general proportions' of his type from Jenson and Rubeus.[33] The chief modifications introduced by Morris were to thicken the strokes and make the serifs more slab-like, producing a printed page that was much darker than Victorian readers were accustomed to. The heavy, bold Golden type represents his reaction to the anaemic typography of late-nineteenth-century England, and it also reflects his search for a typeface that would match the broad lines of wood-engravings cut in the medieval style. 'Crane when he saw it beside Jenson thought it more Gothic-looking: this is a fact, and a cheerful one to me,' Morris said in a letter to Ellis.[34] Another anachronistic quality of Morris's fount is that he did not design an italic alphabet or certain other characters that most modern printers would regard as indispensable (e.g. brackets, dashes, small capitals, bold face). This decision was to produce a grand simplicity and uniformity of texture in the Kelmscott Press pages, but at the same time the lack of an italic made it almost impossible for Morris to print any editorial apparatus such as footnotes or bibliographies (which he was reluctant to include in his books for other reasons as well). One notable departure from the practice of early printers was Morris's avoidance of abbreviations and ligatures – except of course for the usual ones, 'fi', 'ffi', 'fl', 'ffl', 'ff', 'æ', and 'œ'. Writing in September 1890 to Ellis about *The Golden Legend*, which was then envisioned as the first Kelmscott book, Morris declared, 'In one thing I think I differ from you a little, *i.e.*, about the joined letters or queer signs: since our book is to be a *reprint*, not a fac-simile, I do not think that we need reproduce these: indeed I should extend the abbreviations in order to make the book more readable.'[35]

Morris's Golden type evoked a mixed response. Professional printers were quick to observe that, because of the great breadth of the individual letters (not to mention the 14-point size), it was wholly unsuitable for books produced under commercial conditions. The adjective used by De Vinne in writing about Jenson – *coarse* – cropped up frequently in discussions of the Golden type. Typographers in the early decades of this century, influenced probably by the blight of ubiquitous, inferior imitations of Morris's design, tended to offer harsh estimates of it. Stanley Morison, for example, referred to the Golden type in private as 'positively foul', and in a famous essay published in 1924, 'Towards an Ideal Roman Type', he made it one of his chief targets in an attack upon Jenson's large capital letters.[36] Examined individually, the Golden type characters, especially the capitals, do indeed seem heavy and stiff, yet they fit well together (always an important test of any typeface), and their total effect on the page of a Kelmscott book is undeniably attractive. If the Roman type of the Doves Press, created only a few years later and based on the same model, strikes many of us today as a more satisfactory adaptation of Jenson's letterforms, the Golden type must, at the very least, be viewed with the generous tolerance that is accorded a pioneering effort.

Having produced a Roman type more or less to his satisfaction, Morris – even before his Press was in operation – began to lay plans for a Gothic type as well. Morris 'talks of setting to work at once on a blackletter fount!' Cockerell recorded with astonishment in his diary on 1 February 1891. It was not in fact a surprising decision, for Morris's deepest aesthetic instincts always drew him back to medieval forms, and his own library was filled with the finest German and French illustrated books of the fifteenth century, mostly printed in Gothic types. Lecturing on 'The Ideal Book' in 1893, Morris made a strong case for the possibility of reviving blackletter: 'I have hitherto had a fine and generous Roman type in my mind, but after all, a certain amount of variety is desirable, and when you have once got your Roman letter as good as the best that has been, I do not think you will find much scope for development of it; I would, therefore, put in a word for some form of Gothic letter for use in our improved printed book.'[37]

Morris acknowledged that such a proposal would 'startle' some of his listeners; in another lecture delivered a few months later he offered the even more startling suggestion that 'if he had his own way he would have all books printed in it [i.e. Gothic type]'.[38] This may be a characteristic piece of Morrisian exaggeration, but it does reveal his extraordinarily strong preference for Gothic rather than Roman types, a preference based ultimately on their respective connections with the Middle Ages and the Renaissance. Here again Morris was swimming against the current, for, outside Germany, there was an almost universal consensus in the nineteenth century that Gothic typefaces were dead. In *The Printer's Grammar* (1808) Caleb Stower offered this pious graveside homily: 'We may reasonably hope, from the pleasing aspect of the present age, when mankind have dared to burst the fetters of prejudice and antiquity, and have determined to exercise their better judgement, and adopt plans more congenial to true taste, that the Roman character will be universally employed in all civilized states' (p. 35). The assumptions behind Stower's assertation are familiar, because they echo the charges that were still being made in his day against medieval churches and cathedrals: that Gothic types, like Gothic buildings, belong to an earlier, more barbaric period of European history, whereas Roman type, like neo-classical architecture, is associated with order, rationality, enlightenment – with, in a word, civilization. Romanticism, the Oxford Movement, and Victorian ecclesiology were to turn such assumptions on their heads, but English typography was surprisingly late in responding to these new intellectual and aesthetic forces. By the 1890s English cities and towns were crowded with examples of Gothic Revival buildings, and the Pre-Raphaelites had successfully aroused public interest in medieval styles and subjects in paintings; yet, apart from a few isolated experiments by the Chiswick Press and other 'old-style' firms, there had been no serious attempt to restore Gothic types to a central position in the world of printing.

Morris was to describe his Troy type – named after *The Recuyell of the Historyes of Troye* (1892), the first book in which he used it – as a 'semi-Gothic type designed . . . with special regard to legibility'.[39] What he attempted to achieve in the Troy was a modern adaptation of the best fifteenth-century typefaces of Germany that would not seem exotic or illegible to Victorian readers: 'the task I set myself was to redeem the Gothic character from the charge of unreadableness which is commonly brought against it,' Morris declared in the *Note . . . on his Aims in Founding the Kelmscott Press*. 'And I felt that this charge could not be reasonably brought against the types of the first two decades of printing: that Schoeffer at Mainz, Mentelin at Strassburg, and Günther Zainer at Augsburg, avoided the spiky ends and undue compression which lay some of the later type open to the above charge. . . . I entirely eschewed contractions, except for the "&," and had very few tied letters, in fact none but the absolutely necessary ones. Keeping my end steadily in view, I designed a black-letter type which I think I may claim to be as readable as a Roman one, and to say the truth I prefer it to the Roman.'[40]

Morris found pleasure in designing the Troy type. Emery Walker recalled that 'he drew the whole alphabet straight away – more or less "out of his head" '.[41] However, the documentary evidence shows that Morris went through the same painstaking procedure of tracing, drawing, and redrawing photographs that he had used in producing the Golden type, except that in this case Morris's final design was a freer and more skilful translation of his models. Even so, several of Morris's lower-case letters are remarkably close to the examplars he found in the books of Schoeffer ('a', 'd', 'e', 'f', 'm', 'n', 'o', 'p', 'r', 't') and Zainer ('g', 's'). Because of illness, he was unable to give his designs for the Troy type to Edward Prince until late summer of 1891; the

Figure 31

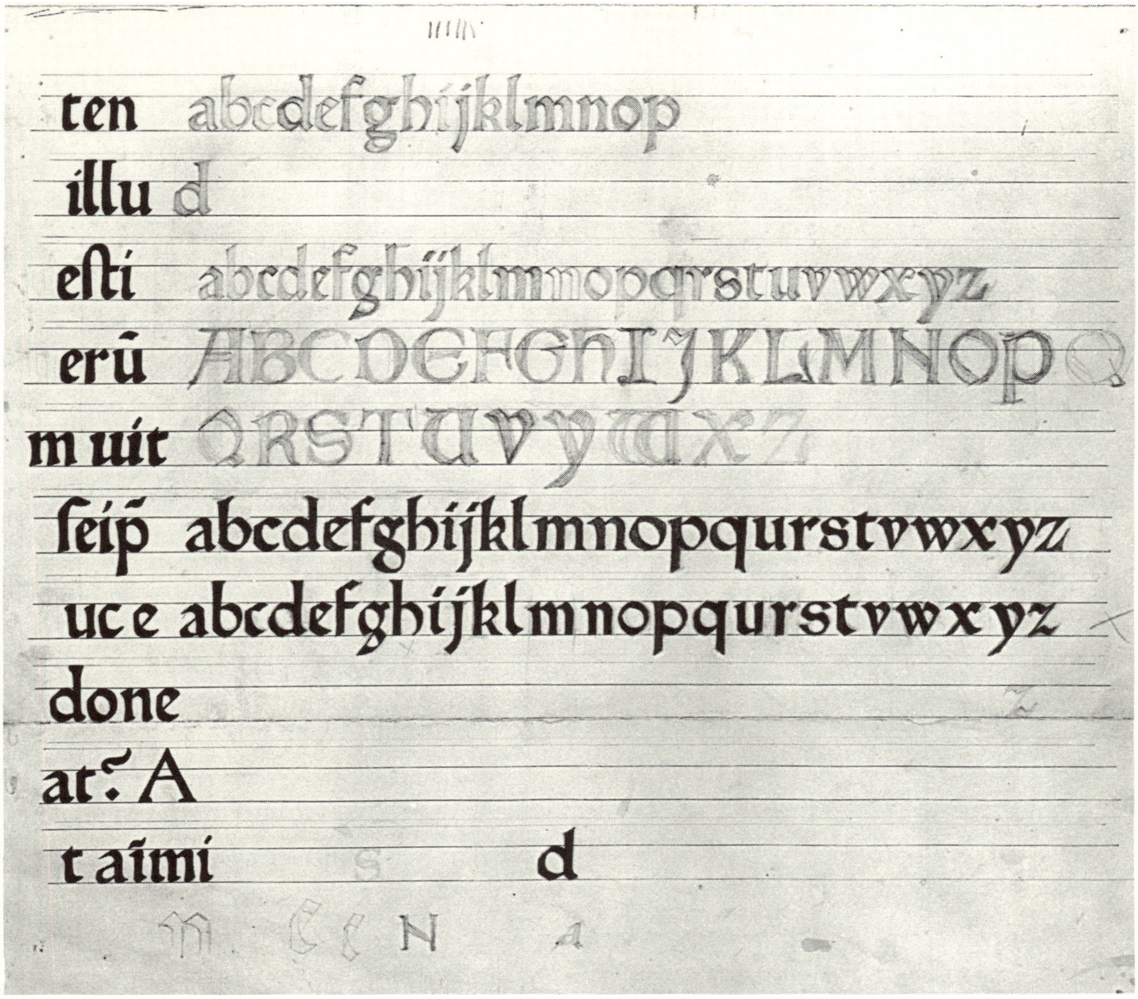

Figure 31 Preliminary designs by Morris for the Troy type.

completed fount was in Morris's hands by the beginning of 1892. 'Morris had trial proofs of pages of prose and verse set up in his new modification of blackletter type,' Cockerell wrote in his diary on 14 January 1892. 'They looked extremely beautiful.' A week earlier (7 January) Cockerell had participated in a lively discussion among some of Morris's friends at Gatti's restaurant (where the officers of the Society for the Protection of Ancient Buildings – known as 'Anti-Scrape' in Morris's circle – normally retired after their committee meetings): 'Had talk as to whether the new Morris type was really medieval in character, I maintaining that it was, and [Philip] Webb asserting that it was nineteenth century.' The Troy type is in reality as characteristic a product of the Gothic Revival in Victorian England as Pugin and Barry's Palace of Westminster or Butterfield's All Saints Church, Margaret Street; like them, it must be judged not merely as a scholarly replica but as a splendid artistic achievement in its own right. It is of course, given its size and character, even more gloriously unsuitable for ordinary commercial use than is the Golden type, but Morris was right in seeing the Troy as his greatest success in type-designing. Printed on handmade paper and accompanied by Morris's lively, graceful ornaments, it produces an unforgettable effect.

One of the two earliest trial pages set in the Troy type was a passage from Chaucer's 'Franklin's Tale', and it is obvious that from the beginning Morris had the type in mind for his great edition of Chaucer. Morris may have been a crusading idealist, but he had a firm grip on financial realities, and he saw that printing the collected works of Chaucer in a type as large as the Troy was out of question. In December 1891, therefore, Emery Walker made several photographic reductions of passages (both prose and poetry) set in the Troy type to help Morris envision the best size for his third fount. Morris wrote to Walker on 20 December: 'Thanks for the photos: which look very well indeed. I must say that I am in some doubts as to which I shall have cut, the English or the Pica; I am almost inclined for the Pica, as it will give us more variety.'[42] Since the Golden type was an English (14 point), Morris finally settled on Pica (12 point) for the Chaucer type, a decision that always left him uneasy. He was 'driven by the necessities of the Chaucer (a double-columned book)' to create this smaller version of the Troy type, Morris said;[43] the Chaucer type was cut by Prince between February and May 1892 and was ready for use in June.

There was another historical model towards which Morris was

drawn, a typeface used by the German printers Sweynheym and Pannartz in three books produced at the monastery of Subiaco near Rome during the 1460s. Morris described it as 'an exceedingly beautiful type' and 'a transition between Gothic and Roman', and when in November 1892 he purchased one of their books, St Augustine's *De civitate Dei*, he designed a lower-case alphabet patterned after the Subiaco fount.[44] It was never completed, but even Morris's discarded ideas created spreading ripples in the typographical waters, for in 1900 Cockerell persuaded St John Hornby to have a private typeface cut for his Ashendene Press: Walker photographed pages of Cicero's *De oratore* printed by Sweynheym and Pannartz, tracings and drawings were made. Prince engraved the punches, and Hornby, to his surprise, found that an investment of less than £100 had made him the owner of the Subiaco type, one of the most successful of all the private-press founts.

The Kelmscott Press began with only one proprietary typeface, but by the second year of operation Morris had at his disposal a 14-point Roman (the Golden type), an 18-point Gothic (the Troy type), and a 12-point version of the same Gothic (the Chaucer type). Because the press was an exercise in deliberate simplification, Morris found these three faces more than adequate for his needs. Likewise, once he had discovered a paper that satisfied him, he had it produced in several sizes and used nothing else thereafter.

As the contract for *The Golden Legend* was being drawn up in September 1890, Bernard Quaritch wrote to F. S. Ellis: 'My business will be to supply the paper, – & if need be – have it specially made for you.'[45] Quaritch's assumption reveals that Morris's plans for his Press were still rather vague at this time, yet we know that he was increasingly coming round to the view that he must control every aspect of production, and that meant he must have his own paper – handmade, needless to say – as well as his own types. 'As to paper,' he declared to Ellis a few days later, 'I have heard of two people who may help us, one whom Walker knows and whose mill I propose to visit with Walker almost at once; and one employed by [George] Allen, Ruskin's publisher. We can do nothing with Whatman but take what he has on the shelves.'[46] Of Allen's papermaker there is no further record in the correspondence, but on 22 October Morris and Walker visited the paper mill of Joseph Batchelor and Son in the village of Little Chart, near Ashford, Kent. Morris took with him a fifteenth-century book as a sample of the kind of paper he wanted,[47] and at the mill he characteristically insisted upon making two sheets himself, with a fair degree of success. That evening after dinner, back in Hammersmith, Morris designed two watermarks, a daisy and a perch, the former of which he sent at once to Batchelor.[48]

What Morris sought was not merely the appearance of handmade fifteenth-century paper but also its permanence and freedom from bleaching chemicals. Once he was persuaded that Batchelor could produce such a paper, Morris seems to have put the subject out of his mind and never visited the mill at Little Chart again. There were, however, minor vexations before the paper problem was completely solved. The first batch of paper, used in *The Story of the Glittering Plain*, was too absorbent because of inadequate sizing; some of the early paper showed variations in shrinkage; and the initial dimensions proved unsatisfactory for the larger books. There was also an abortive experiment with a paper specially made years earlier for Walker which was so hard that its razor-edged deckle cut the fingers of the pressmen.[49] Ten reams of the 'Flower' paper (16 × 11 inches) were delivered on 12 February 1891, and, when Morris realized that he would need a bigger sheet for *The Golden Legend*, Batchelor sent him an additional ten reams of a paper with the same watermark but double the original size (16 × 22 inches). This arrived at the Press on 22 April. Both versions of the paper continued to be used for a time, the smaller size making its final appearance in *The Order of Chivalry* (1893). (That book is rather a bibliographical oddity. The second half of the volume, *L'Ordene de chevalerie*, was printed on the larger Flower paper; hence the first part of the book is a quarto and the second an octavo.) The two additional sizes of Batchelor's paper that Morris eventually adopted are identical to the Flower except for their dimensions, thickness (the smaller sheets being thinner), and watermarks. The 'Perch' paper ($16\frac{3}{4}$ × 23 inches), delivered in February 1893, was designed specifically for the *Chaucer*, and the 'Apple' paper ($18\frac{1}{2}$ × $12\frac{3}{4}$ inches) was an unconventional sheet first used in *The Earthly Paradise* (1896–7). 'I am rather exercised as to its format', Morris confessed of the latter, 'but Cockerell is in favour of this new size – a sort of mild quarto and yet looking like an octavo.'[50] The Press's list of 1 June 1896 described *The Earthly Paradise* as 'intermediate between the large 4to and the 8vo books hitherto issued'.

The superiority of the papers made at Little Chart for Morris did not escape the notice of other printers, especially those in search of stock *Figure 32* suitable for limited or large-paper editions. In 1895 Batchelor wrote to

Figure 32 In 1895 Joseph Batchelor secured Morris's permission to distribute commercially, under the name 'Kelmscott Hand-made', the paper that his mill had been supplying to the Kelmscott Press. It was quickly adopted by some of the leading private presses in England.

Founding the Press

Morris: 'I find that other makers are imitating our Antique Handmade paper. For our protection, and as a means of giving my friends a guaranteed genuine article, I propose calling the paper the Kelmscott Handmade, subject, of course, to your approval. This does not apply to watermarking in any way, but to the wrapping and labelling of the paper.'[51] Morris gave his permission, and thereafter the Batchelor paper was used by many of the major private presses (beginning with the Doves Press) and by a few commercial printers for de luxe editions. The Kelmscott Press also, as Batchelor's letter suggests, inspired a fashion for thick, coarse, machine-made papers which mimicked the most obvious qualities of the Batchelor product, but such palpable imitations misled only the least sophisticated reader, and most have long since discoloured and crumbled. The Kelmscott Press paper, on the other hand, was not a 'quaint', pseudo-antique sham but a serious recreation of the finest papers that had ever been used in books.

Morris's search for a good handmade paper was directly linked to his decision to print books on a hand-press. The *Golden Legend* contract (11 September 1890) specified merely that Morris was to have control over the 'selection of the printer', and Sparling asserts that Morris's original intention was to have only the composition of the Kelmscott books done at Hammersmith, with the presswork to be supervised by Emery Walker at his office, No. 16 Clifford's Inn, near Fleet Street.[52] But by January 1891 Morris owned his first press, a 'Demy Albion Press, complete with Cylinder Inking Table', for which he paid W. H. Bowden, subsequently his foreman, £25.[53] When work on *The Golden Legend* fell behind schedule later in the year, Morris bought 'One Super Royal Genuine Albion Press', for £42 10*s*. on 26 November 1891 from Hopkinson and Cope.[54] Before his death he purchased two additional presses: a small Albion which was used for printing leaflets and proofing, and a full-size Albion which was acquired from Hopkinson and Cope for £52 10*s*. at the end of 1894. Reinforced with steel bands, the latter was used for printing the *Chaucer*, the large ornaments of which put a heavy strain on the press.[55]

With an Albion press, his own typeface, and excellent paper from Batchelor, by January 1891 Morris had gathered together most of the key ingredients he would need at the Kelmscott Press. In the correspondence of Jane Morris with Wilfrid Scawen Blunt during this period, we hear occasional gossip about the Press in gestation. 'My husband . . . does not begin till after Xmas – when he will first print his "News from Nowhere". if it succeeds I believe he contemplates bringing forth quartos and folios in large numbers,' she wrote on 4 December 1890.[56] (*News from Nowhere* (1892) was to be the twelfth book printed at the Kelmscott Press.) And on 29 December: 'he has not started yet, but will soon. he has taken a little place where he can set up the press.' On 30 January she announced: 'The printing is not yet started, some of the letters had to be re-cut, and there were various little delays. the press is in its place at last, and the type has arrived and been pronounced satisfactory so the printing will really begin. the paper is quite lovely, I shall write you next on a sheet of it. . . . we shall call the press the "Kelmscott Press". . . .'

Founding the Press

The name was symbolically important to Morris, for it linked his Press with Kelmscott Manor, the beautiful sixteenth-century house in south-western Oxfordshire of which Morris had been a tenant since 1871 and to which he was deeply attached. Sydney Cockerell, while visiting Kelmscott, recorded in his diary on 10 August 1892: 'W.M. speculating as to whether the Kelmscott Press could not be moved from Hammers[mith] and set up in the meadow here!' The exclamation point always appears in Cockerell's diary when he is reporting Morris's most outrageous proposals: in this instance both he and Morris must have known that shifting the Press to a remote village on the edge of the Cotswolds was not really possible. C. R. Ashbee was to attempt something like this with his Essex House Press, and the result was not encouraging. The Kelmscott Press, then, would have to remain in Hammersmith, on the western outskirts of London, but the pleasant riverside atmosphere of Hammersmith's Upper Mall offered an attractive alternative to rural life which had the important advantage of proximity to the craftsmen who would be essential to the success of the Press. In fact, three of them – Walker, Cobden-Sanderson, and W. H. Hooper (the chief wood-engraver of the Press) – were only a few steps away from Morris's home, Kelmscott House; to no one's surprise, therefore, Morris chose to establish the Kelmscott Press at 16 Upper Mall, a brick Georgian cottage merely a few yards from his own front door, which he rented from 12 January 1891.

Sparling's description of the building makes it clear that, in this earliest phase, the Kelmscott Press was bursting at the seams: 'One upstairs room of the cottage was fitted with racks, cases, imposing-stone, etc.; another housed the single press, bought secondhand; the rooms on the ground-floor were utilized for stores.'[57] The surviving *Figure 33*

Figure 33 A nineteenth-century photograph of Sussex House in Hammersmith, occupied by Emery Walker's firm. The smaller entrance on the left is to Sussex Cottage, which was the chief home of the Kelmscott Press.

rent book discloses that Morris paid a rent of 8*s*. 6*d*. per week and that he gave up his tenancy on 13 June of the same year.[58] In the meantime Morris had acquired the lease of Sussex Cottage, 14 Upper Mall, which was attached to No. 16, Sussex House, used by Walker and Boutall.[59] Cockerell, by coincidence, visited Sussex Cottage for the first time on 4 April 1891, when the colophon for *The Glittering Plain* was being printed, and left this account in his diary: 'Emery Walker took me to see the Kelmscott Press, in which the Glittering Plain, the first book to be issued from it, is now being printed from Morris's new type. . . . Exceedingly interesting. The house a little old fashioned one and the single hand press at the top of a winding corner stair, the compositor being in a little room leading out of that in which the press is. Printed sheets, one on vellum, lying about – all most beautiful, especially the first page with its elaborately designed border.' Sussex Cottage remained the chief home of the Kelmscott Press until it closed, although another cottage, 21 Upper Mall, was leased in January 1895 to house the large Albion that was used for the *Chaucer*.

On 3 January 1891 Morris took the final decisive step in founding the Kelmscott Press. To William Bowden, a retired master-printer who had printed *News from Nowhere* for Reeves and Turner, he wrote: 'Mr. Reeves tells me that you would perhaps engage yourself to me as compositor & press-printer in the little typographical adventure I am planning. I should be glad to see you about the matter: could you call on me here?'[60] Did Morris really think that one man could print the large books he had in mind? Soon Bowden's daughter (Mrs Pine) and son (William Henry Bowden) joined their father in Upper Mall, Hammersmith, and the staff gradually grew as Morris's plans for printing expanded. We have Sparling's word that Morris himself never set a line of type, but, as in his earlier craft revivals, he asked questions constantly, peered over the shoulders of his printers, and familiarized himself with all the technical aspects of printing. Despite increasing ill health, Morris was in exuberant good spirits during the early months of 1891 as he watched his old dream of producing beautiful books begin to take substance. W. H. Bowden has left this lively picture of Morris at the time: 'When the type came in from the founders, he was very anxious to help lay it in the cases; but not having served his time to the business, more often than not put the type into the wrong box. It was very amusing to hear him saying to himself: "There, bother it; in the wrong box again!" But he was perfectly good-humoured, and presently ran off and came back, bustling up the path – and in my mind's eye I can see him now – without a hat, and with a bottle of wine under each arm, with which to drink the health of the Kelmscott Press.'[61]

The first delivery of paper from Batchelor arrived on 27 January, and on the last day of the month the elder Bowden printed for Morris several versions of a trial page of *The Story of the Glittering Plain*. Enclosing one of the trials in a letter to Ellis a few days later, Morris exulted, 'I don't know what you will think of it; but I think it precious good.'[62] He was now impatient to get a book from his Press into print as quickly as possible. Morris had originally intended that *The Golden Legend* should be the first title, but the small sheets supplied by Batchelor seemed more suitable for his *Glittering Plain*, a book of modest size. Unwilling even to wait for some illustrations that Walter Crane had promised for *The Glittering Plain*, he plunged ahead with an unillustrated quarto.

Meanwhile, rumours about Morris's Press began to find their way into print. One of the earliest notices was this item in the 'Literary Gossip' column of the *Athenaeum* on 21 February 1891: 'Mr. William Morris is getting his press into working order. The printing of "The Golden Legend" will be preceded by that of the romantic story first published in the *English Illustrated Magazine* under the name of "The Glittering Plain." A very limited number will be printed as the first issue of the Kelmscott Press, by which name Mr. Morris calls his new enterprise.' There was a further reference to *The Glittering Plain* in the *Athenaeum* on 4 April, with surprising results: 'The fact is on the first announcement in the Athenaeum there was a rush to Reeves & Turner and they sold every copy long before the price was announced,' according to Walker.[63]

Figure 34

So the Kelmscott Press was to be, after all, a commercial success; not even Morris, in his most sanguine moods, had anticipated that. Yet it always remained a small, intensely private enterprise, infused with Morris's own distinctive personality and conducted in a slightly improvisational style. The casual tone of the Press's operations was evident from the beginning. The official publication date of *The Glittering Plain* was 8 May. On the previous day, however, the first copy of the book was actually sold – in Morris's dining-room. Wilfrid Scawen Blunt happened to have lunch with the Morrises that day, and what then took place is recorded in Blunt's diary: 'He [Morris] is full as ever of his work and took me to see his printing press a few doors off, and sold me across the dining table the first copy of

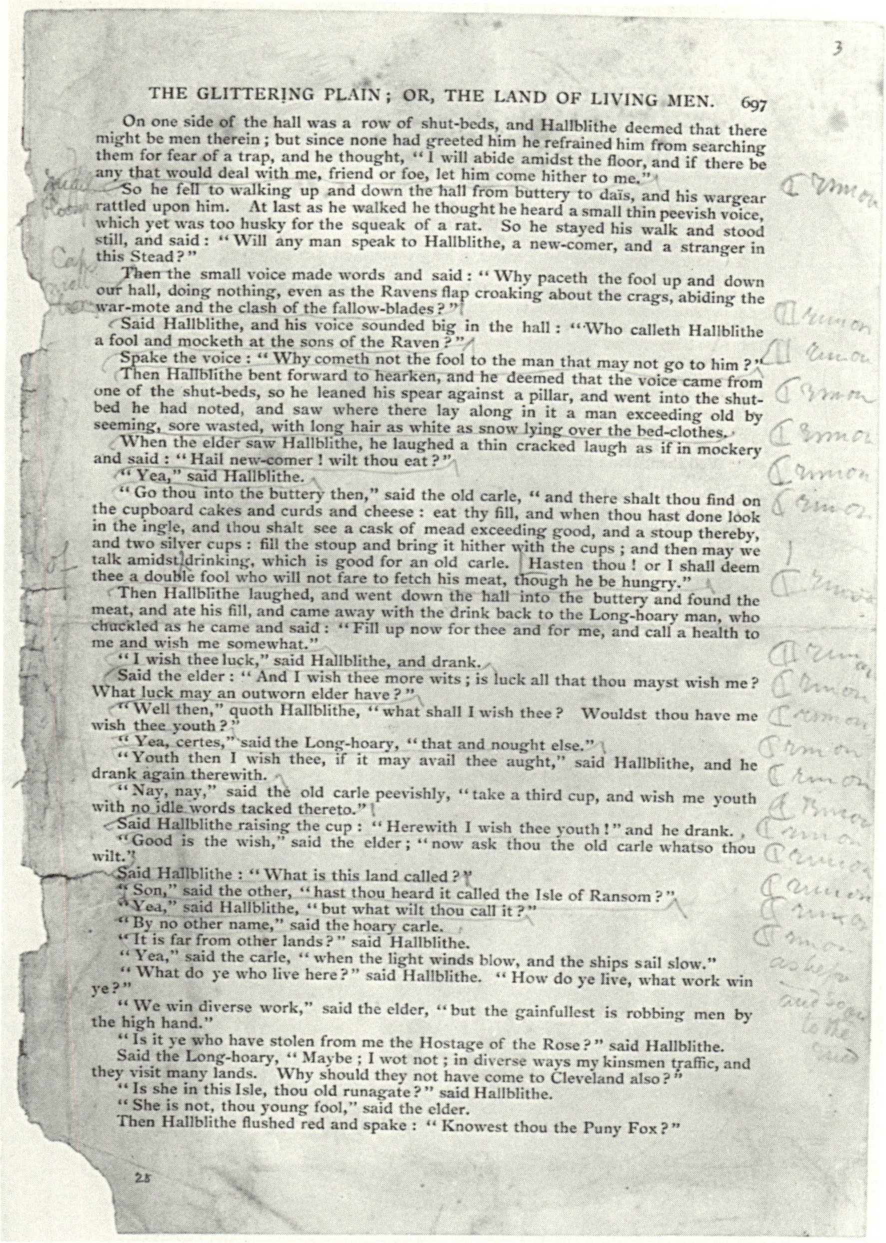
3

THE GLITTERING PLAIN; OR, THE LAND OF LIVING MEN. 697

On one side of the hall was a row of shut-beds, and Hallblithe deemed that there might be men therein; but since none had greeted him he refrained him from searching them for fear of a trap, and he thought, "I will abide amidst the floor, and if there be any that would deal with me, friend or foe, let him come hither to me."

So he fell to walking up and down the hall from buttery to daïs, and his wargear rattled upon him. At last as he walked he thought he heard a small thin peevish voice, which yet was too husky for the squeak of a rat. So he stayed his walk and stood still, and said: "Will any man speak to Hallblithe, a new-comer, and a stranger in this Stead?"

Then the small voice made words and said: "Why paceth the fool up and down our hall, doing nothing, even as the Ravens flap croaking about the crags, abiding the war-mote and the clash of the fallow-blades?"

Said Hallblithe, and his voice sounded big in the hall: "Who calleth Hallblithe a fool and mocketh at the sons of the Raven?"

Spake the voice: "Why cometh not the fool to the man that may not go to him?"

Then Hallblithe bent forward to hearken, and he deemed that the voice came from one of the shut-beds, so he leaned his spear against a pillar, and went into the shut-bed he had noted, and saw where there lay along in it a man exceeding old by seeming, sore wasted, with long hair as white as snow lying over the bed-clothes.

When the elder saw Hallblithe, he laughed a thin cracked laugh as if in mockery and said: "Hail new-comer! wilt thou eat?"

"Yea," said Hallblithe.

"Go thou into the buttery then," said the old carle, "and there shalt thou find on the cupboard cakes and curds and cheese: eat thy fill, and when thou hast done look in the ingle, and thou shalt see a cask of mead exceeding good, and a stoup thereby, and two silver cups: fill the stoup and bring it hither with the cups; and then may we talk amidst drinking, which is good for an old carle. Hasten thou! or I shall deem thee a double fool who will not fare to fetch his meat, though he be hungry."

Then Hallblithe laughed, and went down the hall into the buttery and found the meat, and ate his fill, and came away with the drink back to the Long-hoary man, who chuckled as he came and said: "Fill up now for thee and for me, and call a health to me and wish me somewhat."

"I wish thee luck," said Hallblithe, and drank.

Said the elder: "And I wish thee more wits; is luck all that thou mayst wish me? What luck may an outworn elder have?"

"Well then," quoth Hallblithe, "what shall I wish thee? Wouldst thou have me wish thee youth?"

"Yea, certes," said the Long-hoary, "that and nought else."

"Youth then I wish thee, if it may avail thee aught," said Hallblithe, and he drank again therewith.

"Nay, nay," said the old carle peevishly, "take a third cup, and wish me youth with no idle words tacked thereto."

Said Hallblithe raising the cup: "Herewith I wish thee youth!" and he drank.

"Good is the wish," said the elder; "now ask thou the old carle whatso thou wilt."

Said Hallblithe: "What is this land called?"

"Son," said the other, "hast thou heard it called the Isle of Ransom?"

"Yea," said Hallblithe, "but what wilt thou call it?"

"By no other name," said the hoary carle.

"It is far from other lands?" said Hallblithe.

"Yea," said the carle, "when the light winds blow, and the ships sail slow."

"What do ye who live here?" said Hallblithe. "How do ye live, what work win ye?"

"We win diverse work," said the elder, "but the gainfullest is robbing men by the high hand."

"Is it ye who have stolen from me the Hostage of the Rose?" said Hallblithe.

Said the Long-hoary, "Maybe; I wot not; in diverse ways my kinsmen traffic, and they visit many lands. Why should they not have come to Cleveland also?"

"Is she in this Isle, thou old runagate?" said Hallblithe.

"She is not, thou young fool," said the elder.

Then Hallblithe flushed red and spake: "Knowest thou the Puny Fox?"

25

Figure 34 Morris's *The Story of the Glittering Plain* was originally published in the *English Illustrated Magazine* (June–September 1890); here Morris marks the text for his printers at the Kelmscott Press. In the 1891 edition nearly all indentations were replaced by paragraph symbols, thus creating the solid, black-looking rectangle of type that Morris favoured.

Founding the Press

the first volume printed at the Kelmscott Press, a very wonderful production. . . . While we were sitting talking this copy was brought in to him by the printer, the first that had been bound and made up, and I asked him "what do you mean to do with these books?" He said "to sell them of course[.]" "Will you sell this copy to me?" "Certainly[.]" "How much?" "A pound". So I paid a pound and took it away. . . .'

Two and a half years after Emery Walker's memorable lecture, the Kelmscott Press was successfully launched.

FOUR ❦ MORRIS AS BOOK DESIGNER

EVERYONE knows (or thinks he knows) what a Kelmscott Press book looks like: it is a massive folio printed in dense black type with large, overpowering wood-engraved initials, borders, and illustrations. Such at least is the popular notion of a typical Kelmscott product, a stereotype reinforced by the innumerable photographs of the most heavily ornamented pages of the *Chaucer* that seem inevitably to appear in every history of printing and book-design.

In reality, though Morris did, like most designers, develop a general typographical formula, he applied it to individual volumes with great subtlety and skill, so that there is no such thing as a single completely representative Kelmscott Press book. Notwithstanding their apparent uniformity of manner, one can find evidence of Morris's growing sophistication and self-assurance as a book-designer as the Kelmscott books gradually came off the press. The ornamentation slowly became richer and the pictures better integrated with the words. Morris showed more daring in his use of red ink: at first employed abstemiously, it eventually lent colour to colophons, notes, and his printer's mark. In *Psalmi penitentiales* (1894) the Latin text of each verse in red was followed by the English text in black; in *Sidonia the Sorceress* (1893) the red notes flowed all round the edges of the black text. Even those elements we think of as Morris's typographical signature – the shoulder-note, the engraved title-page, the all-caps, unleaded title (frequently unjustified) – evolved from a surprisingly prolonged process of trial and error.

Because the *Chaucer* is the most famous and widely discussed work of the Kelmscott Press, many readers have assumed that it is also typical. The *Chaucer*'s design represents a solution (a triumphantly successful solution, some would argue) to a particular set of problems of typography and illustration; but had Morris lived to complete the Froissart, another double-column folio of similar proportions, we know that it would have been fundamentally different because of a new style of borders and initials that make their first appearance in the Froissart trial pages. And of course both the *Chaucer* and the unfinished Froissart are, in their amplitude, unrepresentative Kelmscott books: most of the works issued by Morris's Press were octavos, and the majority of them were unillustrated. (Even the *Chaucer* contains many pages, seldom reproduced in photographs, with no wood-engraved illustrations and only a few small initials.) Morris undoubtedly had a great fondness for folios – he argued, rather unconvincingly, that they are easier to read because folios lie submissively and quietly on a table – yet at the other extreme he also produced four of the Kelmscott titles in the format known as 16mo (with a page size of $5\frac{3}{4} \times 4\frac{1}{8}$ inches). These are wonderfully delicate little books which belie the stereotype of ponderous, overweight folios flowing in rapid succession from the Kelmscott Press.

Another way of circumventing the usual misconceptions about the Kelmscott typographical style is to recognize that Morris's principles could be – indeed often were – applied to commercially produced books as well as limited private-press editions. Morris contended that any book, even the most cheaply manufactured shilling reprint, could be improved by the adoption of a few simple rules: the use of old-face types, the reduction of white space between words and lines, and the careful adjustment of margins. As for ornaments and illustrations, it was better to eschew them entirely than to adopt mass-produced or unsuitable ones. Despite his ill-concealed misgivings about the printing press as an instrument for producing universal literacy, Morris was realistic enough to recognize that newspapers, magazines, and the majority of books had to be manufactured as economically as possible. Only the occasional volume could fit his prescription for 'the Ideal Book': 'a book not limited by commercial exigencies of price.'[1]

We can, therefore, see Morris the book-designer in another light when, from time to time, we find him wrestling with the problems of commercial printing in the 1890s. At the very moment when he was about to found the Kelmscott Press, for instance, Morris was also carrying on negotiations with Bernard Quaritch about the 'Saga Library' series that Quaritch was to publish under Morris's editorship. In an undated memorandum to Quaritch (probably written in 1890) Morris emphasized that 'I should wish to have some controul [*sic*] over the printing and general "get-up" of the books. . . . It would, if I may say so, be of some use to the sale of them as *my* public expects from me a certain style.' Specifically he suggested 'that the books should be printed in old-style type', preferably either Caslon or Miller and Richard's Old Style. 'If Caslons were used the size ought to be pica, as his letters have long ascending and descending strokes and consequently look rather smaller for their size. The position of the p[age] of type on the leaf ought to follow ancient precedent, and not be plastered right in the middle as modern printers treat it.' Morris proposed supplying wood-engraved initials if necessary, but 'all "rules" and little twiddledy printers ornaments I should do without'.[2] (Morris's admiration for Caslon's type was not unqualified,[3] but, like Walker, he endorsed it because it was the only typeface then widely available that even approximated his ideal.)

There are other letters from Morris – for example, one written to the publisher Swan Sonnenschein in 1893 about the design of the pamphlet *Socialism: Its Growth and Outcome*[4] – in which he offers remarkably sensible advice about printing books under ordinary trade conditions. In the same vein, when approached in 1894 about producing a syllabus at the Kelmscott Press, Morris dictated to Cockerell the following reply: 'Mr. Morris suggests that you should have the syllabus set up in the plainest possible manner, without rules or ornaments, (& if possible in Caslon's Old faced type) & it will then in his opinion be certain to look well.'[5] As Morris said on another occasion, 'whatever the subject-matter of the book may be, and however bare it may be of decoration, it can still be a work of art, if the type be good and attention be paid to its general arrangement.'[6]

Books have a tendency to become beautiful objects unless we deliberately and perversely make them ugly. 'I lay it down . . . that a book quite un-ornamented can look actually and positively beautiful, and not merely un-ugly, if it be, so to say, architecturally good,' he declared.[7] Yet however much Morris might praise the plain, unadorned book, he in fact produced at the Kelmscott Press volumes that were anything but simple and functional; and this discrepancy between his theory and practice has led to interminable disagreements about the nature of his contribution to modern book-design. Some critics, fixing their gaze on the Kelmscott books, have seen him as the advocate of a regressive, pseudo-medieval style of bookmaking that is irrelevant to contemporary needs; others, emphasizing his writings on design and typography, have identified Morris as the unacknowledged source of the great revival of English commercial printing that took place in the 1920s. Was Morris, then, an escapist who misled several generations of private-press printers into believing that they could turn the hands of the clock back, or was he rather, in Nikolaus Pevsner's famous phrase, one of the 'pioneers of modern design'?[8]

There is no doubt, as Morris never tired of arguing, that choosing an attractive typeface and arranging the block of text carefully on the page add little to the cost of a book. What made the Kelmscott books so expensive to produce, however, was Morris's insistence upon handicraft methods, lavish ornamentation, and high-quality materials. It is not surprising if we are sometimes baffled by the nature of his 'Ideal Book', for Morris himself harbours ambiguous aims: even as he enunciates eminently sane proposals for reforming the commercially manufactured Victorian book, we sense that he is secretly hankering after the illuminated French manuscript of the thirteenth century. In the presence of such myriad-mindedness, we need not be embarrassed by our own confusion when trying to interpret the meaning of the Kelmscott Press and the ideas that lay behind it.

In examining the techniques that Morris as a printer actually adopted, it is important to bear in mind that his primary intention, when he established the Press, was to print a few copies of books for his own pleasure and that of his friends; under these circumstances, he had little choice but to make use of the hand-press, which was ideal even for the somewhat larger pressruns (on an average, between 300 and 500 copies) he eventually settled on for the Kelmscott books. The slow pace of Morris's Albion presses made it possible for his printers to inspect each sheet individually for imperfections, and, by pausing a split second when the forme came into contact with the paper, the pressman could ensure that Morris's heavy ornaments printed solidly and blackly. Indeed, all those wood-engraved initials and borders, so characteristic of the Kelmscott volumes, would have been out of the question had Morris selected a power-driven press.[9] He was of course well aware of this issue: in 1890, writing to Quaritch about the Saga Library, he urged that 'the large-paper copies should be printed with the hand-press, so as to get the impression black enough'.[10]

The adoption of the hand-press was also closely connected with Morris's decision to print on dampened handmade paper, which would have been impossible on a cylinder press. 'At present there is very little good paper made, and most of it is very bad,' Morris declared in a lecture in 1893. 'Our ideal book must, I think, be printed on hand-made paper as good as it can be made. . . .'[11] Joseph Batchelor's attractive (and durable) paper was a primary ingredient in the success of the Kelmscott Press books; one consequence of this is that any photographic reproduction of a Kelmscott page invariably fails to convey the superb tactile quality of the original.

Towards the end of the nineteenth century it became fashionable to produce *éditions de luxe* and facsimile reprints in pseudo-handmade papers, nearly always with ragged, untrimmed edges. The earliest Kelmscott books reflect this 'deckle-fetishism' (as John Carter has called it), but it is significant that Morris assumed his books would be eventually rebound and the deckle trimmed: in a letter written in November 1892 which probably refers to *The Golden Legend*, he explained, 'The edges could be easily trimmed when the book is bound, so as to get more uniformity in the size. Of course the book is issued "done up", and not bound: and under the circumstances it would have been a mistake I think to have trimmed the edges.'[12] But Morris very soon changed his mind, perhaps because he realized that most Kelmscott books were not after all being rebound by their owners, and *Biblia innocentium*, which was issued on 9 December 1892, was the last volume to come from the Kelmscott Press with untrimmed edges. A few collectors of more conventional tastes found this a distressing decision. 'As to the cut edges of the K.P. books W.M. followed the custom of the early printers, & had them trimmed,' Cockerell wrote to Harry Buxton Forman. 'I am sorry that you think this such a blemish.'[13] Even Forman, however, may have found comfort in the fact that the Kelmscott volumes were trimmed only very slightly.

Several of Morris's pre-Kelmscott books had also been issued in large-paper editions, but by the time he set up the Press he realized that this practice ruined the proportions of the margins and introduced too much white space on to the page. 'I am clean against [large-paper editions],' Morris announced in 1893, 'though I have sinned a good deal in that way myself, but that was in the days of ignorance, and I petition for pardon on that ground only.'[14] For the sake of collectors, he printed a handful of copies of most of the Kelmscott books on vellum, though Morris and Walker agreed that they personally preferred paper copies. Finding good vellum – which is made from the skin of young calves – was not easy. Morris in 1891 still owned a small supply of Roman vellum which he had imported for his calligraphic experiments two decades earlier, and this he used to print six copies of *The Story of the Glittering Plain*, the first Kelmscott book; unfortunately he could not obtain more from the same source, because the Vatican bought up all the best Italian vellum. Sparling offers the amusing information that Morris briefly entertained the thought of appealing directly to the Pope for some vellum 'on the ground that the *Golden Legend* was a book in which he ought to be interested';[15] such a violation of Morris's agnostic scruples proved to be unnecessary,

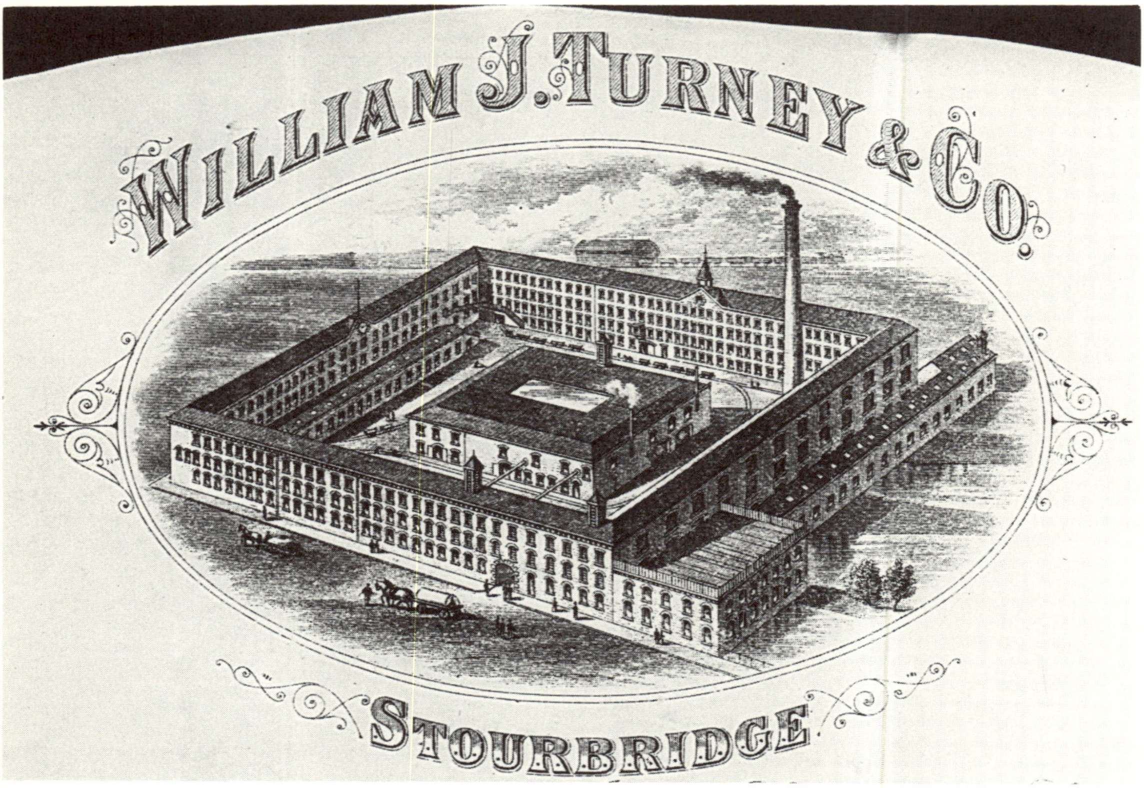

Figure 35 When he discovered that he could not obtain vellum for the Kelmscott Press in sufficient quantity from Italian sources, Morris turned to two English suppliers, one of whom was William J. Turney.

however, when he found an English manufacturer of vellum, Henry Band (of Brentford, Middlesex), who was capable of meeting his exacting standards. Later Morris turned also to a second supplier, William J. Turney and Co. (of Stourbridge, Worcestershire).[16] *Figure 35*

W. H. Bowden has left a detailed record of how the vellum copies were produced at the Kelmscott Press,[17] and his description makes it clear that printing on vellum is a formidably difficult process. Until Morris's printers perfected their procedures, there was a good deal of spoilage. They learned, for example, to print always on the rough side of the vellum first, because that was where problems often arose; and the printing in red ink was done last, after the dampened vellum had dried somewhat, since the moisture caused the ink to run. One can detect also the weary voice of experience in this piece of advice from Bowden: 'Always print *one* more Vellum than the number required, in case of accident; and if there are any heavy borders or larger initials, two should be pulled. This will be found in the end to be a great saving.'

Whatever reservations Morris himself had about vellum as a substitute for paper in printing, collectors have always admired the brilliant contrast of black on white that it produces. A vellum *Chaucer* is undoubtedly a dazzling sight. Of course the Kelmscott books printed on vellum also appeal to the acquisitive instincts of some collectors because Morris sold them for five or six times the price of paper copies and because they are considerably more scarce. 'There are three complete vellum sets in existence,' Cockerell wrote in 1898, 'and the extreme difficulty of completing a set after the copies are scattered makes it unlikely that there will ever be a fourth.'[18] When John A. Saks, an American, succeeded in this quixotic undertaking, his vellum set of the Kelmscott books realized $660,000 at auction in 1983.

In contrast with paper or vellum, ink attracts little attention from readers; even Morris had nothing to say about it in his lectures and essays on the book arts. Nevertheless, ink is a surprisingly important element in printing, and the range of choices available is determined to a large extent by the kind of press being used. As the speed of rotary and cylinder presses increased throughout the nineteenth century, printers were compelled to use progressively thinner inks which produced pages always more greyish-looking. By selecting the Albion hand-press, Morris was able to turn to inks that were much thicker and slower-drying – and presumably closer in consistency and appearance to the inks employed by fifteenth-century printers. Indeed, when an English publisher recently tried to match Morris's ink for a facsimile reproduction of the Kelmscott *Chaucer*, it was discovered that 'the original ink has never completely dried – suggesting that Morris himself never found the perfect answer to the problems which beset him in finding an ink to meet his requirements'.[19]

The ink difficulties proved in fact to be more intractable than the paper problem (which had been solved as soon as Morris found Joseph Batchelor), and, although it constitutes only a minor episode in the history of the Kelmscott Press, the story of Morris's search for an ideal ink is a small but illuminating example of the smoothed-over version of events given to us by earlier historians of the Press. Cockerell, typically, offers an admirably concise but incomplete account: 'The black ink which proved most satisfactory, after that of more than one English firm was tried, was obtained from Hanover. Morris often spoke of making his own ink, in order to be certain of its ingredients, but his intention was never carried out.'[20] Sparling, on the other hand, devotes more than two pages to the matter in his book on the Kelmscott Press, quotes Morris on the 'damned chemists' who have ruined nineteenth-century inks, supplies an excellent description of the ingredients and character of Morris's ink, and in general treats the episode as still another instance of Morris's rejection of modern technology in favour of the sound, simple materials associated with traditional craftsmanship. The German ink which Morris finally adopted, he admits, was rather thick, and 'in the average press-room of those days – or in these? – a little soft soap would soon have got into it . . . ; [but] quickness of working was not asked for at the Kelmscott Press'.[21] (Sparling, in his tribute to the purity of Morris's arts-and-crafts ideals, fails to mention that the German ink was made in an enormous factory with smoke billowing from a multitude of smokestacks.)

Figure 36

Contrast with this the recollections of Emery Walker, who first discovered the superiority of the ink manufactured in Hannover by Gebrüder Jänecke, 'having tried [he wrote] every ink maker in England without success. Many years afterwards when Morris was starting up the Kelmscott Press the question of ink came up and I advised him to use the same ink. It was tremendously stiff and very hard work for the pressmen, and the introduction into the office produced something very near a strike, and Morris[,] anxious to get on having a vast prospect of books before him which he wanted to print, rather weakly gave in and used the best ink he could get in England. This turned out to be unsatisfactory, so Morris read the riot act and said if they wouldn't use the German ink he would close the Press. As his pressmen were very well paid they naturally did not want to lose their jobs and gave way. However, they took it out of Morris another way and the number of sheets turned out per day were extraordinarily small.'[22]

Morris's outburst to his pressmen sounds in character (though the threat of plant closure was perhaps not entirely consistent with his socialist principles), and Walker's version of the Kelmscott Press as a tense, accident-prone operation is a useful corrective to Sparling's lyrical praise of the spiritual harmony of its pressroom. Sparling, Cockerell (though he is generally more cautious in his handling of evidence than Sparling), and other disciples of Morris tended to ignore his failures, to gloss over his unsuccessful experiments and his indecision, and to depict the history of the Kelmscott Press as a series

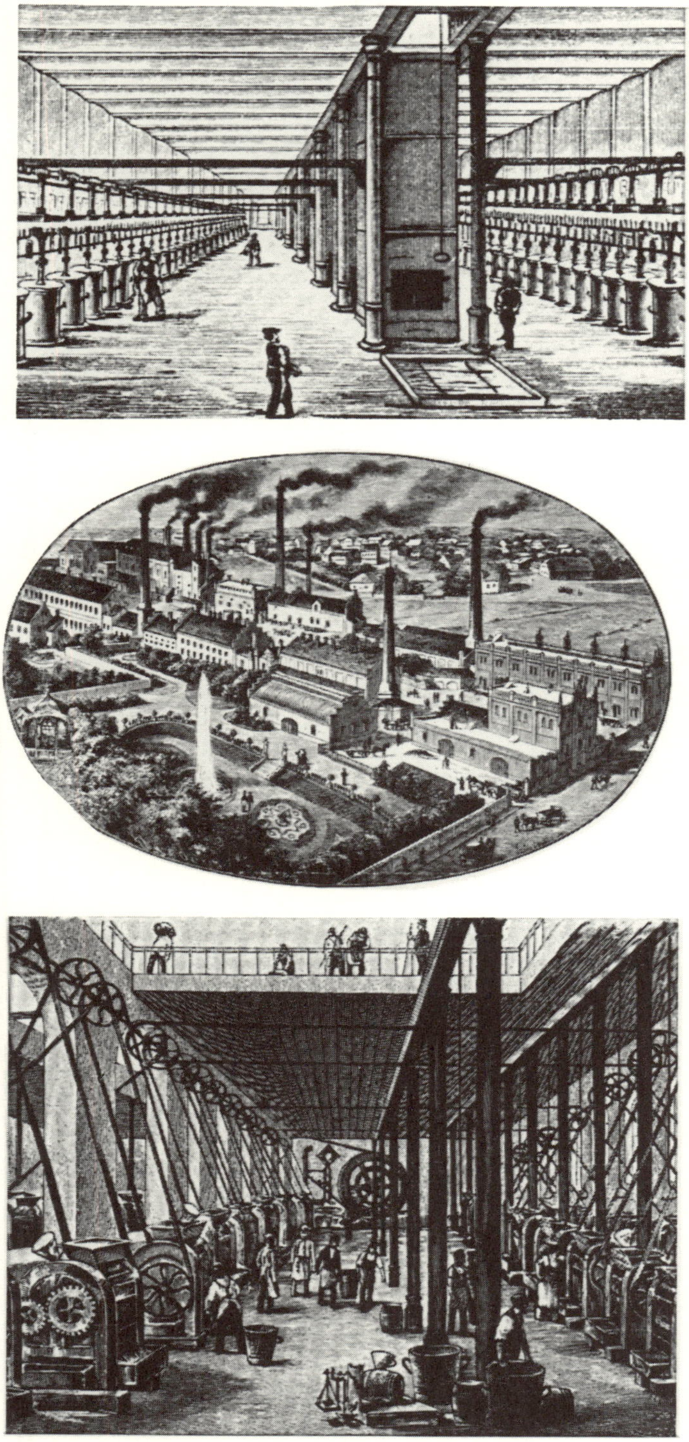

Figure 36 Three views of the Gebrüder Jänecke plant in Germany, the source of Morris's black ink. 'The Hannover ink was tried on Friday and worked well: I shall use no other now,' Morris declared in August 1895.

of unqualified triumphs. In fact, Morris as a printer was guilty of occasional artistic and technical bungling, especially in the early stages of the production of books; Sparling and Cockerell, in their solemn, reverential accounts of the Press, tell us almost nothing about Morris's habitual tendency to muddle through, and to extricate himself in the end from alarming (and sometimes amusing) predicaments.

It was evidently in 1893 that Morris briefly used the German ink which caused so much discomfort to his pressmen. Faced with rebellion in Sussex Cottage, he reverted to the ink supplied by Shackell, Edwards and Co., a London manufacturer. (It is ironic, in view of Morris's problems with Shackell, Edwards inks, that the firm later advertised one of their products as '"Morris" Black Ink'. Like Batchelor, though with less justification, they were attempting to exploit their connection with the Kelmscott Press.[23]) Morris, uncharacteristically, tolerated an ink that seemed to him second-best until a major crisis arose in the spring of 1895, when yellow stains began to appear on some of the pages of the *Chaucer*, then in production. With a large financial investment in the book at stake, Morris was close to panic. On 24 May Cockerell noted in his diary, 'Worry about the ink used in the Chaucer which shows a yellow stain behind some of the heavy borders,' and the same day Morris wrote (in *his* diary): 'Cockerell showed me the p.p. of Ch[aucer] where the ink has gone wrong[;] don't know what to make of it.'[24] There were hurried consultations with the people at Shackell, Edwards (who seem to have suggested tactfully that perhaps the problem lay in Batchelor's paper), and by 8 June Morris was writing to Gebrüder Jänecke and Fr. Schneeman of Hannover: 'Some time ago I tried a sample of your black ink, at the recommendation of Messrs. Walker & Boutall, and found it excellent in all respects, but rather too stiff and difficult to work. I now write to ask whether the quality would suffer if it were made a little softer. It is quite essential to me that it should be sound and lasting and free from any possibility of change or discolouration. I send you a sample page that you may understand the nature of the work for which the ink is required.' Jänecke shipped a softer version of the ink to Morris, who reported triumphantly to Walker on 19 August, 'The Hannover ink was tried on Friday and worked well: I shall use no other now.' All signatures from T onwards in the *Chaucer* were printed with the Jänecke ink.

Figure 37

Meanwhile, there was the problem of the sheets of the *Chaucer* that were already stained. Throughout the spring and summer of 1895

Figure 37 The London plant of Shackell, Edwards and Co., which supplied Morris's black ink until mysterious yellow stains began to appear in the *Chaucer* sheets in 1895. Despite this unhappy episode, the firm later advertised one of its products as '"Morris" Black Ink'.

Morris, Cockerell, and Burne-Jones consulted every expert they could lay their hands on, even including two 'damned chemists'. The most useful advice they received was to bleach the paper in the sun, and for two months Morris's printers were periodically laying out *Chaucer* sheets in the garden of Sussex Cottage. Whenever it rained (and it seems to have been an unusually wet summer, even by English standards), the printers had to dash into the garden and bring the sheets indoors. Finally, in desperation, Morris rented a greenhouse. But the trick worked: 'Re-examined the 6 sheets of Chaucer . . . from which air & light were excluded (after bleaching it in the sun of Aug. 16th'[)],' Morris wrote in his diary on 9 November. 'Examined Sept. 10th & 23rd & Oct. 5. No perceptible return of the yellow stain.' With a collective sigh of relief, the printers resumed work on the *Chaucer*.

Morris never found a convincing explanation of what had gone wrong with the Shackell, Edwards ink, nor (being incurious about theoretical questions) did he seek one once he had found a pragmatic solution of his problem. There are faint yellow stains in several of the early Kelmscott Press books (notably *The Golden Legend*), usually behind heavy ornaments, and, as Morris's friend Jacobi pointed out, similar stains occasionally appear in incunabula: perhaps, Jacobi suggested, the phenomenon stemmed from an imperfect fusion of the two chief ingredients of ink: carbon and varnish.[25] Whatever the reason, Morris was content to have solved the immediate dilemma, yet the price he paid in adopting the Jänecke ink is evident in a Kelmscott Press announcement of February 1896 which informed customers that, 'owing to the strength and slow-drying qualities of the ink used, the Chaucer will not be fit for ordinary binding, with the usual pressure, for at least a year from its issue'. After Morris's death, when the production of *Sire Degrevaunt* (1897) was rushed so that it could be issued at the same time as its companion volume *Syr Isambrace*, his binder found it necessary to insert a flimsy between the frontispiece and the first text page because the ink was not yet dry; one assumes that this was done with reluctance, since it violated Morris's principle that the two pages of an opening must be seen together.

Morris also used red and (less frequently) blue inks at the Kelmscott Press. The red he employed extensively for titles and shoulder-notes, though only rarely for initials; two Kelmscott books, *Laudes Beatae Mariae Virginis* (1896) and *Love is Enough* were printed in three colours. For both the red and blue inks he turned again to commercial sources. The blue (made from ultramarine ash), supplied by Winsor and

Newton to Morris's own specifications,[26] is beyond criticism. On the other hand, the proofs of the Kelmscott books, as well as the books themselves, reveal that Morris experimented with several reds, and the evidence suggests that he was never fully satisfied with any of them. He objected to red inks that were too pink or orange (the ink used in Blunt's *Love-Lyrics* (1892), for instance, is distinctly carroty), and Shackell, Edwards's red, which he adopted for a time, showed a tendency to spread on the damp paper. For this reason, the Kelmscott pressmen always worked the red ink after the black when possible.

Behind the scenes at the Kelmscott Press there were frequent experiments with inks. Proofs of the engraved title-page of *Reynard the Foxe* (1893) were struck off in red ink, a practice Morris never adopted in any of the Kelmscott books; more astonishingly, a set of proofs of the illustrations for *The Earthly Paradise* and *Love is Enough* was printed in gold ink, thus lending new meaning to the phrase 'Golden type'.[27] Morris bought brown inks (which he did not in the end use); he printed blue inks from various manufacturers on both wet and dry papers in attempting to achieve the effects he sought;[28] in 1891 he compared the results from applying ink to the formes with ordinary composition rollers and then with litho rollers;[29] for several years he tried many combinations of hand-coloured and printed ornaments, none, evidently, to his satisfaction.

This sort of restless curiosity did not extend to the Kelmscott bindings: here Morris at once found an answer – admittedly a compromise – to his requirements and never departed from it except in the *Chaucer*, some copies of which were issued in a special pigskin binding designed by Morris and executed at the Doves Bindery. Though all the other Kelmscott books were bound by an ordinary (but conscientious) commercial firm, their structure and appearance were much influenced by Morris's friend Thomas J. Cobden-Sanderson, who achieved fame as both a binder and a printer. Cobden-Sanderson had been an unhappy young barrister in 1883, with suppressed arts-and-crafts longings, when Jane Morris turned to him after dinner and said, 'why don't you learn bookbinding. That would add an Art to our community, and we would work together.'[30] Two days later he apprenticed himself to Roger de Coverley, a leading bookbinder, and, with encouragement from the entire Morris family (including May, who said that his new vocation was 'much nicer than being a barrister'[31]), Cobden-Sanderson soon set himself up in business. 'I think it would be a very good thing for you to get a bookbinding business

Figure 38

Figure 38 The workshop of Thomas J. Cobden-Sanderson. Perhaps the most gifted bookbinder of modern times and the co-founder of the Doves Press, he was a close friend and adviser of Morris, and his views influenced the style of binding adopted at the Kelmscott Press. Cobden-Sanderson was launched on his career in 1883 when Jane Morris asked him, 'Why don't you learn bookbinding? That would add an Art to our little community, and we would work together.'

together,' Morris wrote to him. 'I would be very glad to give you one of my Editions to do if we could get it done at a possible price.'[32] Because the Doves Press was to be the most direct and distinguished successor to the Kelmscott Press, the names of Cobden-Sanderson and Morris have always been linked in the minds of bibliophiles; but in fact their fundamentally differing personalities fixed a great gulf between them. Cobden-Sanderson was visionary, idealistic, and humourless, and it is apparent that Morris, a craftsman of more energetic and pragmatic temperament, often found his neighbour on Upper Mall a source of irritation. Cobden-Sanderson's sober accounts (in his journal) of his encounters with Morris are filled with unintentional humour: in March 1885 Morris launched into an attack upon Cobden-Sanderson's prices and suggested 'that "people would go to the cheapest market," and almost as if he approved; thought my work too costly; bookbinding should be "rough": did not want to multiply the minor arts(!); went so far as to suggest that some machinery should be invented to bind books'.[33] (Though of course Morris was deliberately exaggerating in order to annoy him, there is a ring of authenticity to these remarks: the style of bindings he adopted for the Kelmscott books was exceedingly plain, if not rough.) The tension in their relationship never dissipated. Sydney Cockerell, who usually reflected Morris's attitudes in such matters, attended a lecture by Cobden-Sanderson on bookbinding and confided to his diary (2 February 1894) that it was 'over far fetched and sentimental'. Cobden-Sanderson, for his part, was capable of writing with chilling condescension about Morris after his mentor's death: 'He should have become an architect, a master builder; he became instead – an upholsterer.'[34]

It would have been out of the question for an entire edition of a Kelmscott Press work to be done in the elaborately tooled binding that had become Cobden-Sanderson's speciality, but the Doves Bindery did bind books for a number of Morris's individual customers, and there was close co-operation between the Bindery and the Press. '[Douglas] Cockerell [Sydney's brother] is not exacting in point of finish, but is all the better suited to the rougher, franker work which we propose to develop in association with the Kelmscott Press,' Cobden-Sanderson recorded in his journal in March 1893.[35] (Notice that by now Morris had won the argument about 'roughness'.) The real importance of the Doves Bindery to the Press, however, lay in its proximity, for Morris was able to turn readily to Cobden-Sanderson whenever a technical question about binding arose.

Though in the early stages of planning *The Recuyell of the Historyes of Troye* Morris speculated that he might have it bound in morocco,[36] he worried that goat and calf skins available through the usual trade sources in nineteenth-century England were inferior and not likely to prove durable. None of the Kelmscott books was therefore issued in a conventional leather binding. Morris instead persuaded J. and J. Leighton, an old and respected London firm,[37] to provide for each of his books one of two simple bindings: full vellum or quarter linen (with blue-grey papers on boards), the latter of which he described to Quaritch as 'a satisfactory temporary binding'.[38] *The Story of the Glittering Plain* was issued in a stiff vellum binding, with washleather ties and the title stamped in the Golden type on the spine. 'I have just received from Morris the "dummy",' Cobden-Sanderson wrote in his journal on 4 April 1891; 'it is a dummy binding by Leighton, to show how he proposes to bind *The Glittering Plain*.'[39] Both Morris and Cobden-Sanderson were still searching for an appropriate binding for the Kelmscott books at this point, so that the first appearance of the vellum version displays two features that were later abandoned: Morris soon decided that he preferred limp to stiff vellum (though the stiff vellum binding is in some respects more satisfactory, as it does not wrinkle and curl), and in his second book, *Poems by the Way*, he substituted silk ties for washleather ties.

Despite these subsequent modifications, the Kelmscott Press vellum binding from the beginning possessed a highly distinctive appearance and was widely imitated by other private presses and commercial publishers. One reason it caught the attention of readers and collectors immediately is that vellum as a material for binding books had long gone out of fashion. Joseph W. Zaehnsdorf, in the second edition of *The Art of Bookbinding* (1890), declared, 'This art of binding in vellum seems to be entirely lost at the present day; its imperishable nature is indeed its only recommendation. It has little beauty; is exceedingly harsh; and little variety can be produced even in the finishing' (p. 147). The simplicity and 'imperishable nature' of vellum were precisely what appealed to Morris (who, curiously enough, was especially fond of the occasionally darker vellum binding with untrimmed hairs and always reserved such Kelmscott books for his own library). Another characteristic feature of Morris's books, one that was not generally imitated or admired, grew directly out of a dictum of Cobden-Sanderson: 'The endpapers . . . should . . . *not be pasted on*; they also should be sewn with the rest of the book. And for this purpose additional sections of plain paper should be supplied to the binder by the publisher to be bound up at the beginning and end of the volume.'[40] In a few instances Morris carried this principle a step further by using blank leaves at the beginning or end of ordinary signatures as endpapers, with the result that, if the book were rebound, those signatures would have to be mutilated. There was at least one contemporary protest against this dubious practice.[41] Another deficiency of the vellum-bound books is that vellum hardens with age, making it now nearly impossible to open fully some of the Kelmscott volumes.

The Kelmscott quarter-linen binding, though more unabashedly an expedient, also set a fashion among other publishers, so that for a generation or more after Morris's death English readers were afflicted with a plague of books issued in boards covered with blue or grey paper. The paper used on the Kelmscott books was a laid, handmade stock that varies slightly in thickness and colour from volume to volume; it seems to have been supplied by Leighton,[42] though the advertising of the Kelmscott Press never mentions it, presumably because Morris did not envision the quarter-linen binding as providing anything but short-term protection for a book. As in the case of vellum, the Kelmscott quarter-linen bindings were slightly modified and improved in the later books: the binding of *The Golden Legend*, for example, is unmistakably inferior to its successors.

One small but annoying problem whose solution eluded Morris was the proper form for titles on bindings. As a rule, the title was gilt-stamped in one of his three typefaces (sometimes enlarged or reduced) on the spine of the vellum bindings. On the quarter-linen bindings the title was printed in black ink on the front cover if the book was slender; for larger books it was printed on a paper label which was pasted to the linen spine. Surprisingly, Morris did not always use Batchelor paper for the spine labels, perhaps because it was too thick and inflexible. The label on the *Chaucer*, for example, is printed on the tissue-thin laid paper watermarked 'Silver Linen' which Morris habitually used for stationery and proofs, and as a result the label has disintegrated badly on nearly every copy of the book. Even on the vellum bindings there were difficulties with titles. For *The Defence of Guenevere, and Other Poems* (1892) Morris arranged for Herbert Ellis, F. S. Ellis's son, to hand-letter the title in Gothic script, and, although he praised it in a letter to the younger Ellis ('I think they are very well written & look very well indeed, and for that size and shape of book look decidedly better than the stamped gilded letters'[43]), it is significant that he did not repeat the experiment. Likewise Morris came to realize in 1894 that titles in Gothic capitals were exceedingly difficult to read, and he thereafter used the Golden type rather than the Troy or the Chaucer type for that purpose.[44]

As usual, even though Morris had settled on two standard forms of binding, he found pleasure in carrying on various curious experiments. A few of the vellum bindings were stained: after calling at Kelmscott House on 5 November 1891, Cockerell wrote that evening in his diary, 'Several copies of Poems by the Way on the dining room table with covers stained red, yellow blue & green.' These books, destined for Morris and his immediate circle of friends, were not offered for sale. There were no silk ties on the vellum binding of *The Poetical Works of Percy Bysshe Shelley* (1894–5), provoking an indignant letter from one of Morris's most faithful customers, the Revd Stopford Brooke. 'The ties were omitted from the Shelley as an experiment, as some people don't care for them,' Cockerell explained to Brooke, 'but except in the 2 other volumes of this work, they will appear in all future books that are bound in vellum.'[45] In the summer of 1894 Tregaskis exhibited in London seventy-six copies of *The Tale of King Florus and the Fair Jehane* (1893) that had been sent to the world's leading bookbinders.

Paper, vellum, ink, and bindings were all matters of much concern to Morris, and we have seen that he struggled, sometimes heroically, to find the highest-quality supplies available; Morris's fame as a book-designer, however, rests less on his use of fine materials at the Kelmscott Press than on the typographical theories he enunciated in a series of lectures and articles during the 1890s.[46] Those theories have had an important, albeit indirect, effect on the physical form of the twentieth-century book, and they have provoked much debate among typographers, yet it is essential to recall that Morris was not temperamentally drawn to abstract theorizing. Robert Catterson-Smith, who worked closely with Morris in the preparation of the Kelmscott Press illustrations, offered this analysis of his employer's mind: 'He hated being asked questions which meant an effort at analysis. He appeared to be unable or unwilling to separate his feelings from anything he thought about. The beauty of the outside of things was enough for his mind, he did not wish to probe the unseen.'[47] By instinct Morris distrusted any attempt to codify a set of rules for designing books.

Still, he did formulate and lay down his own laws in splendidly dogmatic fashion. It is obvious that his conclusions were based directly on his earlier attempts at bookmaking (including the calligraphic manuscripts), conversations with Emery Walker (and perhaps secondarily Charles Jacobi), and his firsthand experiences at the Kelmscott Press, which was still in the early stages of its development when most of Morris's typographical lectures and essays were written. Had Morris lived beyond 1896, it is likely that some of his principles of book-design would have been modified, or at least that his *ex cathedra* tone might have softened; for those principles could not always be reconciled with what he actually did at the Kelmscott Press. One wonders, for example, whether he might eventually have realized that no one typographical style is suitable for all books. Morris of course subtly modulated the Kelmscott manner for his individual volumes, but the family likeness is very strong, and even friends most in sympathy with his aims were aware that such a reductionist approach to book-design fails to do justice to the complexity and variety of literary texts. In a characteristic piece of whimsy, Burne-Jones wrote in 1893 that 'Morris vows he will print Dickens (if he lives) at the Kelmscott Press. Oh, if I could see the words "Mrs Gamp" in floriated initials – perhaps in red – thick – [here he draws a witty parody of her name as rendered in Morris initials] . . . for to be printed so finely is equivalent to beatification from the Vatican. . . .'[48] When a journalist put this very question to Morris – whether Dickens might not appear odd in Kelmscott dress – he replied tartly, 'I fail to see the incongruity. That arises simply from preconceived notions of the right thing, established by vicious custom.'[49] He then launched into a slightly irrelevant attack upon shoddy workmanship in the modern world; but the question is a pertinent one, and Morris's response is not persuasive. The Kelmscott formula would *not* work for every book, and only in his less combative moments did Morris allow himself to admit this.

If the chief defect of Morris's views on bookmaking was their intellectual rigidity, his strength resides in a tough-minded pragmatism and sensible grasp of first principles. There can be no greater contrast than that between his typographical writings and Cobden-Sanderson's *The Ideal Book or Book Beautiful*, first published at the Doves Press in 1901.[50] Whereas Morris is interested in how books are actually made now and how they were made in the golden age of printing (the fifteenth century), Cobden-Sanderson constantly loses his way and strays into this sort of cosmic nonsense: 'Finally, in its wholeness, symmetry, harmony, and beauty, without stress or strain, the consummated BOOK BEAUTIFUL is one in principle with that whole of life which is constituted of the world and of ourselves, that complex and marvellous whole which, amid the strife of competitive forces, supremely holds its own, and in the language of the ages writes upon the illuminated pages of the years through the infinitudes of time and space moves rhythmically onward to the completion of its astonishing story, the true archetype of all Books, beautiful and sublime.' Morris, fortunately, was incapable of such rhetorical flights. The tone and underlying assumptions of his writings are in fact closer to those of Stanley Morison's *First Principles of Typography* (1930, revised 1951): both Morris and Morison see typography primarily in historical and ethical terms; both offer closely reasoned principles that are founded, so far as possible, on the best traditional practices of earlier printers; both seem to make annoying claims to omniscience because their attitudes towards book-design are rooted in an interpretation of Western culture as a whole. Morison of course frequently disparaged his predecessor (this appears to be a classic example of what Harold Bloom has called the anxiety of influence), and Morris would no doubt have found Morison irritatingly puritanical. 'Typography', Morison wrote, 'is the efficient means to an essentially utilitarian and only accidentally aesthetic end, for enjoyment of patterns is rarely the reader's chief aim.' One can hear the ghost of Morris sputtering vigorous dissent to that proposition. Yet, if one reads between the lines of Morison's brilliant little pamphlet, there are distinct echoes of the founder of the Kelmscott Press as well. It is possible to trace a direct (if not always visible) line of descent from Morris's exposition of typographical principles in the 1890s to the most influential statement of the twentieth century.

Nowhere is Morris's rigidity more evident than in his insistence upon the reduction of white spaces between words and lines of type. The rich effect of a dense, black, heavily ornamented Kelmscott page is too often achieved at the expense of legibility and light: such a page may create, in some readers at least, a feeling of overwhelming claustrophobia. Morris's explanation of his preference for darker coloration sounds entirely plausible; he praises the sturdier typefaces of the early printers, reminds us that they used leading much more sparingly than their Victorian counterparts, and makes an effective case against the spindly types that disfigured modern books. But beneath this sweetly reasonable line of argument one suspects the presence of an irrational, atavistic yearning for darkness on the printed page that cannot be accounted for in purely typographical terms. The same preoccupation with deeper tones and the exclusion of light is evident in the later stained-glass windows of Morris and Company.[51]

The immediate source of Morris's strong convictions about compact word-spacing was Emery Walker, who already in the 1880s was arguing for a return to the practice of the fifteenth-century printers. Both the theory and practice of Victorian printing favoured placing what were known as 'thick spaces' or (at the very least) 'middle spaces' between words, with even larger spaces after colons and full stops.[52] Insisting that such white patches destroyed the texture of the printed page, Walker advised compositors to remove the thick spaces from their cases completely and replace them with 'thin spaces', which he thought should be adopted as the norm. By the time *The House of the Wolfings* was published in 1889, Morris vowed to 'enter into the conflict on this side also',[53] yet he found it surprisingly difficult to alter the deeply engrained habits of his printers. One can see evidence of this in the series of trial pages of *The Story of the Glittering Plain* which were struck off at the Kelmscott Press on 31 January 1891, a few days after the Golden type had arrived from the foundry. The first of these trial pages[54] is, except for the use of the Golden type, a typical piece of Victorian book-composition: the word-spacing is extremely loose, and indeed after full stops there are em quads (the largest space normally used by a compositor, the width of a capital 'M'). Only in the third and fourth versions does the trial page begin to look like an authentic Kelmscott page with smaller intervals between words. Morris, in over-reacting to the splotchiness of the usual Victorian printed page, sometimes huddled the words together so tightly as to interfere with legibility; English commercial printers, meanwhile, continued in their bad old ways until the 1920s, when there was a gradual but general acceptance in the trade of the Walker–Morris attitude towards word-spacing.[55]

Figure 39

Morris's crusade for the elimination or drastic reduction of leading was less successful. Nineteenth-century books are leaded far more generously than most readers today would wish, but once again Morris went to the opposite extreme by trying to abolish spaces between lines altogether, thus making some (though not all) of the Kelmscott volumes difficult to read. De Vinne later offered the sensible observation that medieval scribes and the first printers probably pushed their lines closely together for the sake of economy.[56] Morris,

The old man answered not a word, and he seemed to be asleep, and Hallblithe deemed that his cheeks were ruddier and his skin less wasted and wrinkled than aforetime. Then spake one of those women : Fear not, young man ; he is well and will soon be better. Her voice was as sweet as a spring bird in the morning ; she was white-skinned and dark-haired, and full sweetly fashioned ; and she laughed on Hallblithe, but not mockingly ; and her fellows also laughed as though it were strange for him to be there. Then they did on there shoon again, and with the carle laid their hands to the bed whereon the old man lay, and lifted him up, and bore him forth on to the grass, turning their faces towards the flowery wood aforesaid ; and they went a little way and then laid him down again and rested ; and so on little by little, till they had brought him to the edge of the wood, and still he seemed to be asleep. Then the damsel who had spoken before, she with the dark hair, said to Hallblithe : " Although we have gazed on thee as if in wonder, this is not because we did not look to meet thee, but because thou art so fair and goodly a man : so abide thou here till we come back to thee from out of the wood."

Therewith she stroked his hand, and with

Figure 39 The first trial page of *The Story of the Glittering Plain* in the Golden type, printed on 31 January 1891. A rejected lower-case 'g' appears on the page, and the word-spacing is much looser than in subsequent versions of the same trial.

however, was not inclined to compromise: 'The only leading I have allowed myself', he wrote in 1895, 'is in some cases a "thin" lead between the lines of my Gothic Pica [i.e. Chaucer] type: in the Chaucer and the double-columned books I have used a "hair" lead, and not even this in the 16mo books.'[57] Since unleaded type does make a visually pleasing block of colour on the printed page, Morris's emphatic preference is an example of the conflict, described by Stanley Morison, between pattern-making and legibility; and Morris, in this instance anyway, unhesitatingly chose the beautiful pattern rather than the comfort of his readers' eyes.

Apart from word-spacing and leading, the question which most engaged Morris's attention as a book-designer was the proportions of margins. The usual practice of Victorian printers was to place the text very nearly in the centre of the page; Morris, with much righteous indignation, argued that the two pages of an opening, since they are always seen together, must be regarded as a single unit, and that therefore the two adjoining inner margins should be narrower than the outer margins. 'I think you will very seldom find a book produced before the eighteenth century, and which has not been cut down by that enemy of books (and of the human race) the binder,' declared Morris, 'in which this rule is not adhered to: that the hinder edge (that which is bound in) must be the smallest member of the margins, the head margin must be larger than this, the fore [i.e. the outer] larger still, and the tail largest of all. I assert that, to the eye of any man who knows what proportion is, this looks satisfactory, and that no other does so look.'[58] As an historical description of the habits of early printers, Morris's statement is more or less accurate; however, real difficulties arise when, on a later occasion, he attempts to formulate the proportions more precisely: 'A friend, the librarian of one of our most important private libraries, tells me that after careful testing he has come to the conclusion that the mediaeval rule was to make a difference of twenty per cent from margin to margin.'[59]

In an article published in 1933, Alfred W. Pollard (who, incidentally, identified Morris's anonymous source as J. P. Edmond, librarian to the Earl of Crawford) subjected Morris's views on margins to close scrutiny and was able to show that the historical evidence is more ambiguous than Morris supposed. Pollard himself, on many winter afternoons, stayed on at the British Museum after the conclusion of his official duties in order to measure the margins of 'fine books of the fifteenth-century and others from modern presses'; he concluded that most printers, including Morris, allowed for less than a 20 per cent difference between the inner and top margins, and that there were many other unexpected variables.[60] Jan Tschichold, one of the most thoughtful book-designers of the twentieth century, had also done a systematic study of the history and theory of margin proportions, and, like Pollard, offers a revisionist interpretation, based this time on the assumption that late-medieval printers arrived at their margins not by trial and error or scribal precedent (as Morris seems to assume) but by the application of the Golden Canon.[61]

Morris's own practice at the Kelmscott Press is not entirely consistent with his public pronouncements on margins. For one thing, Emery Walker had convinced him that headlines (or running titles, as they are sometimes called) weakened the unity of the printed page. Morris therefore always used shoulder-notes instead of headlines. This meant, however, that his outer margin had to be much larger than his theory might indicate, and that in turn necessitated an unusually broad expanse of blank paper at the bottom of the page. On the other hand, the inner margins of the Kelmscott books are frequently so slender that they create problems in rebinding. The meagreness of the inner margin was not accidental: Cobden-Sanderson reports that 'Morris wanted the margins of his borders to meet in the crease leaving no white'.[62] Another source of inconsistency in Morris's application of his formula to margins is that he was reluctant to continue a line of poetry on to a second line, and in order to avoid what printers call a 'turnover' he would allow a long line to drift far into the margin.

Poetry, in fact, provided a severe challenge to several of Morris's typographical theories. Though a poet of some reputation himself, he obviously – as a book-designer – preferred to print prose, because it lent itself more readily to that solid, dark block of type which Morris always sought. However, his habit of placing an enormous decorated initial at the beginning of a poem (or sometimes at the beginning of each section of a poem) produced headaches for the Kelmscott Press compositors. Since Morris preferred to set the opening lines next to the initial in capitals (and yet also wanted to avoid turnovers), his printers faced an insoluble dilemma. Morris's usual formula was to run the opening lines of the poem together, separated only by small leaf ornaments. *Figure 40* In Tennyson's *Maud* (1893) he discarded most of the leaf ornaments and simply printed the first page of the poem as prose, a decision that seems to have satisfied no one, though Morris followed the practice in several later volumes as well.[63]

Nowhere in any poem did he indent turnovers, in disregard to usual typographical convention; for that matter, he was also reluctant to indent paragraphs in prose.

It comes as a surprise to discover that Morris had nothing whatever to say about the designing of title-pages. The explanation of this anomaly is that the fifteenth-century printers whom he most admired did not supply title-pages, or at least nothing beyond a simple label on an otherwise blank page. The Kelmscott Press title-pages are sometimes slightly more elaborate than that, but one can see that Morris's heart is not really in them. With no medieval precedent to guide him, Morris made the brilliant decision to supply, in many of his books, a second title-page (facing the first page of text) on which a hand-lettered title was engraved against a background of floral or vine

Figure 41

Figure 40 The engraved title and first text page of Tennyson's *Maud* (1893). In order to achieve a solid block of text, Morris has set the poem as prose on the first page.

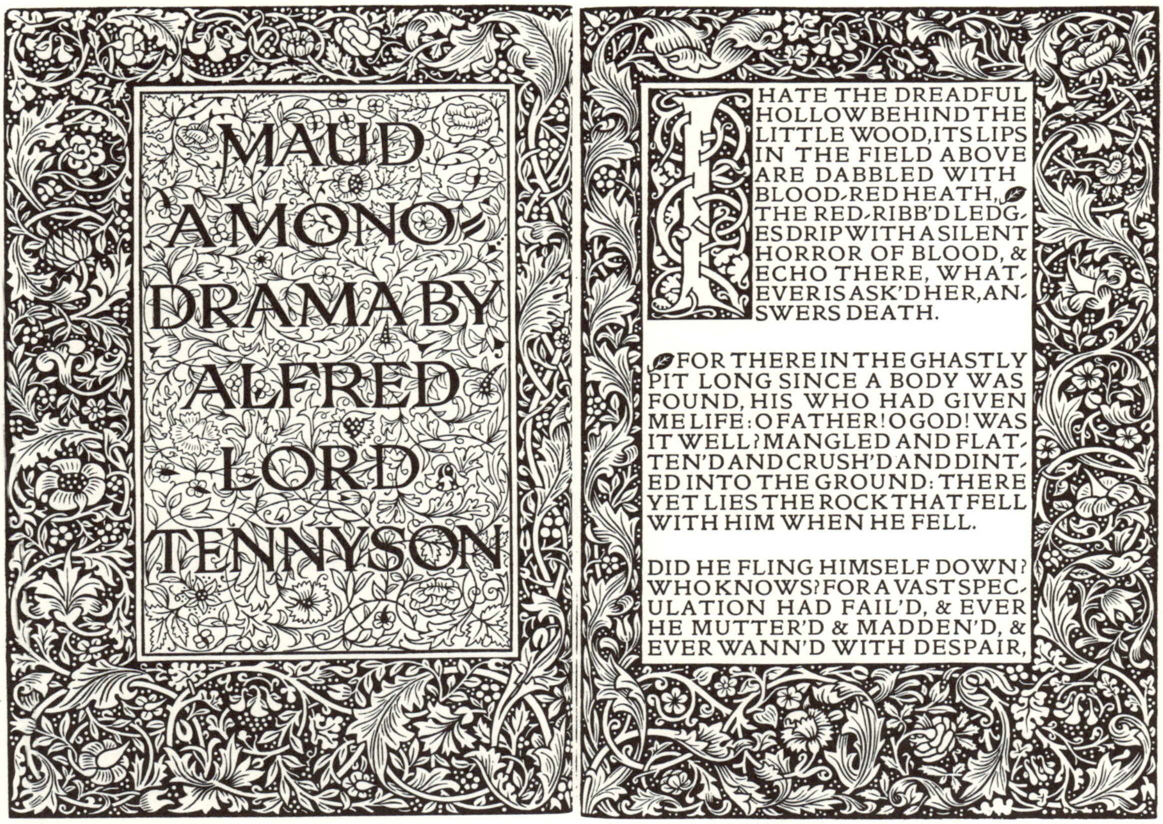
MAUD A MONO-DRAMA BY ALFRED LORD TENNYSON

I HATE THE DREADFUL HOLLOW BEHIND THE LITTLE WOOD, ITS LIPS IN THE FIELD ABOVE ARE DABBLED WITH BLOOD-RED HEATH, THE RED-RIBB'D LEDGES DRIP WITH A SILENT HORROR OF BLOOD, & ECHO THERE, WHATEVER IS ASK'D HER, ANSWERS DEATH.

FOR THERE IN THE GHASTLY PIT LONG SINCE A BODY WAS FOUND, HIS WHO HAD GIVEN ME LIFE: O FATHER! O GOD! WAS IT WELL? MANGLED AND FLATTEN'D AND CRUSH'D AND DINTED INTO THE GROUND: THERE YET LIES THE ROCK THAT FELL WITH HIM WHEN HE FELL.

DID HE FLING HIMSELF DOWN? WHO KNOWS? FOR A VAST SPECULATION HAD FAIL'D, & EVER HE MUTTER'D & MADDEN'D, & EVER WANN'D WITH DESPAIR,

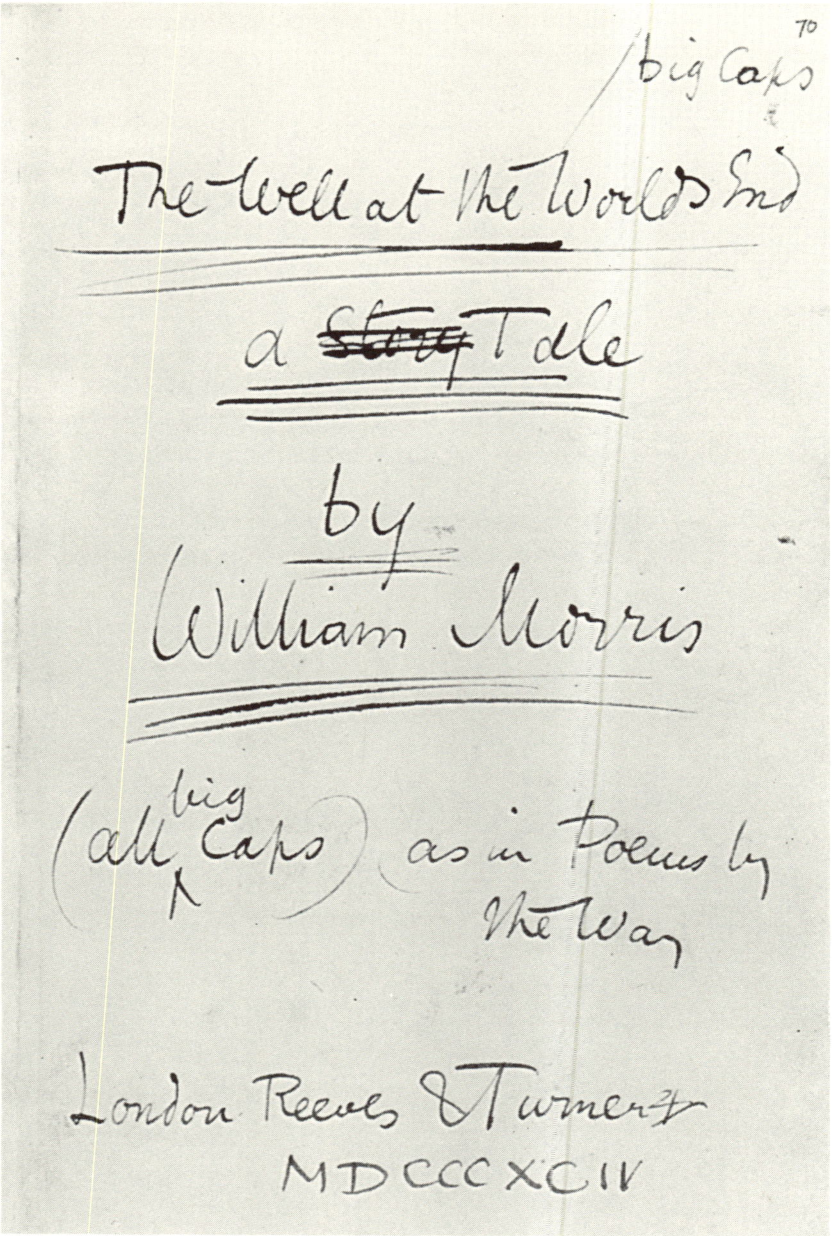
70
/big Caps

The Well at the World's End

a ~~Story~~ Tale

by
William Morris

(all big Caps) as in Poems by the Way

London Reeves & Turner
MDCCCXCIV

Figure 41 Although Morris's designs for ornaments survive in abundance, there are few examples of his sketches for entire pages, probably because he preferred to experiment at the side of the printer. This is the design for the title-page of one of his commercially published books, *The Well at the World's End* (Reeves and Turner, 1894). The instructions to the printer are characteristically terse.

ornaments, all enclosed within a decorated border that mirrored the border on the adjoining page. Here was the ultimate justification of Morris's notion of the opening as a single unit, and here too was a refutation of the charge that he was capable only of imitating early printed books. There can be no doubt that in this case Morris made a distinct improvement upon his models.

Since we have already (in Chapter Three) traced the process by which Morris designed his types, it is not necessary to linger very long over his views on that subject; but it may be useful to touch on a few points briefly. First, Morris's decision to restrict himself to the kind of fount available to fifteenth-century printers imposed limitations on the Kelmscott Press that probably became obvious only after several books had been printed. In particular, the lack of italics made it difficult, if not impossible, for Morris to include any scholarly apparatus with medieval texts, even had he wanted to do so. This may not have troubled Morris, but it was a source of constant frustration to F. S. Ellis. Second, Morris's habit of jamming paragraphs together, separated only by the paragraph symbol or by one of his smaller leaf ornaments, was an unfortunate decision. It did, needless to say, shape the text into the black rectangle which Morris desired, but it also disfigured the page with measles-like spots and made impossible that uniformity of colour which he himself sought. Third, the generous breadth of Morris's individual letters, which made them unsuitable for ordinary commercial purposes, likewise necessitated a broad page that strikes some readers as clumsy: the tall but narrow page which became the trademark of the Nonesuch Press book is unquestionably more gracefully proportioned.

Finally, something needs to be said about one aspect of the Kelmscott books that has aroused much controversy: the considerable depth to which the type is impressed into the paper. Accustomed as we are now to offset printing, which makes no indentation whatever in the page, we may be tempted to conclude that Morris has taken an accidental characteristic of the hand-printed book and carried it to excessive lengths. Certainly some critics of the Kelmscott Press have accused Morris of obsessive archaism in this matter, and in support of their charge they have been able to cite a note signed by him and inserted in *The Golden Legend*: 'If this book be bound the edges of the leaves should only be TRIMMED, not cut. In no case should the book be pressed, as that would destroy the "impression" of the type and thus injure the appearance of the printing.' Yet the true explanation of

Morris's attitude probably lies at a deeper level. The deliberate removal of the type-impression by pressing would have been a dishonest act in his eyes, because it represents an attempt to give an artificial smoothness to a product by concealing the real nature of its material. Furthermore, we need to remember that the unsurpassed presswork of the Kelmscott Press, which has been universally praised, was dependent upon a deep, slow impression of the forme into the fabric of the dampened paper. In short, the 'embossing' of the Kelmscott page, far from being an affectation, was an integral part of a superior system of inking.

Examined minutely, Morris's typographical theories and practices do not display the sort of monolithic uniformity that a merely casual acquaintance with his books might suggest. There are ragged edges to his ideas, and there is some evidence in the later Kelmscott books – and still more in the trial pages of those books – that his views continued to develop until the time of his death. But, even when we have acknowledged the deficiencies and inconsistencies of Morris's work as a book-designer, we are in the end left with a large, impressive body of theoretical writings and fifty-two books which have had a considerable effect upon the books we buy and read today. Exasperating and obstreperous, Morris is none the less still a force to be reckoned with in the world of bookmaking. At the turn of the century Emery Walker felt able to make this large claim for the influence of his friend: 'since the establishment of the Kelmscott Press in 1891, and especially since its close in 1898, the principles which Morris laid down in more than one article and lecture, and which were practically demonstrated in the work of his printing office, have been increasingly accepted throughout the world.'[64] That was a brave assertion in 1901, when Morris's reputation as a printer was temporarily ebbing, but today we are better able than Walker's contemporaries were to recognize the compelling truth of his claim.

FIVE ✿ THE DECORATED BOOK

WILLIAM MORRIS argued with characteristic zest that an austerely simple book, completely unornamented and unillustrated, could please the eye if it were printed in good type skilfully arranged on the page. Yet the Kelmscott books were ostentatiously decorated, with initials, borders, and smaller ornaments creeping sinuously around and within the text, and Morris's tone of voice, when he speaks of such matters, is surprisingly apologetic: 'It was only natural that I, a decorator by profession, should attempt to ornament my books suitably. . . .'[1] What was perhaps a faint misgiving in Morris's mind has become, in many other minds, the most serious objection to the Kelmscott Press books: that they are (according to this view) so overburdened with ornament that they suggest heavily laden ships likely to be capsized by the gentlest breeze. Another, more jocular figure of speech was often invoked after the turn of the century. The volumes of the Kelmscott Press, it was said, bore an unhappy resemblance to his wallpapers.

In fact, the debate about the Kelmscott ornaments has been carried on to a remarkable extent in metaphoric language. Holbrook Jackson, who was outspoken in his reservations about Morris's achievement as a printer, offered this vivid (and, one assumes, admiring) characterization in 1913: 'The Kelmscott books look not only as if letter and decoration had grown one out of the other; they look as if they could go on growing.'[2] Beatrice Warde, on the other hand, suggested in a well-known essay that good printing should be invisible: like a crystal goblet, it is a transparent medium which encloses the wine of literature.[3] Such argument by analogy has had a powerfully damaging effect upon the reputation of the Kelmscott Press in the twentieth century. Of course the 'crystal goblet' figure, though undoubtedly a brilliant image, evades some important questions about the proper relationship between text and ornament in the printed book. Behind it lies the assumption – which by Beatrice Warde's time no longer had even to be expressed – that elaborate ornamentation of any kind is not only ill-advised but also immoral. In the 1920s the new doctrine of functionalism had persuaded many that the lingering influence of Victorianism could be eradicated only by declaring war on all historicist, non-functional ornament, whether in buildings, paintings, household furnishings, or books. The volumes of the Kelmscott Press, so thickly encrusted with initials, borders, and ornamental illustrations, inevitably found themselves caught in the crossfire of this aesthetic battle.

Attitudes towards book-ornament had been very different when the Kelmscott Press was still taking shape in Morris's head. During the final decades of the last century there were influential voices urging a revival of ornament and illustration (wood-engraved, if possible) that would blend harmoniously with the typefaces used on the same pages. Gilbert R. Redgrave concluded an historical survey published in the periodical the *British Bookmaker* (October 1890) with these words: 'It is, we think, a matter of regret that the glory of the early days of woodcut initial letters has been lost to us, and that in the place of the graceful designs which once headed every chapter we have now too often to rely upon the meaningless scrolls and flourishes which the typefounder supplies us. We can only hope that more careful study of the work of the early printers and the growing fondness for all that was best in the days of the masters of the craft may lead to a revival of the use of book-ornament, such as is to be found in the infancy of the art of printing.' In another essay with which Morris was more likely to have been acquainted, published in the catalogue of the second Arts and Crafts Exhibition (1889), Reginald T. Blomfield treated the question 'Of Book Illustration and Book Decoration' by urging the paramount necessity of concord between typeface and decorative elements: he believed, as did Morris, that such harmony could best be achieved if the ornaments and illustrations were wood-engraved. Perhaps, he hinted, the ideal solution was for both text and illustration to be cut on the same block by a single engraver or at least be designed by a single artist.[4]

Walter Crane had adopted this technique in some of his books, drawing lettering and pictures alike, which were then reproduced by photomechanical means; and Morris publicly praised the results.[5] Because he was unwilling to use process engraving, Morris himself turned to sympathetic craftsmen who could translate the decorative designs for the Kelmscott books into wood-engravings that would not clash with his typefaces. His solution was different but the aim was identical: the artistic integration of word and picture. 'This means, in other words,' said Morris, '[that there must be] the designer of the picture-blocks, the designer of the ornamental blocks, the wood-engraver, and the printer, all of them thoughtful, painstaking artists, and all working in harmonious cooperation for the production of a work of art. This is the only possible way in which you can get beautiful books.'[6] Having designed the Kelmscott Press typefaces, Morris also assumed responsibility for all the ornaments used at the Press. The actual wood-engraving was done primarily by W. H. Hooper, assisted by C. E. Keates, W. Spielmeyer, and (for a short time) George Campfield. Both Cockerell and W. R. Lethaby have written lively descriptions of Morris in the act of designing Kelmscott ornaments: sketching faintly in pencil, he then applied India ink to paper with a brush and made occasional corrections with Chinese white.[7] The design was transferred to the surface of the woodblock by photography, though the engraving was done by hand. Under pressure from Walker, Morris allowed the ornaments to be electrotyped by Walker and Boutall after he was persuaded that it was virtually impossible to tell the difference between an electrotyped initial and the original printed on the same proof.[8] (As a result, most of the larger woodblocks containing several Kelmscott ornaments, once they had been electrotyped, remained intact and were never cut apart.) This decision – a grudging concession by Morris to modern technology – had important consequences, for it meant that he had at his disposal a

Figure 42

Figure 42 A proof pulled by Emery Walker (whose note appears on the proof) to demonstrate to Morris that it was difficult, if not impossible, to distinguish between initials printed from the original blocks and the same initials printed from electrotypes.

more abundant supply of initials and borders than if each ornament had been available for use only once on each side of a sheet. A very large book like the *Chaucer* or a book with an unusual text like *Laudes Beatae Mariae Virginis* (which required only initial 'A's, since each stanza began with 'Ave') would have been exceedingly difficult to produce without electrotyped initials.

Morris's desire to ornament his books is scarcely surprising: quite apart from medieval precedent, he was impelled by an innate decorative impulse which revealed itself in every aspect of his career. As a schoolboy Morris was already embellishing his books. Both the aborted *Earthly Paradise* and *Love is Enough* were intended to be decorated as well as illustrated; Morris's manuscripts of the 1870s are at least as notable for their ornament as for their calligraphic distinction. *Figure 43* During the following decade, he was unconsciously drawing floral borders on many of the scraps of paper that lay before him at public meetings (some of the duller socialist conferences were especially fruitful in producing this sort of doodling);[9] what is perhaps more revealing is a dummy copy of *The Roots of the Mountains* in which he had made a pencil sketch of a half border.[10]

Though all the ornaments in the Kelmscott Press books were wood-engraved (or, in a few instances, cast from matrices like type), Morris was fascinated by the possibility of decorative elements added by hand to a printed text, a technique often used in the first century of printing. *The Story of Gunnlaug the Worm-Tongue*, his last pre-Kelmscott experiment in bookmaking, was produced with blank spaces left for initials, as were a few copies of some of the Kelmscott books. There are also two surviving proofs of Kelmscott initials printed in blue with red superimposed by Morris's brush.[11] The person to whom Morris turned most frequently for hand-illumination was Edmond G. Reuter,[12] a Swiss artist employed by an English pottery firm, who decorated books in his spare time. When he submitted an illuminated manuscript to the second Arts and Crafts Exhibition in 1889, Morris was so impressed that he commissioned Reuter to provide ornaments in a copy of *The Roots of the Mountains*, six sheets of which were on display at the next Exhibition the following year (though Reuter did not complete his work on the book until several years later). Reuter, at Morris's invitation, also decorated both a paper copy and a vellum copy of *Syr Perecyvelle of Gales*. By giving him these commissions, Morris hoped to persuade Reuter to remain in England, but in 1895 Reuter returned to Switzerland, convinced that he would find more

Figure 43 An accompanying note by Sydney Cockerell reads: 'These were notes made by Morris during the discussion following a lecture by him at Kelmscott House, Hammersmith, I think in 1886.' In the midst of his socialist activities of that decade, Morris's mind frequently turned to letter-forms and book ornamentation.

demand for his talents there. Had Reuter not emigrated, he might well have played a minor role at the Kelmscott Press, perhaps illuminating a few copies of the Kelmscott books for Morris's special customers.

Compelled, therefore, to use printed rather than painted ornaments, Morris found inspiration especially in the work of an early German printer, Erhard Ratdolt of Augsburg, whose books were adorned with attractive large borders and initials in a matching style of patterns on a black ground. (Morris, however, was uneasy about the comparison: 'I do not profess merely to follow precedent, or to imitate Ratdolt,' he grumbled.[13]) As the Golden type was being cut by Prince in 1890, Morris began working on initials, some of which resemble those in two incunabula, Günther Zainer's sixth German Bible and Johannes Friburgensis's *Summa confessorum* (1476).[14] 'Also I have been designing ornamental letters, – rather good I think,' he wrote to Ellis on 17 October of that year.[15] The first attempt at wood-engraving, however, was not wholly successful. When Morris sent a design for an 'N' to Hooper, he cut it as requested but also engraved a second version (which Hooper preferred) of the letter, with the colours reversed, and supplied a stippled background – of the kind very popular among Victorian 'old-style' printers.[16] One would like to have heard Morris's comment – no doubt explosive – on this odd transformation. Eventually Hopper became more attuned to Morris's style (though disagreements between the two men persisted), and in the following year we find Morris praising him for helping out 'my trembling hand a little' and for cutting the initials 'prettily'.[17] Cockerell, calling occasionally at Kelmscott House, registered his approval of Morris's designs: 'Saw some newly designed letters with vine pattern which I liked as well as any that W.M. has done' (diary, 5 November 1891). 'Morris had . . . some newly designed capitals for his small blackletter type [the Chaucer type] – black with a thin white border [–] very beautiful' (28 July 1892).

Figure 44

Cockerell also records that earlier that year, on 17 March, after his usual dinner with the committee of 'Anti-Scrape' at Gatti's restaurant, Morris, in jovial humour, passed his 'bloomer book' round the table. The book (now in the Bodleian Library) consists of numbered proofs of all his initials and borders, which he used in selecting the ornaments for individual Kemscott books. Of course he continued to design new ornaments until his death; Cockerell remarked astutely that such proofs and the books themselves 'will show the development of Morris's designs for capitals, as he keeps using the new ones as soon as

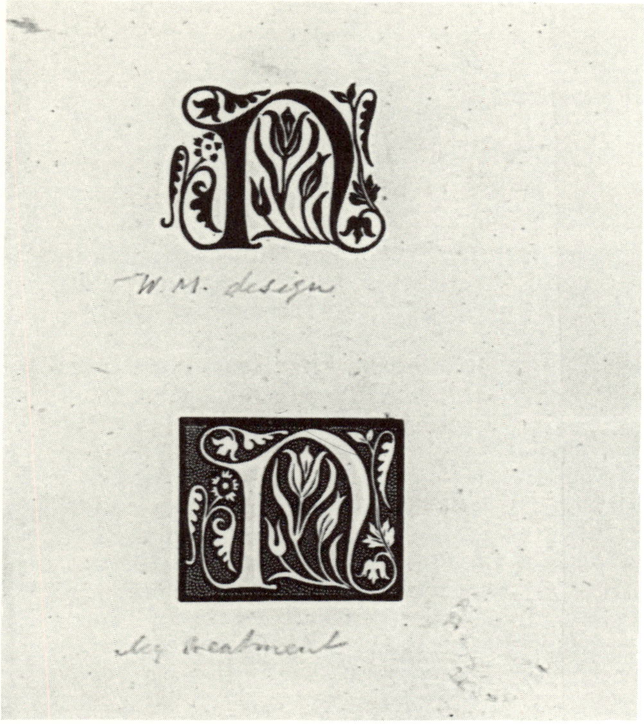

Figure 44 This is apparently the earliest attempt by Morris to design an initial for the Kelmscott Press. Both versions were engraved by W. H. Hooper, who unsuccessfully offered the second initial as a suggested revision to Morris.

they are cut' (diary, 5 November 1891). What the 'bloomer book' also reveals is that, because many of the early initials were too dark for even his own deeply coloured types, Morris used them only once. (This is signalled by their being crossed out in the book.) Another weakness evident in the initials is that Morris sometimes designed them too large for the page. This was notably the case in *The Story of the Glittering Plain*, because the initials used there were originally intended for a book of more ample dimensions; but it is a recurrent problem in other Kelmscott books as well. Occasionally one hears a tactful protest from Morris's compositors. On a proof of the first text page of *The Defence of Guenevere*, one of the printers has scribbled this note about the narrow space to the right of a ten-line initial: 'I tried caps. here, but they fell very awkwardly indeed.'[18] In time Morris became more proficient in matching ornament with text, partly because he adopted

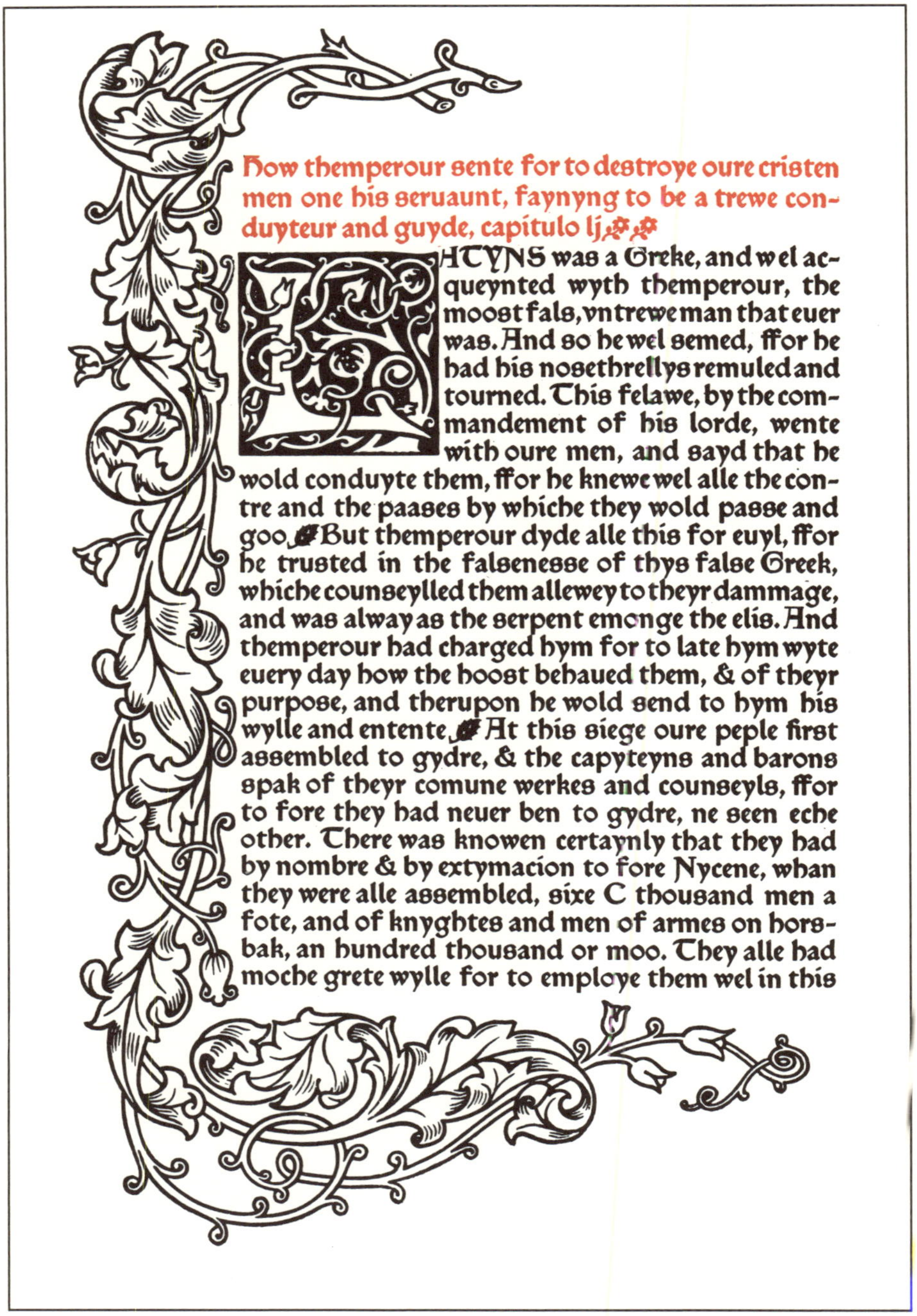
How themperour sente for to destroye oure cristen men one his seruaunt, faynyng to be a trewe conduyteur and guyde, capitulo lj

TATYNS was a Greke, and wel acqueynted wyth themperour, the moost fals, vntrewe man that euer was. And so he wel semed, ffor he had his nosethrellys remuled and tourned. This felawe, by the commandement of his lorde, wente with oure men, and sayd that he wold conduyte them, ffor he knewe wel alle the contre and the paases by whiche they wold passe and goo. But themperour dyde alle this for euyl, ffor he trusted in the falsenesse of thys false Greek, whiche counseylled them allewey to theyr dammage, and was alway as the serpent emonge the elis. And themperour had charged hym for to late hym wyte euery day how the hoost behaued them, & of theyr purpose, and therupon he wold send to hym his wylle and entente. At this siege oure peple first assembled to gydre, & the capyteyns and barons spak of theyr comune werkes and counseyls, ffor to fore they had neuer ben to gydre, ne seen eche other. There was knowen certaynly that they had by nombre & by extymacion to fore Nycene, whan they were alle assembled, sixe C thousand men a fote, and of knyghtes and men of armes on horsbak, an hundred thousand or moo. They alle had moche grete wylle for to employe them wel in this

Figure 45 Contrary to their reputation, not all Kelmscott Press books are extensively ornamented, but *The History of Godefrey of Boloyne* (1893) overflows with initials, side-borders, and half-borders.

the sensible practice of sketching his design for the initial or border directly on a proof of the page where it would appear.

Borders, by their very nature, are used more sparingly than ornamented initials: the same simple border appeared in the first five books of the Kelmscott Press (always on the first text page), and only later did Morris introduce a profusion of full, three-quarter, half, quarter, and corner borders into his books with splendid but sometimes confusing effect. *The History of Godefrey of Boloyne* (1893) must be the most over-decorated of the Kelmscott books, surpassing in this respect even the *Chaucer*; the other end of the spectrum is represented by the earliest books (when Morris did not yet have the resources of a 'bloomer book' at hand) and the latest ones (after Morris's death). Like the initials, the borders consist mainly of stylized floral and vine patterns in white line against a black ground, though Morris also did a few in outline style. Their obvious resemblance to some of Morris's wallpaper designs has struck nearly all readers, and the analogy is not as unflattering to the Kelmscott volumes as some wits have supposed. *Figure 45*

On rare occasions Morris employs a border that reflects the text in some fashion, as when a rose border surrounds the opening of Chaucer's *Romaunt of the Rose*; but for the most part he believed that borders should be ornamental rather than illustrative. (Hence there is not one historiated initial or border in any Kelmscott book, even though Morris could have found many examples of them by going no further than his own library.) Various sources have been suggested for his borders, including the Windmill Psalter (which he owned), sculpture in the porch of Beauvais Cathedral, and of course Ratdolt's books. In fairness to Morris, it must be said that Ratdolt's borders, while vaguely similar, are much less successful and more static than those in the Kelmscott books. This appears to be yet another instance of Morris improving upon, rather than directly imitating, a medieval model.

Even the most severe critics of the Kelmscott Press have acknowledged that Morris's borders are things of extraordinary beauty. The borders have inspired prolonged debate, however, about whether they distract and dazzle the eye and thus call too much attention to themselves. Morris's full borders, like Ratdolt's, reflect the classic proportions of the book page – that is to say, the border progressively increases in breadth from the inner margin to the lower margin. R. B. McKerrow justly complained that such a pattern produces an unpleasantly heavy concentration of black ink at the bottom of the page.

It is a pattern that ultimately derives from the medieval scribes and illuminators, but they had the advantage of being able to produce borders of lighter texture.[19] These broad borders also produced technical problems for Morris: they were in part responsible for the discoloration of the *Chaucer* sheets, and in fact the press on which the *Chaucer* was printed had to be reinforced with steel bands because of the unusual size of the ornaments.

We have already seen that one of the chief decorative elements in the Kelmscott books, the engraved title facing the first page of text, was apparently Morris's own invention, since the first two generations of printers used, at most, label titles that allowed no scope for ornamentation. Morris's distinctive engraved title evolved slowly, perhaps accidentally. The first five books from the Press had only simple titles set in the Golden type, but the sixth Kelmscott volume, *A Dream of John Ball* (1892), included a frontispiece by Burne-Jones (the first illustration in the Press's books) across from the opening text page, with large Roman lettering underneath designed by Morris. Probably this gave Morris the idea of placing an engraved title (within a border, like the *John Ball* illustration) in this position, because the first such title appears in the next Kelmscott book, *The Golden Legend*.[20] Of the fifty-two books printed at the Kelmscott Press, twenty had wood-engraved title-pages, though naturally there were no more after the *Chaucer* because of Morris's death in the autumn of 1896.

Figure 46

Morris took pains with the lettering, even consulting Arrighi, one of the sixteenth-century writing masters represented in his library,[21] when he was designing the *John Ball* frontispiece caption; yet the Roman titles which he created leave much to be desired. Rather than resembling any of the graceful hands recommended by Arrighi, Morris's hand-lettered titles suggest a crude version of Caslon or, at worst, the wood-types used on Victorian posters. They are nevertheless redeemed by Morris's consistently fine floral and vine backgrounds and by the borders that surround them. The Gothic titles, not surprisingly, are much more satisfying. Walter Crane writes that Morris 'was asked why he did not use types after the style of the lettering in some of his title-pages, but he said this would not be reasonable, as the lettering of his title was especially designed to fit into the given spaces, and could not be used as movable type'.[22]

Figure 47

Following the example of his medieval predecessors, Morris designed three printer's marks for the Kelmscott Press, all of which he regarded as ornamental as well as informative. The first, reading

Figure 46 The engraved title of the Kelmscott Press *Chaucer*, designed while Morris was seriously ill, is a classic example of the overpowering yet graceful ornamentation that he created for many of his books.

Figure 47 The two printer's marks designed by Morris for the Kelmscott Press books; the lower one (reduced) was used in larger volumes.

'Kelmscott | William Morris', was contained within a small rectangle and was designed for use in the octavos and 'small quartos' (i.e. quartos printed on the first Flower paper, which produced a book the same size as an octavo printed on the later version of the Flower paper). It made its début in *Poems by the Way*, the second Kelmscott book, and one senses that Morris was pleased with it, for he used it twice in the same volume. The second printer's mark – reading 'Kelmscott' against a leafy background rendered in outline style – was first printed in *The History of Godefrey of Boloyne* and generally used thereafter in quartos; Morris's drawing[23] discloses that he originally placed his name at the bottom of this design but then changed his mind and covered it with Chinese white. A third printer's mark ('Kelmscott') is a long rectangular device which appears only in the *Chaucer*. Each of the three designs blends admirably with the texture of Morris's types, borders, and initials.

All the ornaments described so far were cut on wood; but there was also a smaller category of ornaments, very useful to Morris, that were engraved on punches and cast like type. Morris first revived the paragraph symbol, which he employed frequently in the early years of the Press, as a substitute for a paragraph indentation. After *News from Nowhere* he abandoned it, no doubt wisely, though Charles Fairfax Murray persuaded him to revive it briefly in *Epistola de contemptu mundi* (1894). Morris in effect replaced the paragraph symbol with three other small ornaments, which produced the same unfortunate effect of destroying the even colour of the page. Morris's passion for filling every white space can also be seen in the considerably larger 'line-fillings' (as Cockerell calls them) which made their initial appearance in *The Life and Death of Jason* (1895). Curiously, Morris never made use of the simple ornament that is the favourite of most printers, the rule – either plain or swelled – though in the proofs of *The Nature of Gothic* (1892)[24] there is a rule between a border and the enclosed text that was eliminated before the book was printed.

Book-ornamentation and illustration may seem to be two separate subjects, but Morris believed that illustrations – except in the most baldly functional books such as reference works and textbooks – should be primarily ornamental and must, like ornaments, be designed in reference to the book as a whole and to the typeface in particular. This was a subject that much exercised the minds of Morris's contemporaries, because the advent of zincos and halftones, which were printed most effectively on glossy paper, posed a new threat to the structural unity of books in the late-nineteenth century. Alfred Dawson, whose firm had introduced photomechanical engraving to England, offered a radical but plausible-sounding solution: in a paper entitled 'The Most Artistic Book', read at the Annual Meeting of the Library Association in 1891, he suggested that the illustrations and text should be produced on separate presses and brought together only at the binding stage. The illustrations, under this scheme, would be printed on a copperplate press with dry paper; the text, set in old-face type, would be printed on dampened paper, if possible handmade.[25]

But this was an admission of defeat in the struggle for unity, and Morris would have none of it. His answer was to bring harmony to a book by returning to the older method – printing wood-engraved illustrations locked in the same frame with the type, and thereby using the same paper and ink. Morris, in a lecture given a year after Dawson's, summarized his conception of the ideal book with pictures: 'An illustrated book, where the illustrations are more than mere illustrations of the printed text, should be a harmonious work of art. The type, the spacing of the type, the position of the pages of print on the paper, should be considered from the artistic point of view. The illustrations should not have a mere accidental connection with the other ornament and the type, but an essential and artistic connection. They should be designed as a part of the whole, so that they would seem obviously imperfect without their surroundings. The designs must be suitable to the material and method of production, and not offer to the executant artist a mere thicket of unnatural difficulties, producing no result when finished, save the exhibition of a *tour de force*. The executant, on his side, whether he be the original designer or someone else, must understand that his business is sympathetic translation, and not mechanical reproduction of the original drawing.'[26]

It was perfectly natural for Morris, given his passion for the medieval book, to prefer wood-engravings to photomechanical reproductions; as early as the 1860s he had made an ambitious effort in *The Earthly Paradise* to restore a broad, simplified style of engraving to its former pre-eminence. Both he and Walker were convinced that an engraved line was somehow 'sweeter' than one reproduced by a camera. And of course Morris's insistence upon the artistic freedom of the wood-engraver, who is not to obey the dictates of the designer servilely, is wonderfully Ruskinian and consistent with Morris's own view of the craftsman's role. It is a seductively attractive theory of how book illustrations should be produced. In practice at the Kelmscott Press, alas, all was not so simple.

Part of the difficulty lay in the temperament and training of William Harcourt Hooper, who engraved most of the Kelmscott illustrations. Described by Walter Crane (in 1897) as 'almost the only first-rate fac-simile engraver on wood left – and a good artist and craftsman besides',[27] Hooper had engraved the drawings of artists like Fred Walker, Millais, du Maurier, Tenniel, and Leighton for the *Illustrated London News* and *Punch*. A neighbour of Walker's and a committed socialist (and, in earlier days, a pupil of Rossetti at the Working Men's College), he was persuaded by Morris to come out of retirement to work for the Kelmscott Press; but almost immediately there were disagreements between Hooper and his new employer. Morris, like Ruskin before him, believed firmly that the craftsman ought not to be merely the tool of an artist–designer. In practice this meant that the engraver should participate in the creative process by adapting and modifying the artist's design as he transfers it to the surface of the woodblock. Hooper's entire career, however, had been devoted to producing exact facsimiles of the drawings given to him. This fidelity to the design was in fact a matter of pride to engravers like Hooper: he was delighted on one occasion when Ruskin was unable to distinguish one of his proofs from the drawing on which it was based.[28]

During the first year of the Kelmscott Press's existence, Hooper engraved only initials and borders, but there were even then occasional outbursts of indignation from Morris about his work, as in this letter to Walker of 4 February 1892: 'Thanks for note and proof of Hooper's cuts. . . . I think these not at all good: in fact I think them downright bad, and chiefly of use as illustrating every fault that woodcut can have. Of course I know that he can do better, and that he is doing this out of perversity. Anyhow, anybody can do them as well as this, as they are quite *un-intelligent* in cutting; and Leverett [who was to engrave Walter Crane's illustrations for the second version of *The Story of the Glittering Plain* (1894)] may as well cut them at a lower price. But I daresay with a little talking to I can bring Hooper round and get him to do better: if not I must in future go to someone else.'[29] Morris's anger was usually short-lived, and Hooper was not replaced, but it is clear that the discord between Morris's ideal of the creative craftsman and Hooper's ideal of facsimile engraving never diminished. In another letter from Morris to Walker, written three months later, we hear the same lament: 'Looking again at the proof I see one place which will have to be altered, where in the pursuit of paradox H[ooper] has cut through a line which ought to have joined: however 'tis not in the "Corner" [probably one of Morris's corner borders] which is the one I want most. – Ah I see he has altered my drawing to its disanvantage [*sic*] in the "Corner"; so his principle of exact copying has gone by the board, and I can chaff him.'[30]

Hooper's 'principle of exact copying' was in conflict with more than Morris's Ruskinian doctrine of the dignity of the workman; it was also incompatible with the habits of Edward Burne-Jones, the usual illustrator of the Kelmscott books. In all the designing that Burne-Jones had done for Morris and Company he exhibited a reluctance to provide finished drawings to scale, and in an interview published in the *Studio* (1894) Morris explained how his firm had got round this problem when producing textiles. Morris's remarks are worth quoting at length, because they shed considerable light on both Burne-Jones's technique as a draughtsman and Morris's use of photography: 'The original studies are not above 15 inches high. The figures are grouped and drawn from carefully prepared studies: for the rest there is but little minuteness of detail; and they are only slightly tinted. That is the form in which they come into our hands. We have to have them enlarged by photography, in squares varying in size and number according to the full dimensions required. The enlarged sections are then fitted together, and the whole, now of the proper size, submitted, together with a small coloured study, to the artist for his revision and approval; and on these enlargements he does a great deal of work, especially to the heads and hands. The ornamental accessories, the patterns of brocades in the draperies, the flowers and the foliage, are left to us, and are drawn, for the most part, by Mr. H. Dearle. . . .'[31] We know that from the mid-1870s the same procedure was used at Morris and Company for translating Burne-Jones's small sketches into suitable designs for stained-glass windows; and in the end something not unlike this had to be done to his Kelmscott Press illustrations as well.

The most distinctive characteristic of a wood-engraving is that it can render only blacks and whites; intermediate tones must be suggested by parallel lines (which Morris believed should be used sparingly) or cross-hatching (which he thought should not be used at all). But the designs which Burne-Jones provided for the Kelmscott illustrations were pencil sketches filled with delicate nuances of shading. Obviously a great deal of 'translation' had to be done before they could be made into wood-engravings. According to Morris's understanding of craftsmanship, Hooper should have been the person to adapt the sketch to the limitations of the engraving process, yet in practice a veteran 'facsimile engraver' like Hooper was incapable of bridging the gap. Nor was there much rapport between Hooper and Burne-Jones, the latter of whom wrote in September 1894: 'Sometimes men come at half past eight in the morning for instructions. . . . I have the Chaucer engraver man one day in every week, and have to begin at the beginning, at the words of one syllable with them. . . . But I am so dejected that I have a mind to say to them when they come "Good morning to you, I hope you are well and your wife and family, for, judging by your impoverished appearance, and the lack of interest in your toilet, I take it for granted you have both wife and family. I hope you are well. Now to begin about this engraving. Once upon a time darkness was everywhere, and matter mingled with water. Then came a great wind." – and so on by slow degrees leading them to the Tower of Babel – to the beginnings of the Egyptians, and so through the Greek pots to Albert Durer. It[']s always like that.'[32]

If Hooper was too uninstructed and too rigid in his loyalty to the ideal of facsimile engraving (not to mention his shabby clothes which were so offensive to Burne-Jones), then someone else would have to be found who could convert the Burne-Jones sketches into a form suitable for wood-engraving. One obvious candidate was Charles Fairfax Murray, an artist, collector, and connoisseur who had done similar copying tasks for Rossetti and Burne-Jones in the past,[33] and in fact Murray 'touched up' photographic copies of both Burne-Jones illustrations in *The Golden Legend* before Hooper engraved them.[34] Whether he adapted Burne-Jones's other early Kelmscott illustrations is not known; but when work began on the *Chaucer* it became evident that Murray alone was not equal to the task of translating eighty-seven drawings. Walker, perhaps exaggerating, claimed that it took him nearly three months to complete one illustration.[35] An early state of the picture which appears on page 9 of the *Chaucer* has survived,[36] and it is a remarkably inept piece of drawing, the faces and flowers are lifeless, and the shading is unsatisfactory. This is probably not the engraving that was adapted by Murray,[37] but its lacklustre quality does at least suggest the seriousness of the problem facing Morris and Burne-Jones.

The first Kelmscott Press illustration, the frontispiece of *A Dream of John Ball*, was an adaptation of a gravure reproduction which had appeared in the first edition (1888) of the book, and it is significant that at about the time of publication Robert Catterson-Smith, a young *Figure 48*

Figure 48 Burne-Jones's frontispiece for *A Dream of John Ball* (1892) was the first illustration to appear in a Kelmscott volume and may have suggested the engraved titles that were used in many of the later books.

artist who later became Headmaster of the Birmingham School of Art, redrew the frontispiece on a larger scale for reproduction in the *Daily Chronicle*.[38] In the same year Catterson-Smith adapted – by means of an ink tracing on a photograph of the pencil sketch – Burne-Jones's design for the frontispiece of the Earl of Lytton's *King Poppy*, published by Longmans in 1892 (though the frontispiece was printed from a line block rather than a wood-engraving).[39] Here, then, was at last a possible solution to the impasse in producing illustrations at the Kelmscott Press.

Walker and Catterson-Smith have given detailed accounts of how the latter mediated between Hooper and Burne-Jones;[40] and an ingenious procedure it was. Once Walker and Boutall had photographed the pencil sketch, Catterson-Smith covered the print with a thin wash of Chinese white so that the lines were just barely visible. He then traced over the drawing 'with a fine round sable brush and very black Chinese ink', thickening the strokes as he did so and converting Burne-Jones's grey tones into engraving-style shading. 'When difficulties arose in the treatment of passages I consulted B-J,' Catterson-Smith recalled, 'Sometimes work by Albert Dürer was consulted to see how he had dealt with passages. Some of the drawings were done over several times.' Since most of the communication between Catterson-Smith and Burne-Jones was oral, we have few written records of the negotiations between them; but we catch an interesting glimpse of the difficulties faced by Catterson-Smith in a letter written many years later when he was redrawing some unfinished designs by Burne-Jones once intended for the Kelmscott edition of *Biblia innocentium*. 'One of my greatest problems', he declared, 'is to keep the compositions unchanged, to keep line for line – and the second problem is to get rid of the rawness which arises from the change from pencil to black line: without introducing too much of my *own*.'[41] In this instance Catterson-Smith also struggled with angels that had too many wings and a sword's edge that was awkwardly depicted by a line which simultaneously represented a wing.

Catterson-Smith's ink tracings of Burne-Jones's designs were transferred photographically to the surface of the woodblocks and engraved by Hooper. Throughout the process the artist played an active role, and both the redrawn designs and the proofs of Kelmscott illustrations are heavily marked and annotated by Burne-Jones. 'Slight alteration in position of base of second column,' he writes on one. Or: 'Have taken out a buttress between Troilus's leg and post.' 'Ought there to be a nimbus on angel?' 'Please engrave as far as possible the accidental changes of tone in wicker-work.' 'Mail on left arm | Sword hilt | alter proportion of soldier's head | Change hand on sword hilt | Increase size of child | Best for general work yet | Perspective of stone stairs wall not true.'[42] Proofs were first pulled on India paper (Burne-Jones and Morris, however, declined to look at these earliest impressions), then on thicker stock, and finally on Batchelor paper.

The relationship between Burne-Jones and Catterson-Smith appears, on the whole, to have been amicable, though Burne-Jones always adopted a patronizing tone in speaking of the Birmingham artist. 'I have had to give a drawing lesson to the lad,' he wrote. 'The lad is about 35, but still a lad – as in art one is a lad at a hundred.'[43] Catterson-Smith, for his part, professed great admiration for Burne-Jones but complained that he was out of touch with the ordinary realities of life.[44] The nature of Catterson-Smith's contributions to the Kelmscott *Chaucer* was not widely known or publicly discussed until 1898 (since his name did not appear in the book's colophon), when an anonymous contributor to the London *Daily Chronicle* – probably Joseph Pennell, the American artist–journalist – suggested that Morris had unfairly (and perhaps dishonestly) failed to acknowledge Catterson-Smith's role as collaborator of Burne-Jones in the *Chaucer*.[45] Such a claim elicited an angry letter to the editor of the *Daily Chronicle* from Cockerell (15 November 1898) and a somewhat milder epistle from Catterson-Smith himself (25 November), who wrote: 'Fingers, eyes, and sympathy I brought, but Sir Edward was responsible for every line and dot in the eighty "Chaucer" drawings which I did under his guidance. I worked at his very elbow for months, often spending whole days seeking out a simple and expressive treatment of a passage, and in many cases doing drawings over and over again, until he was satisfied that the treatment or convention (he used to call it his "shorthand") expressed *him*.' Burne-Jones, adopting a similar but less flattering metaphor, described Catterson-Smith as an efficient tool in his hands.[46] That it was an extraordinarily successful collaboration there can be no doubt; yet at the same time Morris must have been uncomfortably aware that the use of a craftsman as a mere 'tool' of an artist undermined the ideological basis of the Kelmscott Press. It illustrates once again the chasm that sometimes appeared between Morris's theories of bookmaking and what actually took place at the Press.

The wood-engravings created by the joint endeavours of Burne-Jones, Catterson-Smith, and Hooper are a curious hybrid of the characteristics of engravings and woodcuts. They are unquestionably engravings, because they are cut on the end grain of boxwood with gravers' tools; yet they display little of the delicacy of effect usually associated with an engraving. The lines are broad, the wood is engraved deeply, and there is no hint of the white-line technique used by most engravers after Bewick. The methods of shading are similar to those of a woodcut. What Morris and Burne-Jones produced, through their craftsmen, was an engraved version of the woodcut as it appeared in the fine illustrated books of the fifteenth-century printers of Ulm and Augsburg.

Once they were completed, Morris found that wood-engraved illustrations are technically difficult to print. They require exceptionally careful presswork, with overlays and underlays, and, because they bite deeply into dampened paper, problems arise when large engravings appear on the verso of a page of text. In fact, not until the *Chaucer* was published in 1896 did Morris succeed in fully integrating words and pictures. In the illustrated Kelmscott Press books published before the *Chaucer* the wood-engravings always appeared alone on leaves that were blank on the verso; some were printed in an entirely separate operation and the leaves merely inserted when the book was bound, a procedure perilously close to the formal dichotomy of text and illustration that Morris always condemned in the books of others. (Another reason for Morris's adoption of this unhappy expedient is that Burne-Jones was often so slow in producing his designs that they were engraved only after the rest of the book had been printed.) There were occasional proposals that the Kelmscott Press illustrations might be issued by themselves, in a form suitable for framing, but here Morris drew the line. Writing to the bookseller Frederick H. Evans in November 1894, Cockerell declared: 'I have spoken to Mr. Morris about the proposal in your letter . . . & I am sorry to say that he does not see his way to printing the pictures [of the *Chaucer*] apart from the book. He thinks that if this were done it might injure the sale of future issues, as some people would abstain from buying the books themselves in the hope of being able to secure the illustrations separately.'[47] Cockerell's letter was expressed in appropriately commercial terms, yet Morris was no doubt more troubled by the theoretical implications of Evans's proposal. He might be willing to compromise in his search for unity of word and picture, but to publish the *Chaucer* illustrations separately would have made a mockery of his ideal of integration.

Another symptom of the obstacles that Morris encountered in

furnishing illustrations for the Kelmscott books is the surprisingly large number of false starts and abandoned plans. Probably the most spectacular example is *The Well at the World's End* (1896): Morris first announced that Charles Fairfax Murray would illustrate it, then assigned the task to Arthur J. Gaskin, and finally – but only after Gaskin's designs had been engraved and several leaves printed – persuaded Burne-Jones to rescue the book. A map was prepared for *Utopia* (1893) and a line block made, but it was scrapped. During 1893 and 1894 Charles M. Gere produced a series of drawings for a projected Kelmscott edition of *The House of the Wolfings* which was not in the end published. In 1892 Aymer Vallance visited Morris at Kelmscott House, accompanied by Aubrey Beardsley, who had prepared an illustration for *Sidonia the Sorceress*. Morris, while conceding that young Beardsley had a good eye for drapery, declined the drawing – and not very politely, according to Vallance.[48] Beardsley was so deeply offended that for the rest of his short life he nursed a grudge against Morris and sought revenge by creating a witty parody of the Kelmscott style in his ornaments and illustrations for the Dent edition of Malory's *Le Morte Darthur* (1893–4). Beardsley's encounter with Morris appears to have been an exceptionally unhappy one, but the truth of the matter is that every book-illustrator found Morris nearly impossible to please. It is a significant fact that no artist except for Burne-Jones illustrated more than one Kelmscott Press book.

When he was planning the first Kelmscott volume, *The Story of the Glittering Plain*, Morris hit upon the notion of enlisting the services of his friend, Walter Crane, a fellow-socialist and one of the leading illustrators in Victorian England. In the autumn of 1890 he wrote out a formal proposal specifying that 'The illustrations [are] to be drawn by Walter Crane, & cut on wood by Mr Leverett or others to Walter Crane[']s satisfaction.'[49] It was agreed that they would share any profits from the book, an arrangement Morris never again offered to an illustrator. But Morris was so impatient to get his first title into print that he was unwilling to wait for Crane's designs, and *The Glittering Plain* was issued in 1891 in unillustrated form; to atone for his haste, Morris published another version of *The Glittering Plain* at the Kelmscott Press in 1894 with Crane's pictures. After such a long delay, the result was anticlimactic, for the *Glittering Plain* illustrations are, quite simply, the most feeble that appeared in any Kelmscott book.

Figure 49

THE STORY OF THE GLITTERING PLAIN OR THE LAND OF LIVING MEN
Chapter I. Of those Three who came unto Hallblithe to the House of the Raven
THAS been told that there was once a young man of free kindred and whose name was Hallblithe: he was fair, strong, and not untried in battle; he was of the House of the Raven of old time. This

Figure 49 *The Story of the Glittering Plain* was the only title issued twice at the Kelmscott Press. The second version (1894) was illustrated with wood-engravings designed by Walter Crane, which Morris found unsatisfactory.

What had gone wrong? It is evident that Crane, whose own books are characterized by great delicacy and grace, found Morris's dark, medieval page uncongenial; an examination of his designs[50] reveals that he worked mainly with white ink on a black ground, thus producing a suffocatingly black picture which was then engraved in the most mechanical facsimile fashion by A. Leverett. 'I think your woodcuts look delightful,' Morris wrote to Crane, trying to put the best face on the matter; but Crane later doubted 'if I was ever quite Gothic enough in feeling to suit his [Morris's] taste'.[51] Privately Morris offered a more blunt assessment of Crane's work. The *Glittering Plain* illustrations, he told the artist Edmund New in October 1895, 'are poor in execution'; to Cockerell he acknowledged that he 'thought this volume his one Kelmscott Press failure'.[52] Philip Webb fired off a letter to Morris filled with indignation at the inadequacies of Crane: 'do not be so rash in trusting the picture makers in future,' he implored.[53]

But if not even the distinguished Walter Crane could be counted upon to come up to what Webb called 'the Thames-side mark', where in England was Morris to find the artists who could satisfactorily illustrate the Kelmscott books? The answer, surprisingly, was Birmingham, where the arts-and-crafts movement flourished more vigorously than anywhere else in the provinces, mainly through the influence of the municipal School of Art.[54] Three youthful artists associated with the School – Arthur J. Gaskin, Charles M. Gere, and Edmund New – were doing attractive book illustrating during the 1890s, and Morris was aware of them and their work. 'Gaskin, a young Birmingham artist, called in the afternoon [at Kelmscott House] with a number of very pretty drawings for an edition of Hans Anderson which Geo Allen is going to publish,' Cockerell noted in his diary on 3 December 1892.[55] Nine days later this entry appears in the diary: 'C. M. Gere, another nice young Birmingham artist, called in the aft[ernoo]n with some drawings, and W.M. asked him to have a shot at illustrating the [House of the] Wolfings.' Both men also visited Kelmscott Manor on later occasions, as did Edmund New, though in New's case Morris offered him no assignment for the Kelmscott Press.

Morris's attitude towards Gaskin and Gere was ambivalent: he was grateful for their loyalty to the arts-and-crafts ideals, yet his praise for their work was often cautious and qualified. In an interview in the *Daily Chronicle* (9 October 1893), he was quoted as saying that 'there is a great quantity of excellent art, but the only thing that is new, strictly speaking, is the rise of the Birmingham school of book decorators. . . . these young men of the Birmingham School of Art – Mr. Gaskin, Mr. Gere, and Mr. New – have given a new start to the art of book decorating.' In another interview, published in *Bookselling* two years later, however, Morris remarked: 'I think they have, in Gaskin and New, two very good men, who have ideas and originality. For the most part, however, they follow too slavishly the opposition to conventionality. . . . but you must remember that the Birmingham people have not yet found their feet. They will do good work yet, I am sure.'[56] This was faint praise indeed. The same ambiguous note is struck by Morris in the introduction he wrote for *Good King Wenceslas* (Birmingham, 1894), illustrated by Gaskin: 'they [the illustrations] have given me very great pleasure, both as achievements in themselves and as giving hopes of a turn towards the ornamental side of illustration, which is most desirable.'[57]

As Cockerell's diary indicates, Gere was the first of the Birmingham artists to be asked to illustrate a planned Kelmscott Press book, *The House of the Wolfings*. Gere – who in later years achieved a reputation as a painter as well as an illustrator for the Ashendene Press[58] – made a number of designs for the book during 1893, submitted them to Morris, and for his pains received a bundle of letters from Morris[59] filled with the most severe criticism. The execution was wrong, said Morris; the lines were too thin. 'I am sorry to say that I do not think that you have done yourself justice in them' (1 July 1893). This went on for some months, and finally the correspondence faded away. Though Morris had two trial pages printed at the Press, *The House of the Wolfings* was stuck forever in limbo.

But in the meantime Gere had undertaken a more modest task for the Kelmscott Press that proved to be a considerable success. At Cockerell's suggestion, Morris asked Gere to do a drawing of Kelmscott Manor to serve as a frontispiece for *News from Nowhere*, and in this instance he was able to supervise Gere's work much more directly. (The engraving, in fact, seems to have permanently fixed the image of Kelmscott Manor in the public mind: some years ago the trees and bushes in front of the house were rearranged and in some cases replanted so that they would appear exactly as they do in Gere's illustration.) Gere's style – simple, unfussy, honest, and linear – is markedly similar to that of New and Claude Napier-Clavering, the director of the Birmingham Guild of Handicraft, whose work also appeared frequently in the Guild's intensely Morrisian magazine, the *Quest*; and in the November 1895 number of the *Quest*, the *News from*

Nowhere frontispiece was reprinted from the original block (something Morris normally did not allow) to accompany an essay by Morris on Kelmscott Manor, 'Gossip about an Old House on the Upper Thames'.

Gaskin's relationship with Morris was, if anything, even more turbulent than Gere's. Subsequently well known as a designer of jewelry,[60] Gaskin was scheduled to illustrate a Kelmscott Press edition of *The Roots of the Mountains* that never materialized; Morris also arranged for him to design the pictures for *The Well at the World's End* and *The Shepheardes Calender* (1896). The twelve designs for the latter book are impressive – Gordon Ray has called them 'perhaps the most successful of Kelmscott Press illustrations', a judgement in which Colin Franklin concurs[61] – but one suspects that they were not entirely to Morris's liking, and Morris would probably also have disapproved of the use of process blocks furnished by Walker and Boutall. *The Shepheardes Calender*, however, was printed shortly after his death, when Walker and Cockerell felt a desperate urgency in trying to finish Morris's uncompleted books at the Press and thus were inclined to adopt short cuts.

Figure 50

The story behind Gaskin's illustrations for *The Well at the World's End* is longer and more complicated. The earliest reference to Gere's connection with the book appears, as usual, in Cockerell's diary (23 April 1893): 'Gere to illustrate The Wolfings, Gaskin The Roots & The Well at the World's End.' Gaskin produced some designs in July 1893, but Morris confided to his daughter Jenny that 'I am sorry to say that I did not think they would quite do'.[62] Morris sent back detailed criticisms, Gaskin offered more drawings which Morris in turn criticized, and so it went for month after month: *The Well at the World's End* was longer in production than any other Kelmscott book. Altogether Gaskin submitted nineteen designs, some of which were actually engraved by Hooper, and trial pages incorporating them were pulled at the Press. Finally in February 1895 Morris realized he could temporize no longer. After consulting Philip Webb, who confirmed his own unfavourable opinion of the designs, Morris wrote to Gaskin that he would be unable to use them. The artist protested and begged to be allowed to do four new illustrations; Morris declined. *The Well at the World's End*, when it eventually came off the press in 1896, was embellished with four designs by Burne-Jones. What is puzzling about all this is that Gaskin's drawings are in fact beautiful. One can only guess that Morris objected to them because he feared their graceful delicacy would not harmonize with his sturdy Chaucer type.

Figure 50 At Morris's request, Arthur J. Gaskin produced wood-engraved illustrations for *The Well at the World's End* (1896), but Morris was unhappy with them and substituted four designs by Burne-Jones.

And so, once again, Morris fell back on the support of Edward Burne-Jones. In reality, Burne-Jones's illustrations were the only ones that satisfied him, and Morris's attempts to recruit other artists to work for the Kelmscott Press were always rather half-hearted. Whatever the merits of the designs by Crane, Gere, and Gaskin, it is the name of Burne-Jones that will forever be associated with the Press, for not only did he illustrate more Kelmscott works (twelve) than any other artist, but his visual imagination also matched perfectly the typography and design of Morris's books. Reflecting their deep personal rapport, the aesthetic harmony of those volumes serves as another reminder that Morris's and Burne-Jones's ideas about book-illustration had evolved slowly, and in unison, since their Oxford days. In the evening at Exeter College they read Chaucer aloud to each other; during the day they *Figure 51*

Figure 51 This photograph of Edward Burne-Jones and William Morris was taken in 1890, shortly before they began their memorable collaboration at the Kelmscott Press. Though Morris occasionally turned to other illustrators, their work failed to satisfy him as fully as Burne-Jones's did.

examined illuminated manuscripts at the Bodleian. The Oxford experiences established a psychological pattern for life: until his death Morris spent Sunday mornings in Burne-Jones's studio, reading and talking, while the artist worked at his easel. It was the high point of the week for both of them. Burne-Jones sent to Lady Horner this account of one of those Sunday communions: 'To-day at breakfast came Morris, and he talked hard all morning whilst I drew – why the mediaeval world was always on the side of the Trojans. We quarrelled a good deal and often he said "Look here, Ned, that's all gammon, you know," and I said, being more polite, "Pardon me, you are quite wrong there – the fact of the matter is," and sometimes we were both polite, and he would say, "There's a good deal in that," and I would say "Yes, you are quite right." . . . All morning through we never mentioned one living person. We never got to speak of anyone later than the ninth century.'[63]

As Burne-Jones's words suggest, he and Morris hardly thought alike on all subjects. There were in fact two great sources of tension between them. The first was that, whereas Morris was drawn to the art, literature, and values of the Gothic North, Burne-Jones (for all his medievalism) felt an equally strong attraction to the Mediterranean world. 'I think Morris's friendship began everything for me, everything that I afterwards cared for,' Burne-Jones wrote to J. Comyns Carr. But he added: 'I quarrel now with Morris about Art. He journeys to Iceland and I to Italy, which is a symbol.' When Morris returned from Iceland in 1873, Burne-Jones made a delightful drawing of his friend seated in the snow and gnawing a fish. 'Mr. Morris has come back more enslaved with ice and snow and raw fish than ever,' Burne-Jones declared in a jesting tone that only partially conceals the seriousness of his complaint – 'I fear I shall never drag him to Italy again.'[64] This painful division between their loyalties to North and South is sometimes reflected in the Kelmscott Press illustrations, most conspicuously in the *Chaucer*.

Morris's commitment to socialism also threatened at times to divide him from his closest friend. Burne-Jones acknowledged the corruptions of the capitalist system, but unlike Morris he did not rejoice in its impending collapse: violent change, which seemed to him a bad thing in itself, would have separated him from what he described as 'the dream country where my real home is'.[65] On his side, Morris expressed strong disapproval when Burne-Jones was made a baronet in 1894. 'W.M. much vexed thereby. A great pity I think also,' Cockerell recorded in his diary on 5 February. Jane Morris has left a more vivid record of the incredulous response in the Morris household to the news: 'my husband refused to believe it at first, but afterwards when the plain fact was known, he said "Well a man can be an ass for the sake of his children".'[66]

At a more profound level the relationship between Morris and Burne-Jones could have been undermined by Burne-Jones's conviction that his friend was too emotionally self-sufficient. 'I suppose he [cares?] for me more than anyone, yet the day I go he will lose nothing, only he will have to think to himself instead of thinking aloud – no more than that,' Burne-Jones wrote with a trace of bitterness.[67] But this was only a passing mood: more characteristic is his comment about Morris recorded in T. M. Rooke's diary on 24 May 1853: 'Oh! he's most beautiful, he's a *darling*, he's the queerest thing that ever existed on this earth, there's nobody like him, there isn't anyone that *exists* that in the least *bit* resembles him.' Contrary to Burne-Jones's expectations, Morris preceded him in death; the experience left him broken in spirit, and within two years he died as well.

Apart from his Kelmscott Press work, Burne-Jones's achievements as an illustrator were slender. In 1854 he was asked to provide pictures and ornaments for Archibald Maclaren's book, *The Fairy Family*, which was to be published by Longmans. Burne-Jones's designs, rather in the style of Dicky Doyle, do not at all suggest his mature work, and he never completed them, apparently because of his own dissatisfaction with any work begun before he was exposed to Rossetti's illustrations; *The Fairy Family* was eventually issued without his drawings.[68] There was much talk during the 1860s about Burne-Jones's great promise as an illustrator, yet little came of it. Holman Hunt in 1861 sent the Dalziel Brothers a strong endorsement, for, he said, 'I feel very strongly [Burne-Jones] might be of great value to you in the illustrating of *Good Words*. . . . He has yet, I think, made but few if any drawings on wood, but he has had much practice in working with the point both with pencil and pen and ink on paper, and so would have no difficulty with the material. I have not seen him lately, but remember that he has sometimes said that he should like to try his hand at drawing on wood. . . .'[69] Though Burne-Jones did a handful of designs for magazines during the decade, his heart was not in it, and he was not temperamentally or aesthetically drawn to the 'sixties' version of wood-engraved illustrations; only Morris was able to provide the proper environment for his talents.

One reason Burne-Jones was able to produce so many illustrations for the Kelmscott Press of such consistently high quality within a short time span is that he borrowed heavily from his own earlier work. A large proportion of the Kelmscott pictures are derived from paintings on similar or identical themes: this ought not to surprise us, for Burne-Jones was the sort of obsessive artist who returned constantly to certain recurring motifs, which Penelope Fitzgerald has ably summarized as 'the enchantment of the willing victim, sleep, waiting, imprisonment, loneliness, guiding, rescue, the quest, losing and finding, tending the helpless, flying, sea-crossing, clinging together, the ritual procession and dance, love dominant and without pity, the haunting angel, the entry into life'.[70] Yet he did not offer Morris mere mechanical adaptations of earlier designs. The Kelmscott illustrations based on specific paintings by Burne-Jones are always transposed into a new, more haunting key. Human figures are shown in graceful but static poses against sometimes disturbing, surrealistic backgrounds, gazing past each other (and past the reader) with strangely hypnotic expressions. It may be significant that most of Burne-Jones's Kelmscott Press illustrations, except for the *Chaucer* and *The Well at the World's End*, are upright rectangles, often quite narrow, filled with emphatically vertical and slender figures (usually female), for the pattern reminds one immediately of the scores of stained-glass windows that he had designed for Morris and Company. As in the windows, the human beings seem divided from each other, though here it is an invisible barrier rather than a mullion that separates them. The darker side of Burne-Jones's imagination is much in evidence in these illustrations.

His distrust of deep perspective, in the paintings as well as these designs, has often been commented on, and of course the comparative flatness of Burne-Jones's drawings goes well with the markedly shallow depth of Morris's ornaments. What is perhaps less commonly noticed about the Kelmscott illustrations is that Burne-Jones, in accordance with the theories that he and Morris had evolved, made certain that his pictures were above all ornamental. Whether they told a story was beside the point; he confessed that he always evaluated his illustrations by looking at them upside down.[71] The real tests, in the end, were the decorative quality of the pictures and the extent to which they contributed to the *mise-en-scène* of the printed page.

Burne-Jones's view of his own role as illustrator of the Kelmscott Press books is nowhere more beautifully expressed than in this eloquent letter to Charles Eliot Norton in 1894: 'And so you don't like

Chaucer – that is very sad – for I am beside myself with delight over it. I am making the designs as much to fit the ornament and printing as they are made to fit the little pictures – and I love to be snugly cased in the borders and buttressed up by the vast initials – and once or twice when I have no letter under me, I feel tottery and weak; if you drag me out of my encasings it will be like tearing a statue out of its niche and putting it into a museum – indeed when the book is done, if we live to finish it, it will be like a pocket cathedral. . . .'[72] This was the splendid vision that beckoned Morris and Burne-Jones onwards at the Kelmscott Press, and, whatever else may be said against their books, they did indeed find a way of reconciling type, border, initial, and picture. It was no mean achievement.

SIX ✾ THE BUSINESS OF THE PRESS

NOT larks!' Morris would exclaim to his daughters as he plunged into another of his adventures. The play impulse was strong in his temperament, and one is constantly amazed by the degree to which he was able to combine pleasure and work in all that he did. It is useful, and no doubt essential, to see Morris as a serious social reformer who sought a transformation of Victorian life and labour; but we overlook a vital clue to his mind if we fail to take into account the sheer zest he brought to every task at hand. The Kelmscott Press itself was begun in this joyous spirit. Even though Morris had brooded over the problem of how to print beautiful books for several decades, he made the decision to design his own types on impulse, in a moment of enthusiasm, when he heard a stimulating lecture by Emery Walker. Throughout those final years of his life, Morris's pleasure in his typographical adventure never diminished. Wilfrid Scawen Blunt, visiting Kelmscott Manor, noted in his diary on 5 July 1892: 'Morris was in fine spirits and inexhaustible energy over his new hobby of the printing press.' 'I do so like seeing a new book out that I have had a hand in,' Morris commented to Jenny.[1] The Kelmscott Press was pleasantly intertwined with Morris's family life; snatches of conversation with Walker and others in the library floated up the stairway into the drawing-room, proofs and trial pages lay scattered about Kelmscott House, and in time the books themselves drifted into the house in great piles.

But the Kelmscott Press, which began as a lark, turned into a large and complex business operation as the months went by. There are signs of this change in the growing number of titles printed each year. In 1891 only two books were issued, whereas in the following year that figure increased to seven; in 1893 it became twelve. Though thereafter the number of books produced each year tapered off slightly, that is a misleading statistic, because the works actually printed were in some instances very ambitious undertakings, such as the *Chaucer* and the eight-volume *Earthly Paradise*. Even at the beginning the Kelmscott Press was never really a one-man operation, since Morris had the assistance of Walker and various craftsmen. It must have soon become evident, however, that Morris needed to build up a staff to help him in business and editorial matters as well as the printing itself.

Figure 52 Morris's daughter May and her husband Henry Halliday Sparling, who served as the first Secretary of the Kelmscott Press until 1894, when he and May separated. The testimony of his contemporaries suggests that he was not an effective manager.

The Business of the Press

Figure 52

The person who initially bore much of the responsibility of the day-to-day operation of the Press was Henry Halliday Sparling, the husband of May Morris. Sparling assumed the title of Secretary of the Press, a position he held until July 1894 when he was succeeded by Sydney Cockerell; but, whereas Cockerell was a born archivist who left behind enormous collections of papers recording his activities at the Press, Sparling is a far more elusive figure about whom we know little. Everything we can learn about him comes from a few scraps of correspondence, a sketchy outline of his career supplied by newspaper obituaries, and the passing comments of his contemporaries. These observations by those who knew him, it must be added, are uniformly unflattering. Sparling seems to have struck everyone who met him as a weak, unremarkable young man. Cockerell, who was normally discreet in writing about the Morris circle, described Sparling as 'a rather second-rate Socialist'. Joseph Pennell and his wife discovered Sparling talking 'like a romantic schoolgirl' at the meetings of the Hammersmith Socialist League; Mrs Pennell declared that Morris's future son-in-law was 'disagreeably young and pleased with himself'. Yeats, who met Sparling at the British Museum, recorded his fatuous remark about May Morris, to whom he was then engaged: ' "She is very beautiful," he would say, "Morris says so." '[2]

Sparling kept himself occupied during the 1880s as a socialist lecturer, literary journalist, and co-editor of Morris's *Commonweal*. Exactly what functions he performed at the Kelmscott Press in the following decade is not clear, because the documentary record is so skimpy. We know that he edited three of the Kelmscott books (*The Recuyell of the Historyes of Troye*, *The History of Reynard the Foxe*, and *The History of Godefrey of Boloyne*), did most of the proof-reading, and handled some of the correspondence and perhaps the bookkeeping as well. There is no evidence that he carried out his clerical duties especially well. In fact, when Cockerell took over, he found the records in some confusion and had to exercise his formidable organizational skills to put the Kelmscott Press on sounder footing as a business.

Sparling's marriage to May Morris (which took place in June 1890) seems to have been fragile from the start, and it was finally broken up, if Bernard Shaw's account can be trusted, by the prolonged presence of Shaw as a guest in their home.[3] 'May's married life has come to an end,' her mother announced to Blunt in May 1894, 'and although we always expected some catastrophe in that direction, the blow is no less heavy now it has come.' Writing again two days later, she added that

May 'has been seeing a good deal of a former lover [Shaw?], and made her husband's life a burden to him'.[4] On 25 May Cockerell noted laconically in his diary: 'W.M. asked me to take up the Secretaryship of the Kelmscott Press, which Sparling is resigning.' Sparling officially stepped down as Secretary in July when the following note was inserted in a Kelmscott Press circular: 'Mr. Halliday Sparling, having been compelled by the pressure of other work to resign the Secretaryship to the Kelmscott Press, has been succeeded by Mr. S. C. Cockerell. Correspondents are asked to note the consequent change of address.'

Thereafter Sparling fades mysteriously out of the picture. He fled to France, returning to England only for Morris's funeral, and a divorce was granted in 1899. He continued to live on the Continent in obscurity and possibly also poverty, though in later years we hear of his delivering lectures on 'William Morris as Prophet of Reform' and 'The Glory that is France' to wealthy ladies in Pasadena, California. His last act before his death in 1924 was to complete a volume entitled *The Kelmscott Press and William Morris Master-Craftsman*, published in the same year. Though Emery Walker dismissed it as 'not of much value',[5] the book is a useful, detailed account of the Press which supplies some factual information not preserved elsewhere. It is weakened, however, by poor Sparling's need to protest his constant loyalty to Morris's ideals despite the collapse of his marriage to Morris's daughter: 'an adoring and eager disciple throughout, I may claim to be especially qualified as an interpreter of his teaching,' he declared in the preface. It seems fair to assume that the tone of unctuous piety which pervades Sparling's history has had a detrimental effect upon the reputation of the Kelmscott Press in our century.

It is difficult to imagine a greater contrast than that between the first and second Secretaries of the Press. Sydney Cockerell possessed a keen, inquiring mind, was an efficient administrator, and in later years achieved a position of considerable eminence as director of the Fitzwilliam Museum, Cambridge. What is especially remarkable about Cockerell is that he sensed at once the historical importance of the Kelmscott Press and from its founding collected every scrap of paper, every piece of ephemera, associated with the Press. If it were not for Cockerell's acquisitive instincts, our knowledge of Morris's work as a printer would today be greatly impoverished. One particularly valuable document preserved by Cockerell is the series of small pocket diaries in which he recorded his daily activities from the age of eighteen

Figure 53

Figure 53 'I haven't a spark of imagination,' Sydney Cockerell once complained, but in face he was a brilliant administrator who held the Kelmscott Press together after Morris's death.

until his death; thanks to these diaries, the great masses of correspondence that he accumulated, and his large collection of Kelmscott Press proofs and trial pages, it is possible to reconstruct – at least after July 1894 – the activities of the Kelmscott Press in vivid detail. Cockerell claimed that his diary, straightforward and simple as the entries are, 'has not only been of daily service to me for casual reference to events that may only have been a few days or a few weeks old, and for verifying names and dates, but, after a longer interval, has formed a sort of ladder on which recollection could climb with sure feet'.[6] Cockerell's later published recollections of Morris and the Kelmscott Press were in fact based directly on his diaries (as early as November 1896 he was copying out the references to Morris).

Left fatherless and not well provided for, Cockerell reluctantly joined the family firm of George Cockerell and Company, Coal Merchants, in 1884, and when he received a legacy of £2,000 from his grandfather in 1889, he invested it in the company only after much soul-searching; two years later he decided to sell his share in the business and seek his fortune in the world. But what was he to do? Though he considered the possibility of Cambridge, London seemed to hold more attractions, for by now, with his socialist and artistic leanings, he had drifted into Morris's circle. He temporarily postponed the hard decision about his future, concentrating instead on reading (especially Ruskin, with whom he was corresponding), visiting museums and art galleries, and attending lectures sponsored by various socialist and arts-and-crafts groups. It was a heady experience for a coal merchant's son still in his early twenties. In 1892 he applied for the curatorship of the Soane Museum; in 1894 he was offered (but declined) the editorship of the *Studio* magazine. And regularly, sometimes several days each week, Cockerell dropped in at Kelmscott House, where Morris showed him the latest proofs and books from the Press and they talked about Morris's exciting plans for printing. In August 1892 he visited Kelmscott Manor for the first time.

At the beginning of his diary for 1892 Cockerell wrote the following note (dated 10 January 1893): 'Out of my stays at Kelmscott [Manor], and thanks also to Philip Webb's influence, arose my present work of cataloguing W.M.'s books, a truly enviable task, on which I have been engaged for 2 months. It will take me some time longer, and having done it I shall the more readily obtain other employment. I already begin to know something of the various 15th century types.' Morris had engaged Cockerell's services for 'two guineas a week and a copy of the Golden Legend!' (diary, 3 November 1892), and, aided by the advice of F. S. Ellis and Robert Proctor, Cockerell, though inexperienced, found his way with surprising agility through the bibliographical thickets of incunabula. The catalogue of Morris's library was never completed, but one can see an indication of the form it would have taken in one of the last of the Kelmscott Press books, *Some German Woodcuts of the Fifteenth Century* (1897), where Cockerell printed, as an appendix, the result of this painstaking analyses of the woodcuts in Morris's collection of early books. 'I haven't a spark of imagination, and am only good for dry-as-dust cataloguing,' Cockerell protested; but it was clear to many of Morris's friends that this unusual young man possessed 'the excellent organ of orderliness' (as Ellis called it) to an amazing degree.[7]

Figure 54

The lesson was not lost on Morris, who increasingly gave Cockerell other clerical assignments, some of which were connected with the Kelmscott Press. His name appears occasionally, for example, on receipts for the Kelmscott books during the period when Sparling was still nominally the Secretary. Once it became obvious that Sparling's marital arrangements were crumbling and that he could no longer serve as Secretary, Morris turned without hesitation to the studious, meticulous young man already working in his library. Cockerell at once made himself indispensable. He was able to relieve Morris of administrative responsibilities at the Press much more efficiently than Sparling had ever done, and Morris's standard response to questions now became 'Ask Cockerell'. As Philip Webb remarked to Cockerell, 'you know every body, almost tripping up Walker in that capacity. . . .'[8] After Morris's death he was effectively in charge of the Press during the last year and a half of its existence.

What Cockerell brought primarily to the secretaryship was resourceful intelligence rather than experience, though for several years before 1894 he had been discussing typefaces and early printed books with Morris and Walker, and wood-engravings with Walter Crane and W. J. Linton. About printing, as about everything else, Cockerell displayed an insatiable curiosity. Consider, for example, this typical entry in his diary (10 June 1890): 'Called on Mr Jowett, the manager of Hazell Watson & Viney's [the firm that printed Ruskin's books], and was shown over the works, where I learned a great many things – and saw the processes of type casting, electrotyping, handpress and machine printing, sewing, binding etc etc which I had been very anxious to be acquainted with.'

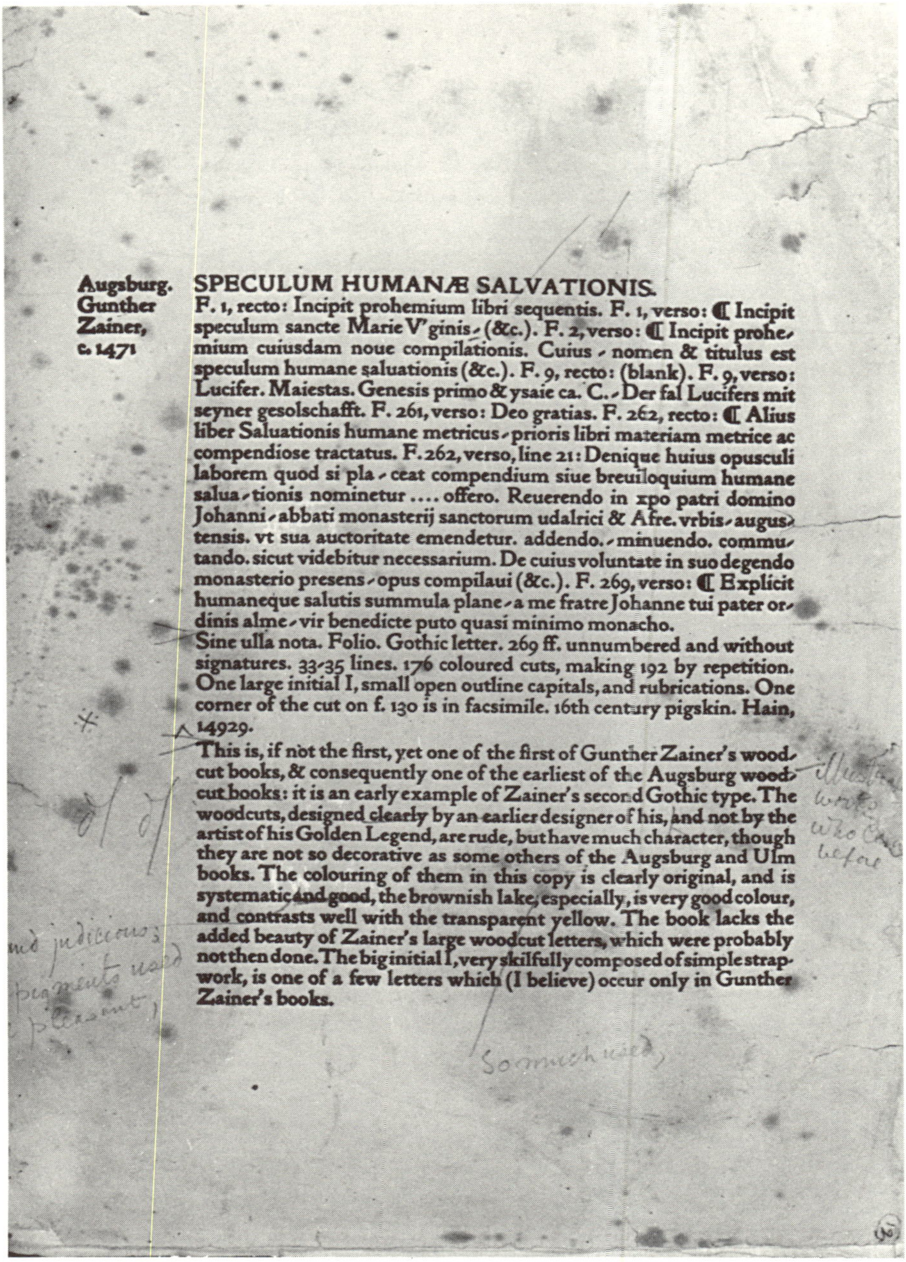

Augsburg. Gunther Zainer, c. 1471

SPECULUM HUMANÆ SALVATIONIS.

F. 1, recto: Incipit prohemium libri sequentis. F. 1, verso: ℭ Incipit speculum sancte Marie V'ginis- (&c.). F. 2, verso: ℭ Incipit prohe-mium cuiusdam noue compilationis. Cuius - nomen & titulus est speculum humane saluationis (&c.). F. 9, recto: (blank). F. 9, verso: Lucifer. Maiestas. Genesis primo & ysaie ca. C. - Der fal Lucifers mit seyner gesolschafft. F. 261, verso: Deo gratias. F. 262, recto: ℭ Alius liber Saluationis humane metricus - prioris libri materiam metrice ac compendiose tractatus. F. 262, verso, line 21: Denique huius opusculi laborem quod si pla - ceat compendium siue breuiloquium humane salua - tionis nominetur .... offero. Reuerendo in xpo patri domino Johanni - abbati monasterij sanctorum udalrici & Afre. vrbis - augus-tensis. vt sua auctoritate emendetur. addendo. - minuendo. commu-tando. sicut videbitur necessarium. De cuius voluntate in suo degendo monasterio presens - opus compilaui (&c.). F. 269, verso: ℭ Explicit humaneque salutis summula plane - a me fratre Johanne tui pater or-dinis alme - vir benedicte puto quasi minimo monacho.

Sine ulla nota. Folio. Gothic letter. 269 ff. unnumbered and without signatures. 33-35 lines. 176 coloured cuts, making 192 by repetition. One large initial I, small open outline capitals, and rubrications. One corner of the cut on f. 130 is in facsimile. 16th century pigskin. Hain, 14929.

This is, if not the first, yet one of the first of Gunther Zainer's wood-cut books, & consequently one of the earliest of the Augsburg wood-cut books: it is an early example of Zainer's second Gothic type. The woodcuts, designed clearly by an earlier designer of his, and not by the artist of his Golden Legend, are rude, but have much character, though they are not so decorative as some others of the Augsburg and Ulm books. The colouring of them in this copy is clearly original, and is systematic and good, the brownish lake, especially, is very good colour, and contrasts well with the transparent yellow. The book lacks the added beauty of Zainer's large woodcut letters, which were probably not then done. The big initial I, very skilfully composed of simple strap-work, is one of a few letters which (I believe) occur only in Gunther Zainer's books.

Figure 54 Proof (corrected by Morris) of an entry for a catalogue of Morris's library that he hoped to publish at the Kelmscott Press. The technical bibliographical description in the first two paragraphs is by Sydney Cockerell; the third paragraph is by Morris. The catalogue was never completed, but material from it was absorbed into *Some German Woodcuts of the Fifteenth Century* (Kelmscott Press, 1897).

Another conspicuous trait of Cockerell was his loyalty to the Morris family: after 1896 he became a confidant, financial adviser, and surrogate son of Jane Morris, and in later years stepped into a similar role (though now more like a kindly older brother) in his relationship with both Jenny and May Morris. There was something about his manner that inspired confidence; he appeared to be everyone's choice for executor or trustee. In old age, laden with honours, a bedridden invalid in Kew, he became a formidable character indeed. With undiminished force of personality, he held court in his bedroom for those who came to pay their respects, surrounded by the most astonishing relics of Morris, Rossetti, Ruskin, and the other great Victorians. The slightest disparagement of Morris in print brought him roaring out of his den with a fierce letter to the editor. When he died in 1962, he seemed to many the last vestige of the Victorian era, an improbable survivor of an inconceivably distant age.

Apart from Cockerell and Walker, the person who exercised the most direct influence on the Kelmscott Press was Frederick S. Ellis, a retired bookseller and publisher, who edited no less than fifteen Kelmscott books and became Morris's chief literary adviser at the Press.[9] Morris was introduced to Ellis by Swinburne in the 1860s, found him a pleasant, knowledgeable man (as did everyone), agreed to let him publish most of his books from 1869 onwards, and was joint-tenant of Kelmscott Manor with Ellis for a decade. By the time the Kelmscott Press was in operation, Ellis had sold his business and was living in Torquay, but he frequently visited his son at 7 Hammersmith Terrace and dashed in and out of Kelmscott House in his inimitable, breezy fashion.

Figure 55

One of Ellis's major contributions to the Kelmscott Press was his untiring insistence upon the importance of accurate proof-reading. There was an amusing rumour at the time that Morris was unaware of the existence of printers' readers; this of course was not true, but for the first three years proofs seem to have been read very haphazardly at the Press. Morris's attitude towards such matters was decidedly casual. When *Poems by the Way*, the second Kelmscott book, was in production, he wrote to Walker from Folkestone (where he was convalescing from an attack of gout): 'Yes, I think it would be best to print that sheet again; a misprint of that kind, though it does not otherwise matter, is such a defacement of the book, – and 'tis a 2

Figure 55 F. S. Ellis, a retired bookseller who had once been Morris's publisher, became a key figure at the Kelmscott Press, where he served as an editor and proofreader. He brought pressure on Morris to improve the textual accuracy of the Kelmscott books.

guinea book let us remember.'[10] Sparling, Morris, and the printers themselves at first handled the proof-reading, but this was soon seen to be an unsatisfactory arrangement. Morris appears not to have read against copy, and many of his alterations in proofs were intended to improve the word-spacing of loose-fitting lines rather than to eliminate typographical errors. Ellis certainly had no illusions about Morris's capabilities as a proof-reader. While he was supervising a reprint of *The Defence of Guenevere* in 1900, he wrote to Cockerell, 'I am following the K.P. [edition] but strange to say though dear old Morris corrected that much as to punctuation – he has not only left the old misprints of 1858 but perpetuated others [–] "sleepby" for "sleepy" &c &c.'[11]

Sparling reports that 'as time went on and the work grew in volume an ordinary professional reader was employed'[12] – whose identity he unfortunately does not record. An article by 'W.H.H.' (a proof-reader at the Chiswick Press) in the *Printers' Register – Supplement* (December 1897), however, offers a few more tantalizing details: the Kelmscott Press reader, he says, though he was 'treated in a most liberal manner by Mr. Morris', resigned when he contracted a serious illness resulting from the dampness of the riverside cottage in which the Press was housed. Morris must have been immensely relieved to discover that Cockerell, in addition to his many other talents, was also an excellent proof-reader. 'Mr. Cockerell will read for literals,' Morris wrote in the proofs of *The Life and Death of Jason*,[13] and that short sentence serves as a summary of the solution he eventually adopted at the Press. After her father's death May Morris also gave Cockerell some assistance.

Nevertheless, the Kelmscott Press was bedevilled by careless proof-reading throughout its history (though the standard of accuracy certainly improved after Cockerell became Secretary), and Ellis frequently reminded Morris of this sad state of affairs. In January 1894 he sent him a stern letter about *Sidonia the Sorceress*: 'there are certainly more misprints than there ought to be,' declared Ellis, offering his own services as a reader; the same was true, he added ominously, of *Godefrey of Boloyne*.[14] It was in 1893, during the last full year of Sparling's tenure, that sloppy proof-reading became a problem of major proportions. Three books published consecutively in 1893 were disfigured by serious errors. In a copy of *Utopia* Ellis inscribed the following plaintive note: 'I cannot help feeling vexed at my name as Editor [in the colophon] opposite a palpable mistake of the first printer's name.

I did correct it, but my proof was lost and the misprint left.'[15] *Maud* contained so many typos that Bertram Dobell, a leading London bookseller, complained, and Morris issued four cancel leaves.[16] Jacobi marked the errors in his copy of *Gothic Architecture*, which corrections Morris then incorporated in the second and third impressions of the book.[17]

The surviving Kelmscott Press proofs indicate that Morris was not very receptive to editorial suggestions from his proof-readers, and when he made changes he did so with bad grace. In the proofs of *Beowulf* (1895), responding to a reader's query, he protested indignantly, 'I cant have my best lines spoiled!'[18] Another reader, probably Sparling, offered a more persuasive argument in the proofs of *A Dream of John Ball*: '? to W.M. Is there not a better word than "surplus"? It smacks of Fabian essays!'[19] Morris substituted the word 'overplus'. Even Cockerell, in whom Morris had more confidence than in any other assistant, occasionally provoked the famous Morris temper: 'When I read the proofs for him', he recalled, 'I often used to point out things, which he altered (as a rule) as a matter of course. On one occasion I remember an outburst & a grinding of teeth, & then the alteration after a few minutes, & a confession at Gatti's the same evening by way of amends to myself. But this was over a sentence that he had designed on the first page of the Chaucer, & his irritation was natural as some words had to be taken out & designed again.'[20]

Ellis was aware that he was treading on dangerous ground when he disagreed with Morris about textual matters, but he nevertheless did so frequently, though always with self-deprecating good humour, as in this comment on the proofs of *The Tale of King Florus*: 'I think this may go for press. There are singularly few typographical errors. I daresay you will despise some of my queries & questions.'[21] The notes written by Ellis and Morris in the proofs of the books edited by Ellis constitute an extended debate about such matters as old-spelling and textual apparatus. The fundamental difference between them was that Ellis, a man of scholarly and antiquarian outlook, thought the Kelmscott books should include introductions, notes, bibliographies, glossaries, and the like, whereas Morris wanted to print the texts with as little apparatus as possible. Ellis, for instance, persuaded Edward Gordon Duff to prepare a list of English-language editions of *The Golden Legend* to serve as an introduction to the Kelmscott Press version, yet at the last moment Morris balked at printing it, and Ellis found himself in the embarrassing position of having to explain to

Duff, 'The author–printer kicks much against introductions, and there is some practical difficulty as to bibliographical description in that there are but two sets of type[:] caps. for headings & that used in the text.'[22] The most that Morris would permit was a brief essay by Ellis entitled 'Memoranda' at the end of *The Golden Legend*.

The text of *Psalmi penitentiales*, which was written in Middle English, provoked a good deal of wrangling between Morris and Ellis. When Ellis suggested in the proofs[23] that the phrase 'caris colde' was a corruption, Morris replied, 'Of course it is caris colde[,] a most common expression.' They disagreed about the form of the glossary. At times they both had to admit defeat when confronted by a puzzling text. 'I cannot make very much sense of this line,' wrote Ellis, 'but it is exactly to MS. & I daresay the old boy who wrote it knew what it meant.' 'All right,' responded Morris wearily. But Morris was unbudging in his refusal to admit footnotes to the Kelmscott books. In the proofs of *The Nature of Gothic*[24] one can see him experimenting with side-notes, then placing notes within brackets, and finally relegating Ruskin's annotations to an appendix.

One editorial issue on which Morris and Ellis were in substantial agreement was that the original spelling of medieval and Elizabethan texts should be preserved. In his foreword to the Kelmscott edition of *The Poems of William Shakespeare* (1893), Ellis called attention to the question: 'After a lapse of about three hundred years Shakespeare's Poems are in this edition printed with the original spelling. That the orthography of an author of the 16th or 17th century cannot without detriment be translated into that of the 19th, seems a proposition too evident to leave room for controversy. The fact of being able to study these poems in the language in which they were written, ought we think, to add materially to the pleasure of the reader.'[25] This was a point of view long advocated by Frederick J. Furnivall, the founder of the New Shakspere Society (which Shaw wittily described as the Old Spelling Society) and numerous other literary and philological groups; Furnivall was a friend of Morris – it was at his suggestion that Morris planned an edition of Shakespeare's plays – and Ellis was active in his societies. Like Ellis, Morris believed that to alter the spelling of a word was to conceal its history: he no doubt saw modern-spelling editions as the literary equivalent of insensitive Victorian restoration of medieval churches.

Yet, if Morris was generally scrupulous in his fidelity to an author's orthography, he was in other respects remarkably free in his treatment of literary documents. He never hesitated to alter punctuation, and he offered emendations when he believed the text was corrupt. The liberties that Morris took with the Caxton books in particular were described by Sparling: 'Those edited by F. S. Ellis, the *Golden Legend* and the *Order of Chivalry*, in deference to the editor's tastes and desires, were, as nearly as might be, textual and literal reproductions of Caxton's editions. The three others, the *Historyes of Troye*, *Reynard the Foxe* and *Godefrey of Boloyne* [all edited by Sparling], were differently treated, as Morris wished them to be regarded as Kelmscott Press editions, and therefore to be amended where this was desirable. Caxton's text was to be taken as a basis, but not looked upon as archaeologically sacrosanct. It was to be collated with Caxton's originals, and corrected where need was, mistranslations being put right and omissions filled in, care being taken to preserve the style and flavour of Caxton in doing this. When we came to the *Godefrey of Boloyne*, Morris decided that the original spelling need not be rigorously adhered to, as Caxton was an erratic speller, following no discernible rule, and that we were consequently free to retrench or add a letter where the justification of a line could be improved or a "river" avoided thereby.'[26]

On the other hand, when it came to modern writers, Morris worried a good deal about textual accuracy. Following his usual practice of carefully inspecting the first sheets that came off the press, Morris was horrified to discover that *The Poems of John Keats* (1894) included a slightly garbled version of 'La Belle Dame sans merci' and insisted with some passion that the poem be reprinted at once.[27] What is especially fascinating about this episode is its revelation that Morris had not inspected the text before it was set up in type, and this in fact was frequently the case at the Kelmscott Press. Depending heavily on the literary advice of Ellis, Cockerell, and others, Morris focused his own attention mainly on the designing of the books.

The basic question that faced Morris when he founded the Kelmscott Press was what books he should print, and his initial short answer was 'Caxton's books'. The Press was created, in part, as an outlet for Morris's antiquarian enthusiasm for England's first printer (which of course did not extend to Caxton's typography): *The Golden Legend* was the first book planned, and it was only its length and minor technical problems that prevented it from being the first book printed at the Kelmscott Press. Morris never completely lost his interest in Caxton, yet as time went by he came to speak of the Caxton reprints as books of mainly archaeological interest – surely a somewhat dismissive phrase – and, encouraged by Ellis and Cockerell, turned his attention to writers of the nineteenth century as well. 'The books that I would like to print are the books I love to read and keep,' he explained in an interview in 1895.[28] That meant, not surprisingly, his own books above all; as early as November 1891 he was expressing the hope that he could reprint all his works at the Kelmscott Press, beginning with *The Story of Sigurd the Volsung*[29] (which in fact was to be one of the last books issued at the Press, in 1898). Morris also wrote especially for the Press several prose romances with vaguely medieval settings. These struck Bernard Shaw as a lamentable form of escapism that failed to advance the cause of socialism: 'Morris, to fill the maw of the Kelmscott Press, began to pour out tale after tale of knights in armour, lovely ladies, slaughterous hand-to-hand combats, witches in enchanted castles changing humans into beasts, lovelorn heroes going mad in the mountains, haunted woods, quests after impossibilities, all under medieval conditions. . . .' Shaw claimed that he attempted without success to 'check the voracity of the Kelmscott Press' by suggesting to Morris that he turn his attention to manufacturing musical instruments to the old standards of craftsmanship.[30]

About some of the other writers he published at the Press Morris displayed a conspicuous lack of enthusiasm. He did not particularly care for Tennyson's *Maud* and would have preferred to print his earlier poems, but Macmillan, the publisher, insisted upon *Maud*. 'Coleridge', wrote Morris to Ellis, 'was a muddle-brained metaphysician, who by some strange freak of nature turned out a few real poems amongst the dreary flood of inanity which was his wont.'[31] Of Herrick he remarked languidly, 'I like him better than I thought I should: I daresay we shall make a pretty book of it.'[32] The truth is that Morris's deepest literary preferences lay in the Middle Ages, and medieval texts, whether by Caxton or others, were the real *raison d'être* of the Kelmscott Press, despite his efforts to enlarge the range of titles he printed. Something of the intensity of Morris's passion emerges in a note he composed about *The Recuyell of the Historyes of Troye* for Quaritch's catalogue: 'For though written at the end of the Middle Ages and dealing with classical mythology, it has in it no token of the coming "Renaissance" but is purely mediæval. . . . Surely this is well worth reading, if only as a piece of undiluted Mediævalism.'

The Business of the Press

Morris received editorial advice at the Kelmscott Press from a small inner circle of friends – Ellis, Cockerell, Furnivall, Bernard Quaritch, and perhaps Sparling – and one suspects that he would have been pleased if the books could have been printed by an equally small group of craftsmen. It is typical of Morris that he initially hired three members of one family (the Bowdens) to work at the Press, for it reflects the atmosphere of intimacy and familiarity that characterized the Press, particularly in its early years. William Bowden, the master-printer brought out of retirement by Morris,[33] found the work of the Kelmscott Press too burdensome for one pair of hands and enlisted the aid of both his son W. H. Bowden and his daughter Mrs Pine, who appear to have joined their father in February and April respectively (see Appendix C). W. H. Bowden soon became Morris's foreman; Mrs Pine, who was paid the same wages as the men (contrary to the usual practice at the time), achieved the distinction of becoming the first woman to join the London Society of Compositors.[34]

Mrs Pine, her father, and a pressman named Giles soon retired from the Kelmscott Press, but Morris's staff grew as the output of the Press increased. A photograph taken in September 1895 shows eleven printers working for Morris at that time; the size of the establishment obviously fluctuated according to how many books were in production, and it seems probable that Morris's staff reached its maximum size in 1895 and early 1896 when work on the *Chaucer* was approaching completion. By the time the final sheet of the last book was printed in March 1898, there was only one pressman, Stephen Mowlem, still employed at Sussex Cottage.

Figure 56

Though W. H. Bowden himself was not a socialist, several of Morris's printers – John Tippett, Stephen Mowlem, H. Howes, and Thomas Binning – had been associated with either (or, in Binning's case, both) the Hammersmith Socialist League or C. R. Ashbee's Guild of Handicraft.[35] Binning, who had been the foreman printer of the *Commonweal* and became father of the chapel (the printers' equivalent of shop steward) at the Kelmscott Press, was a stern, unbending old trade-unionist who intimidated even Ashbee, for whom he later worked at the Essex House Press. 'I never met a man so solid, so upright, so honest, so reliable, so obstinate,' said Ashbee. 'For the 10 years he has been with me he has never missed a day[,] not an hour, he has never shirked or loitered at his work, never made a mistake that I know of, he hates the world and its institutions and he loves his craft to its minutest details.'[36] It was a point of honour with Binning never

Figure 56 The staff and friends of the Kelmscott Press, in a photograph taken on 13 September 1895: (*front row, from left*) Stephen Mowlem, William H. Bowden, William Morris, May Morris, W. Collins, W. L. Tasker; (*back row, from left*) H. Howes, Carpenter, F. Collins, Emery Walker, R. Eatley, Halliday Sparling, J. Tippet, Thomas Binning, G. Heath.

to touch his hat to anyone ('least of all to me', added Ashbee), but he was loyal to Morris and his ideals to the point of fanatical intensity.

It was probably Binning who, through the London Society of Compositors, arranged for the unionization of the Kelmscott Press. Though the presence of Mrs Pine at first created obstacles, in time all were enrolled as members of the LSC. When initially approached by the union authorities, Morris was able to reply truthfully that his printers were already well treated. Precisely how much they were paid is not known: Emery Walker implied that their wages were high, and H. Gage-Cole, who was an apprentice at the Kelmscott Press before working at the Doves Press, claimed that the Kelmscott pressman were paid five shillings a week above the standard rate.[37] W. H. Bowden testified that 'Mr. Morris never tied his men down to any certain number of hours. He was most generous in the matter of holidays. On every Bank Holiday they had two days instead of the usual one day. When any of his men were ill, if it was only a short time, they always received full pay; if it was for a long time they had full pay for a month and half pay afterwards.'[38]

The Kelmscott printers must also have enjoyed the odd sensation of being the centre of attention in the world of English book-lovers. Journalists and other curious visitors descended upon Sussex Cottage regularly to see the celebrated Kelmscott Press. 'Press invaded in the aft[ernoo]n by 60 members of the Librarians' Association,' Cockerell wrote in his diary (22 May 1895) on one such occasion. 'Went off all right however.' In the autumn of 1893 one of the Kelmscott books, Morris's *Gothic Architecture*, was actually printed at the Arts and Crafts Exhibition in front of crowds that, according to Sparling, 'imposed a severe strain upon the pressman Collins's Celtic modesty'.[39] And there were pleasant social occasions. On the day of the University boat-race in 1893 Morris invited his printers to tea at Kelmscott House (Cockerell diary, 22 March 1893), while once a year he and May joined them on the outing known in the trade as a 'Wayzegoose'. After a substantial dinner at a hotel or inn near London (the Dumb Bell Hotel at Taplow, Berkshire, in 1894, for example), the printers celebrated with a lengthy programme of toasts, songs, and recitations. In 1894 they heard, among others, Binning toast the Kelmscott Press, F. Collins sing 'At Trinity Church I Met My Doom', and Binning (once more) recite 'Cassius instigating Brutus'. The bill, paid by Morris, came to a total of £9 15*s*. 4*d*. and included both luncheon and dinner for seventeen, many bottles of wine and port, and a good supply of cigars.[40]

Figure 57

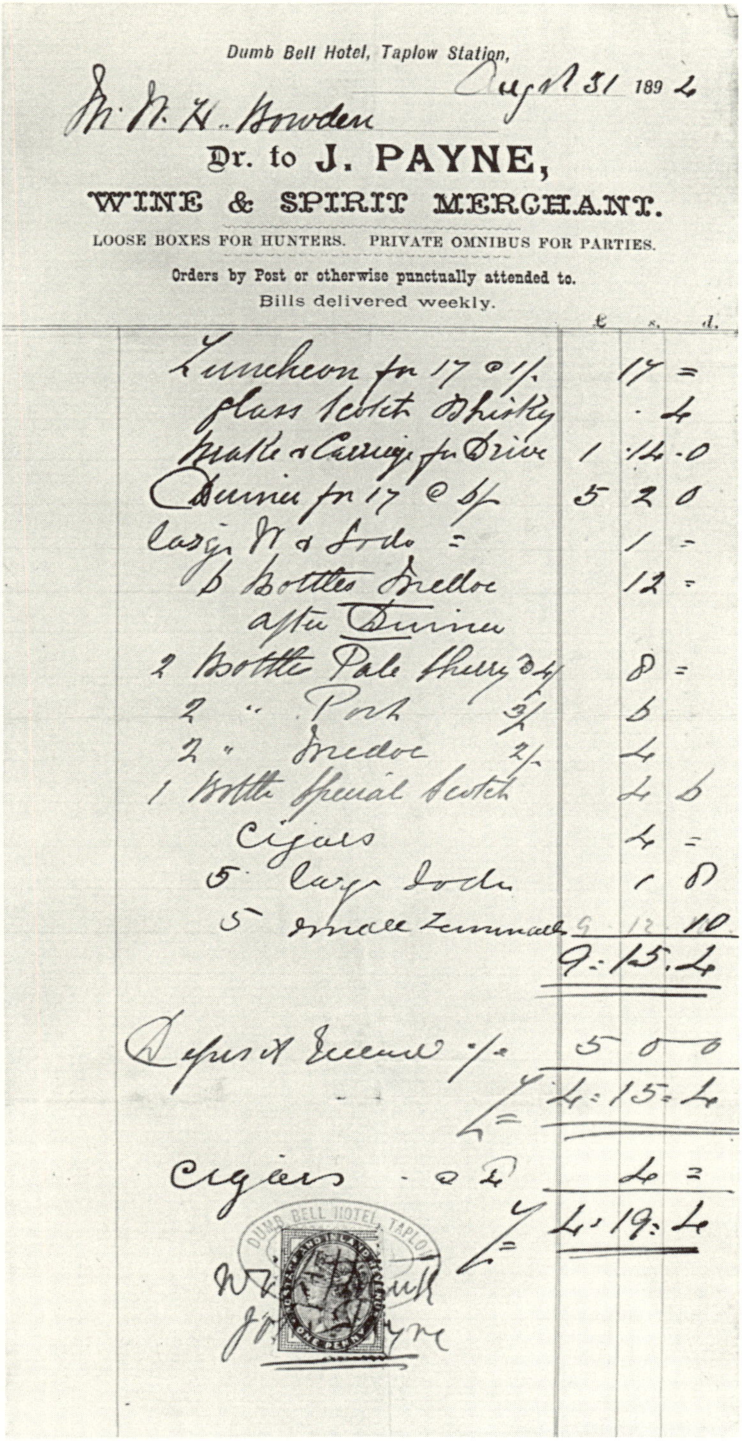

Dumb Bell Hotel, Taplow Station, Aug 31 1892

Mr W. H. Bowden

Dr. to J. PAYNE,

WINE & SPIRIT MERCHANT.

LOOSE BOXES FOR HUNTERS. PRIVATE OMNIBUS FOR PARTIES.

Orders by Post or otherwise punctually attended to.

Bills delivered weekly.

	£	s.	d.
Luncheon for 17 @ 1/-		17	=
glass Scotch Whisky		·	4
Brake & Carriage for Drive	1	14	0
Dinner for 17 @ 6/-	5	2	0
Large W & Soda		1	=
6 Bottles Medoc		12	=
after Dinner			
2 Bottle Pale Sherry @ 4/-		8	=
2 " Port 3/-		6	
2 " Medoc 2/-		4	
1 Bottle Special Scotch		4	6
Cigars		4	=
5 large Soda		1	8
5 small Lemonade	9	12	10
	9	15	4
Deposit Received	5	0	0
	£4	15	4
Cigars @ 2		4	=
	£4	19	4

DUMB BELL HOTEL, TAPLOW

Figure 57 Morris took his printers on an annual outing known in the trade as a 'Wayzegoose', where they enjoyed luncheon, port, and cigars. After dinner there were recitations and songs.

This all sounds rather jolly, and no doubt was, yet it is important to bear in mind that Morris's relationship with his printers was not always serene. The conflict over the use of the Jänecke ink, described in Chapter Four, resulted in a serious rebellion by the pressmen in Sussex Cottage, and only by threatening to close the Press altogether was Morris able to persuade them to adopt the ink he wanted. Afterwards 'the number of sheets turned out per day was extraordinarily small', according to Walker's sarcastic summary of the episode.[41] The perspective that Walker is able to offer on Morris's printers is helpful in this case because, although he became strongly identified with the private-press movement, Walker, unlike Morris, was also accustomed to producing books under normal commercial conditions. The Kelmscott Press, notwithstanding its modest financial success, represented a special kind of experiment that would not have been possible in any ordinary printing establishment. When Morris was unhappy with the appearance of a trial page, he sent it back to be redone for the sixth time. When he discovered an error on the first sheet off the press, he had the passage reset while the pressmen stood idly by.

Figure 58

This pursuit of perfection both impressed and horrified visitors to the Kelmscott Press. Sparling, rather complacently, tells the story of a head of a large commercial printing firm, who watched Morris's compositors 'carefully setting, and still more carefully justifying, line after line; looked with a discontented eye at the pressmen heedfully pulling sheet after sheet, minutely examining each one to see whether it were up to the mark; and as he left, summed up his impressions: "We-e-ll? That's all very well for Mr. Morris, but there isn't a man here that would be worth a penny an hour to me after he'd been here for a week!" '[42] Sparling finds obvious pleasure in reciting this anecdote, which he sees as a symbol of the iniquities of the printing trade; yet it may strike the more impartial reader as a very sad tale indeed, revealing as it does the profound incompatibility of Morris's ideal of painstaking craftsmanship and the actual conditions under which printing had to be done in the world outside Sussex Cottage.

There was another person employed at the Kelmscott Press, not mentioned by Sparling, who also played an important role: Mrs Sarah A. Peddie, Cockerell's secretary. In the later years she coped with much of the voluminous correspondence which flowed incessantly into the Press, and Frank Colebrook (editor of the *Printing Times and Lithographer*) claimed that she transcribed all the Caxton books reprinted by Morris, except for *The Golden Legend*, on a typewriter at

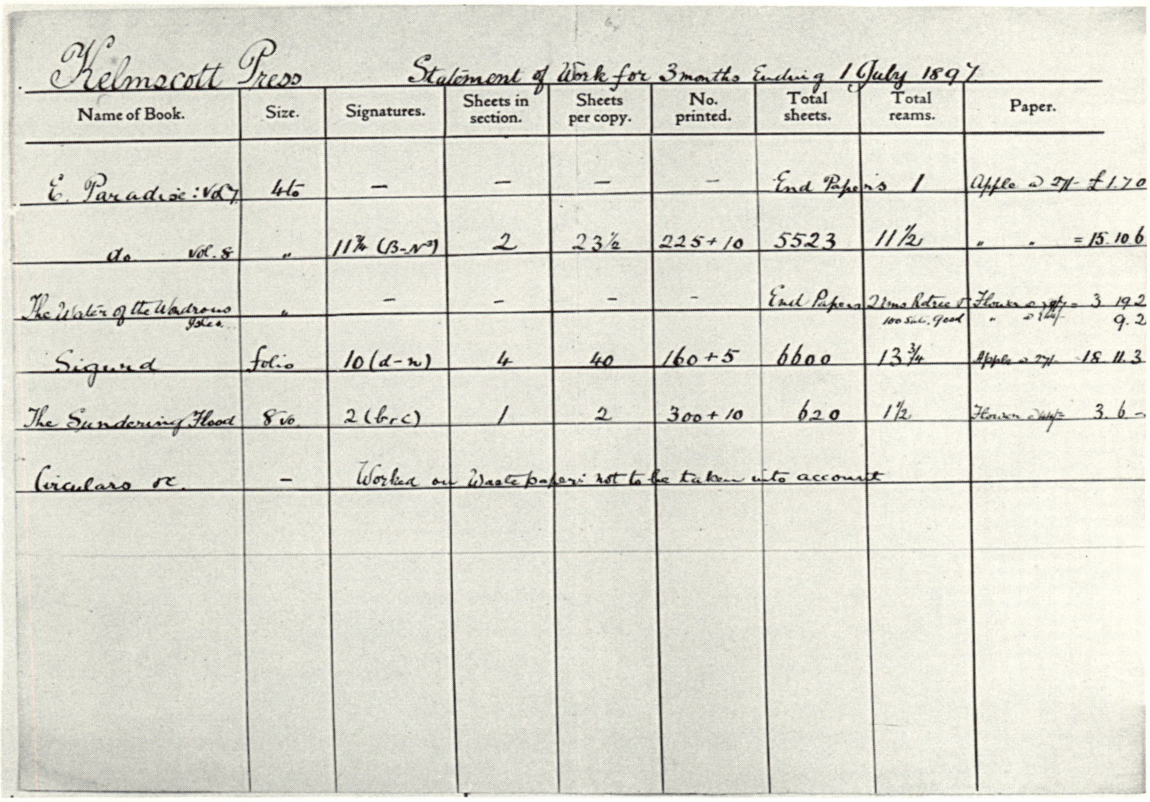

Kelmscott Press   Statement of Work for 3 months Ending 1 July 1897

Name of Book.	Size.	Signatures.	Sheets in section.	Sheets per copy.	No. printed.	Total sheets.	Total reams.	Paper.
E. Paradise: Vol. 7	4to	–	–	–	–	End Papers	1	Apple @ 27/- £1.7.0
do Vol. 8	"	11¾ (B–N³)	2	23½	225 + 10	5523	11½	" " = 15.10.6
The Water of the Wondrous Isles	"	–	–	–	–	End Papers	2 Rms Retree & 100 Sht. Good	Flower @ [illegible] = 3.19.2 " @ [illegible] 9.2
Sigurd	folio	10 (d–n)	4	40	160 + 5	6600	13¾	Apple @ 27/- 18.11.3
The Sundering Flood	8vo	2 (b, c)	1	2	300 + 10	620	1½	Flower @ [illegible] 3.6.–
Circulars &c.	–	Worked on Waste paper: not to be taken into account						

Figure 58 A form used for record-keeping at the Kelmscott Press in 1897. It shows four books in progress: *The Earthly Paradise, The Water of the Wondrous Isles, Sigurd the Volsung,* and *The Sundering Flood.*

the British Museum.[43] The story may be apocryphal, since the Museum's records do not indicate that Mrs Peddie held a reader's ticket during the 1880s or 1890s; but if it is true, it lends wonderful irony to an irrelevant attack upon typewriters made by Sparling in his history of the Kelmscott Press.[44] (Sparling's surviving letters, incidentally, are all typed.)

Journalists, after visiting the Kelmscott Press, often rhapsodized about the extreme simplicity of the equipment used by Morris's printers: 'It was like stepping into Caxton's shop' (or words to that effect) was the usual response. One writer, so overwhelmed by the medieval splendour of the Press, even made Kelmscott House four hundred years older than it was.[45] In fact, as we have seen, Morris was using nineteenth-century hand-presses – Albions – that had been commonplace fixtures in nearly every English printing office until the middle of the century. The hand-press did of course seem slightly archaic by 1891, but it was still widely used for pulling proofs, printing posters, and producing limited editions; hence we must avoid the vulgar error of admiring Morris's reversion to medieval technology. It should be observed that Morris had no great difficulty in finding printers who were skilled in the use of the hand-press, for what he revived at the Kelmscott Press was a handicraft whose survival had been threatened only within the previous three or four decades.

Nevertheless, Morris was engaged in a deliberate simplification of the printing process when he founded his Press. But, unless one returns to scribal methods (which Morris always longed to do), books cannot be manufactured without machinery of some sort; even an Albion press is a machine, simple though it may be. The choices that Morris made in acquiring equipment for the Kelmscott Press were complicated by his well-known aversion to mechanical devices, yet his hostility appears to have been directed primarily against what he called 'the great intangible machine of commercial tyranny' rather than 'this or that tangible steel and brass machine'.[46] His types were cast mechanically; his inks were manufactured in large factories; he used photography constantly as a labour-saving device at the Kelmscott Press; Frank Colebrook records that Morris even visited and admired the modern printing establishment of R. and R. Clark in Edinburgh, and that a Linotype salesman who called at the Press 'positively escape[d] with a whole skin'.[47] Morris's attitude towards Victorian technology was altogether more ambivalent than some of his doctrinaire disciples supposed.

The small cottages strung along Hammersmith's Upper Mall in which Morris housed his Press were as unpretentious as the equipment the printers used. The rent book[48] for 16 Upper Mall, the first home of the Kelmscott Press, shows that Morris's tenancy began on 12 January 1891; it ended in June of the same year, because in the meantime (from 20 May[49]) Morris had begun leasing Sussex Cottage, 14 Upper Mall, which remained the primary building of the Press until its closing. In 1893 Morris persuaded Cobden-Sanderson to lease 15 Upper Mall, opposite Sussex Cottage, for the Doves Bindery, with two rooms to be reserved for the Kelmscott proof-readers: one of Cobden-Sanderson's workers recalled that 'this brought Morris in and out many times a day, and with his marvellous interest in the detail of all craft-work, he made himself a familiar figure to the employés . . .'.[50] This, however, was only a temporary arrangement, for on New Year's Day 1895 Morris took possession of still another small house, 21 Upper Mall (Sunderland Cottage), which was used for the large Albion press printing the *Chaucer* and for Cockerell's office. W. Irving Way, the American publisher, visited the Kelmscott Press in April 1895, and the vivid (albeit slightly sentimentalized) description he offers is probably of Sunderland Cottage: 'No sign marks the Kelmscott Press, . . . but . . . the right one is finally reached, and here Mr. S. C. Cockerell . . . greets the visitor with a smile, and at once puts him at ease. A quiet, tidy, orderly place it is, but with nothing modern about it. No noise of machinery, escaping steam, or hum of electric motor, distracts one. No mahogany furniture, Axminster rugs, or click of type-writer reminds one of nineteenth century progress. The furniture is plain deal like the floor, and with a "gray goosequill" the visitor registers his arrival in the guest-book.'[51]

The modesty of the Kelmscott Press's machines and buildings, though it had an ideological basis, also reflected Morris's limited goals. His original intention had been to print only a small number of books – just enough of each title, perhaps, to distribute to his friends – and he did not at first envision becoming his own publisher. Morris's publisher during the 1870s and early 1880s was F. S. Ellis, but in 1888 he turned to Reeves and Turner, who operated a bookshop and small publishing establishment in the Strand.[52] Thus there was nothing unexpected about Morris's decision to allow Reeves and Turner to publish most of the early Kelmscott books.

Nevertheless, those close to Morris grumbled occasionally about the inefficiency they detected in his publishers. 'You don't see [Morris's] books on bookstalls and I think it's for want of a businesslike publisher,' Burne-Jones commented; Jane Morris, in a similar vein, described them as 'the most unbusiness-like publishers but extremely honest'.[53] Reeves and Turner published eleven of the Kelmscott Press books, the last of which was More's *Utopia*, issued in September 1893, and Morris's connection with them was officially severed in June 1896.[54] His relationship with Reeves and Turner, while it lasted, seems to have been happy, but he was uneasy about any suggestion that the Kelmscott Press books were less than completely his own productions. Reeves and Turner were merely the distributors for the Kelmscott Press, not the publishers in the usual sense, and Morris emphasized this in the formula he adopted in the Kelmscott colophons: 'Sold by Reeves & Turner.' (In the proofs of *The Defence of Guenevere* that line read 'Published by Reeves & Turner' but was changed.[55]) 'I don't mind having a publisher', Morris declared, 'so long as he has nothing whatever to do with any question except purely business ones.' One possibility that he considered briefly was to print a book and then sell the entire edition at auction, thus eliminating the need for a publisher at all.[56]

Figure 59

Four of the Kelmscott Press books were published by Bernard Quaritch, who, following his arrival from Berlin in 1842, had rapidly made himself London's leading bookseller. To his first English employer, Henry G. Bohn, Quaritch announced, 'Mr. Bohn, you are the first bookseller in England, I mean to become the first bookseller in Europe.'[57] Energetic, aggressive, learned, Quaritch was a very different person from nice, inoffensive old William Reeves, and there were serious misunderstandings between him and Morris from the beginning. Since the price of *The Golden Legend* was not specified in the contract, Quaritch raised it on the day of publication and eventually doubled it. 'The trade, being dissatisfied with the terms which Mr. Q. offered to supply them, did not push the book at the time, & the result was that a good many copies remained on Mr. Q's hands,' Cockerell wrote two years later.[58] There was other evidence of what Morris took to be sharp dealing by Quaritch, and early in 1893 he reached a decision that infuriated Quaritch. 'Mr. W. Morris wishes to have *all* the profits of the Kelmscott press,' Quaritch wrote to his

Figure 59 Bernard Quaritch, London's leading bookseller, commissioned *The Golden Legend,* the most ambitious book issued by the Kelmscott Press during the early years of its existence. Later he and Morris quarrelled about publishing arrangements for some of the other Kelmscott books.

son in America on 8 February; 'he will *retail* himself the ["]Godfrey of Bouloyne" – he will be a loser by it, but it will be a nuisance to me on account of the Chaucer.'[59] Morris was at first conciliatory: 'It is true that I shall probably publish the Sidonia myself, as I have done with the Godfrey, but I do not suppose I shall generally publish the Kelmscott books myself,' he declared in a letter to Quaritch the following July.[60] Yet this was precisely what he did do. Quaritch, who had obviously hoped to be the publisher of the *Chaucer*, was furious. Morris's decision therefore had the unfortunate effect of poisoning the later negotiations with Quaritch about the number of copies of the *Chaucer* – and on what terms – he would be supplied. 'Mr. Wm Morris will call on me tomorrow,' Quaritch informed his son on 20 August 1894; 'he is laying down the law to me about the Chaucer, – and I must submit.

'This has cooled my ardour to obtain the transfer of his publications from Reeves & Turner. Mr. Morris's success has turned him into a despot.' The following day Quaritch reported that 'W. Morris did not dare to show his face' but instead sent Cockerell ('a very sharp fellow') as his representative. 'My friendship with Morris is at an end,' lamented Quaritch. In the autumn they reopened negotiations, and at last Quaritch felt satisfied with the trade discount he was to receive on the *Chaucer*. 'Mr. William Morris & I are on the road of being reconciled, he sees now how hastily he acted,' Quaritch announced triumphantly to his son on 12 November 1894. But it was a partial victory at best, because Quaritch was never allowed to publish another Kelmscott book.

Several publishers – Ellis and Elvey, George Allen, Macmillan, and Way and Williams – commissioned books to be printed at the Kelmscott Press, and, at F. S. Ellis's suggestion, Ellis and Elvey even proposed that they be allowed to succeed Reeves and Turner.[61] Yet there was little inducement for Morris to seek the services of another publisher, for J. and J. Leighton, his binder, performed some of the same functions as Reeves and Turner (the books were sent directly to buyers and book dealers from Leighton), and Morris was eager to replace, as much as possible, the middlemen between the craftsman and his customers. Acting as his own publisher, as Sparling pointed out, also gave Morris 'exact information with regard to the real and immediate sale of each book, and thus placed him in a position to guard against over-printing'.[62]

Many of the Kelmscott Press books, especially the 16mos and

octavos, were completely sold before they were off the press. Any advertising of the books was at first done by Quaritch, and Reeves and Turner, but in May 1892 Morris issued a simple, two-page list indicating the titles already and soon to be printed at the Kelmscott Press, and periodically thereafter he announced his books by means of these lists (which grew longer and more elaborate) and similar circulars describing individual volumes, some of the latter containing specimen pages. 'We only send out special announcements of books which are not nearly sold out when they go to the binder,' Cockerell explained.[63] They proved to be an effective, low-key method of advertising: in all fifty-one such circulars were issued by the Kelmscott Press, and they are prized by collectors today.

Figure 60

Because the Kelmscott Press was operated in the glare of publicity – the *British Bookmaker* (July 1892) summed up a popular attitude when it declared simply, 'The doings of the Kelmscott Press are of interest' – Morris did not have to advertise his books widely. Every bibliophile, bookseller, and librarian in the English-speaking world seemed to be aware of the existence of the Kelmscott Press. Even when the Press was attacked in print, its fame increased. And strong attacks there were: someone writing in 1895 under the pseudonym of 'Arthur Pendenys' addressed an open letter to Morris which began, 'I presume that the Kelmscott Books are published for your own amusement, because I have inquired extensively and find they do not amuse any one else. They are very dull books, much more dull than Law books and Medical books, and there are only two on your list, the *Shakespeare Sonnets* and the *Keats*, which I should prefer to a Post Office Directory.'[64] The letter went on to condemn Morris's prices, types, and every aspect of the Kelmscott books. Morris was presumably enraged when he saw this abusive yet amusing document, but he might have found consolation in the commonsense principle – subsequently elevated to a divine truth in Hollywood – that bad publicity is better than no publicity at all.

None the less, it was necessary for the Kelmscott Press to communicate with its potential customers in some systematic fashion. The Press's list of 'Addresses to which announcements were sent [in] 1897' has survived (in the Morgan Library), and it makes instructive reading. The document includes more than 800 names – mostly book dealers – of whom twenty-one were Americans ('We have at present very few private customers in America,' Cockerell wrote in 1894[65]), with several other nationalities represented; among the names of individuals

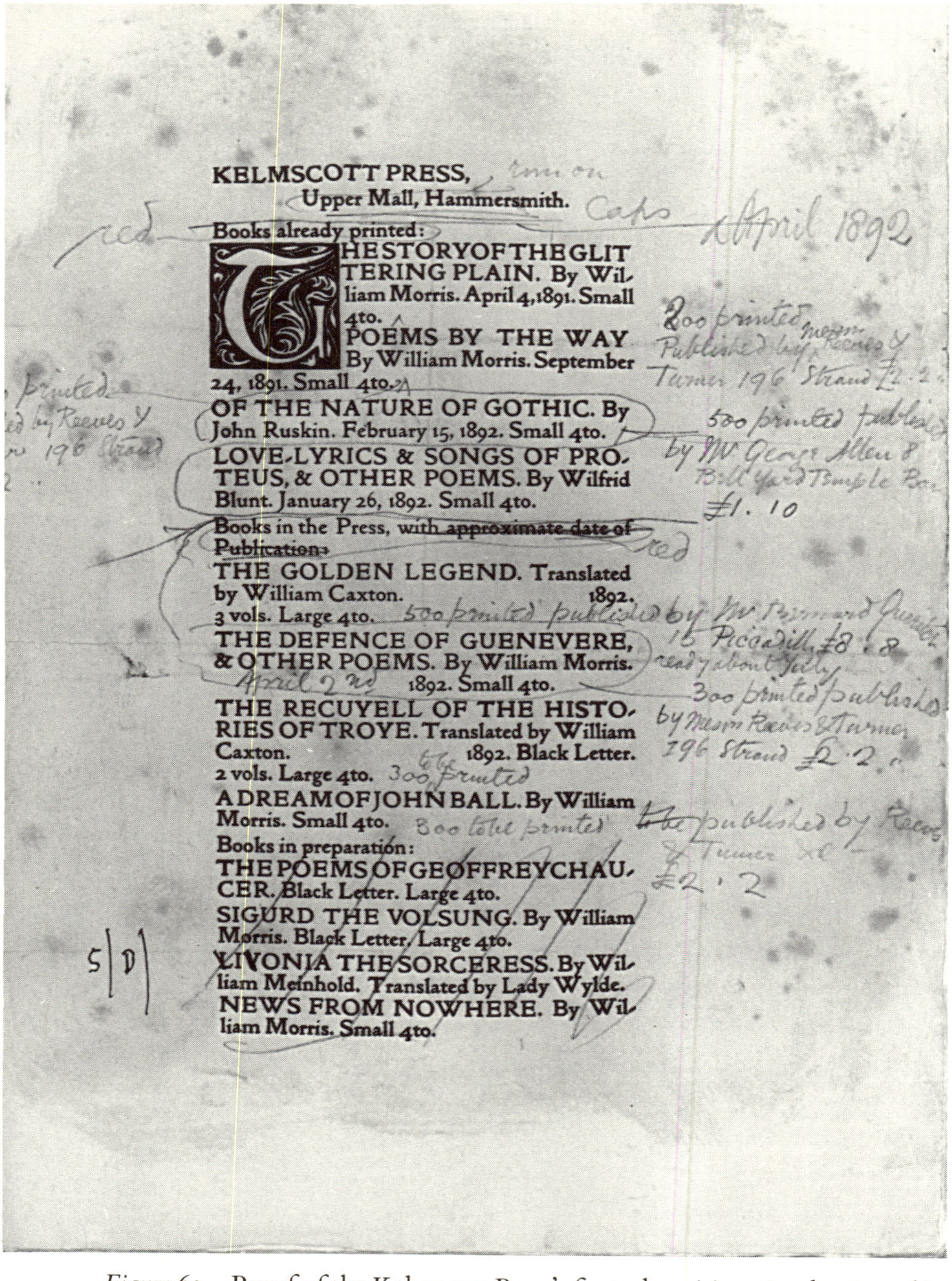

KELMSCOTT PRESS,
Upper Mall, Hammersmith.

Books already printed:

THE STORY OF THE GLITTERING PLAIN. By William Morris. April 4, 1891. Small 4to.

POEMS BY THE WAY By William Morris. September 24, 1891. Small 4to.

OF THE NATURE OF GOTHIC. By John Ruskin. February 15, 1892. Small 4to.

LOVE-LYRICS & SONGS OF PROTEUS, & OTHER POEMS. By Wilfrid Blunt. January 26, 1892. Small 4to.

Books in the Press, with approximate date of Publication:

THE GOLDEN LEGEND. Translated by William Caxton. 1892. 3 vols. Large 4to.

THE DEFENCE OF GUENEVERE, & OTHER POEMS. By William Morris. 1892. Small 4to.

THE RECUYELL OF THE HISTORIES OF TROYE. Translated by William Caxton. 1892. Black Letter. 2 vols. Large 4to.

A DREAM OF JOHN BALL. By William Morris. Small 4to.

Books in preparation:

THE POEMS OF GEOFFREY CHAUCER. Black Letter. Large 4to.

SIGURD THE VOLSUNG. By William Morris. Black Letter. Large 4to.

LIVONIA THE SORCERESS. By William Meinhold. Translated by Lady Wylde.

NEWS FROM NOWHERE. By William Morris. Small 4to.

Figure 60 Proof of the Kelmscott Press's first advertising circular, issued in May 1892, with corrections and additions in Morris's hand. The *Chaucer*, in a deleted passage, is described as being 'in preparation', though it was not to be published until 1896.

are Bernard Shaw, Ellen Terry, George Meredith, A. Edward Newton, Holman Hunt, C. R. Ashbee, F. J. Furnivall, and Luther A. Brewer (identified as 'State Inspector of Illuminating Oils, Cedar Rapids, Iowa'). Walker was later to make use of this same list at the Doves Press.[66] Morris and Burne-Jones often protested that only a handful of people had any interest in the Kelmscott books ('I do the books mainly for you and one or two others; the public does not really care about them a damn,' Morris confided to Philip Webb[67]), but the size of the address list and the actual sales of the books belie this claim. Morris offered a fairer appraisal when he wrote to Quaritch before the publication of *The Golden Legend*: 'for my part I feel sure there will be a moderate public who will like it very much.'[68]

There was a small group of customers who were prepared to buy anything Morris printed at the Kelmscott Press. In March 1898, when he was closing down the Press, Cockerell sent Jane Morris a list of twelve names (ten British, two American) of persons who owned either all the Kelmscott books or all those written by Morris.[69] And Morris always took care to see that his closest friends received complimentary copies: a letter from Cockerell specifying those who were to be given the second volume of *The Poetical Works of Percy Bysshe Shelley* mentions Burne-Jones, his wife Georgiana, Philip Webb, Emery Walker, Elizabeth Burden (Jane's sister), Theodore Watts, Swinburne, Kate Faulkner, May (Morris) Sparling, and F. and R. Smith (the last two employees of Morris and Company).[70] Review copies were, according to Cockerell, posted to less than half a dozen of the London newspapers, though the *Daily News* and the *Daily Chronicle*, whose political views Morris approved, were consistently on this short list.[71]

Despite the commercial success of the Kelmscott books in general, there were naturally variations in this pattern from one volume to another, and Morris never ceased to worry about the possibility of printing too many copies. The first Kelmscott book, *The Story of the Glittering Plain*, stunned everyone when the entire edition was disposed of long before publication – indeed, long before even the price was announced by Reeves and Turner – and Morris found himself short of copies for his friends.[72] Another of the early books, *The Poems of William Shakespeare*, was probably the most popular volume ever issued by the Press. At the other extreme, *The History of Godefrey of Boloyne*, as glumly predicted by Quaritch, sold so badly that Jane Morris had to donate the overstock to British public libraries in 1897.

Morris abandoned one of his announced books, *Vitas patrum*, when the advance subscriptions proved to be too meagre. The last circular announcement distributed by the Kelmscott Press – dated 25 February 1898 – reveals that, as the Press was winding down its operations, it still had on hand copies of the following books: *The History of Godefrey of Boloyne* (two), *The Well at the World's End* (eleven), *Sidonia the Sorceress* (six paper, two vellum), and *The Tale of Beowulf* (thirteen).

Morris was distressed to discover that many of the Kelmscott books were being bought by dealers who held them in stock until their value rose – as it did almost immediately. A contributor to the *Contemporary Review* in 1898, praising the *Chaucer*, adopted an unfortunately apt figure of speech when he wrote that the book's paper was 'firm and crisp to the touch like the paper of a Bank of England note'.[73] Works printed in the Golden type generated real gold in the sale rooms, though none of it found its way into Morris's pocket. Periodicals such as the *Connoisseur* and the *Scottish Review* began to publish league tables in which the escalating prices of Kelmscott books were charted in stupefying detail.[74] J. H. Slater, an experienced observer of the book trade, commented in 1899 that the Kelmscott books 'will soon be unprocurable except at a ruinous price'.[75] Though, like other examples of fine printing, the Kelmscott Press volumes have fluctuated in market value from time to time (especially during the first two decades of this century), the trend to this day has been upward: one could not reasonably expect in 1987, for example, to pay less than £6,500 for a paper *Chaucer* or £13,000 for a vellum copy. (And these figures continue to escalate unpredictably, for as this book went to press, newspapers throughout the world were trumpeting the astonishing price of $330,000 realized by a vellum *Chaucer* sold at Christie's, New York, in May 1989.) P. H. Muir once remarked that 'anyone well well informed on book-collecting history could venture a shrewd guess at the state of the money-market from the current price of a Kelmscott Chaucer'.[76]

As an old bookseller, Ellis professed to be bewildered by this development but could scarcely conceal his professional excitement. 'Chaucer £27 – at Sothebys last week. We shall (some of us) live to see it bring £100 – . . . ,' he wrote to Cockerell in 1899. And in another letter later that spring: 'The K.P. books keep up in a wonderful way! Shelley £23 – & Keats £26! And we were told they were dear at £3.15 – & 30/-. What a comedy we live in.'[77] While Ellis secretly rejoiced, Morris – who saw this happening in his lifetime – was frustrated by his inability to control the book speculators. 'I am afraid that the "forestallers and regraters" got most of the benefit,' he said in an interview in 1895; 'but although I don't like that, I don't see what's to be done to alter it as things are.'[78] The two unfamiliar terms of condemnation that Morris used are medieval words meaning, respectively, those 'who bought up produce to hold it for a rise' and those 'who bought and sold in the market or within five miles of it'.[79] But although Morris would not tolerate the freewheeling entrepreneurial style of Quaritch (a 'vulture', Cockerell called him in private, no doubt reflecting Morris's opinion[80]), he could not prevent his books from falling into the hands of other, even more unscrupulous, forestallers and regraters.

Another form of capitalistic exploitation of his goods that Morris resented was the pirating of the Kelmscott Press types and ornaments, particularly in the United States where he had no legal protection. Morris insisted that he was concerned not about preservation of private property but about the improper use of his designs, which often appeared in books printed in weak modern typefaces on cheap paper. To G. W. Jones he remarked, 'Men have a perfect right to use my designs, they have a perfect right to copy me *if they can use my designs rightly*, but not if they cannot use them. Every man has a right to everything he can use well in this world, but only to what he can use well.'[81] In practice, these high-principled sentiments translated into a consistent refusal by Morris to allow his types and ornaments to be employed by another printer. He reluctantly authorized the periodical *Bookselling* to reproduce several of his borders but grew agitated when E. F. Strange requested permission to use them as an illustration in a book as well. Morris 'allowed them to appear there as it is an ephemeral publication, but he does not think them very satisfactory, & would not like to see them in a permanent book', Cockerell declared in a letter to Strange.[82] Even when George Allen reprinted three of the Kelmscott Press books under the title *Old French Romances Done into English by William Morris* (1896), Morris declined to let him make use of the engraved title-pages which had appeared in the Kelmscott editions.[83]

But the Kelmscott Press had scarcely issued its first book before trouble began to brew in the United States. Roberts Brothers of Boston, who in 1890 had published an apparently unauthorized edition of *News from Nowhere*, wrote to Morris in 1891 ordering some copies of the Kelmscott *Glittering Plain* and requesting permission to produce an American edition of *Poems by the Way* from early sheets. 'I will send them sheets on the definite understanding that they are not to attempt to imitate the style of the typography,' Morris informed Reeves and Turner.[84] What Roberts Brothers did instead was to create a photographic facsimile of *The Glittering Plain*, which they issued in a limited edition of 500 copies in October 1891.

News of still more alarming developments came from across the Atlantic. In early 1894, while Morris was preparing an edition of Rossetti's *Sonnets and Lyrical Poems* for Ellis and Elvey, the latter unexpectedly received from another Boston publisher, Copeland and Day, a version of *The House of Life* done in elaborate imitation of the Kelmscott manner. With three borders and 114 initials designed by Bertram Grosvenor Goodhue, *The House of Life* sold for $2.50 (ordinary edition) and $5.00 (Michallet paper, rubricated initials). 'However much we may disapprove of this Publication, & the bad imitation of the Kelmscott Press we fear we cannot help it,' Ellis and Elvey wrote to William Michael Rossetti, who was seeing the Kelmscott edition through the press. 'They do not tell us what the price is but from their letter we infer it is comparatively cheap.' The only consolation Ellis and Elvey could offer him, a few days later, was that 'it is not likely to interfere with the Kelmscott edition'.[85] The reaction at Kelmscott House was less restrained. Morris himself was 'very scornful' of the American imitation but temporarily tricked his wife into believing it was one of his own books; Jane, once enlightened, described it as 'an impudent swindle'.[86]

The Roberts Brothers and Copeland and Day volumes were merely the first trickle of the great flood of Kelmscott-style books to be produced in the United States during the last decade of the century, most of them printed in types closely patterned after Morris's. 'Have you heard of the wrong they've [i.e. the Americans] just done him. . . ?' Burne-Jones demanded of his studio assistant in January 1896. 'Some wicked wretch over there has been photographing and copying all his letters and ornaments, and those he reproduces exactly he calls the Jenson letters, and those he mommocks [*sic*] and alters and spoils he calls Morris letters. *That is* irritating, isn't it? He's in a great rage, and it's no wonder.'[87] The fame of the Kelmscott Press had travelled rapidly across the ocean, and photoengraving techniques made it absurdly easy for American typefounders to reproduce – and modify to any extent they wished – Morris's designs. Either Joseph W. Phinney of the Dickinson Type Foundry (Boston) or H. L. Bullen

THE JENSON OLD-STYLE SERIES, MADE BY AMERICAN TYPE FOUNDERS' CO. (DICKINSON TYPE FOUNDERY BRANCH), BOSTON.

IN THIS specimen the Dickinson Typefoundery (Boston) exhibits a variation of the types of Nicholas Jenson, a Frenchman, who at Venice, in 1470, founded the true Roman, which thereafter dominated European typography. William Morris, of the Kelmscott Press, has modified these shapes from the originals, but in all the essentials they belong to the Jenson font. The form is round and clear, with great boldness and dignity; the lines are full of strong black and white effects, adapted to old-style work. Jenson derived his Roman characters from a mixture of alphabets of various countries, modelling these letters into marvellous perfection. The original font is on about a 14-point size, and consists of twenty-three capital alphabets (J, U and W then not being in use); twenty-three lower-case characters, fifteen contractions, six double letters and three points (period, colon and interrogation), making a total of seventy-three punches. Aldus Manutius finally inherited these punches, matrices and types, and later became famous for the fine alphabets designed by himself. The attempt to lessen Jenson's fame as the originator of the present forms of Romans by crediting their use to the Weidenbach monks, at their monastery, in 1465 (five years before Jenson's publication), has not succeeded with thoughtful students.

Figure 61 Jenson Old-Style type, patterned after Morris's first typeface, had a considerable vogue in the United States and elsewhere during the last decade of the nineteenth century.

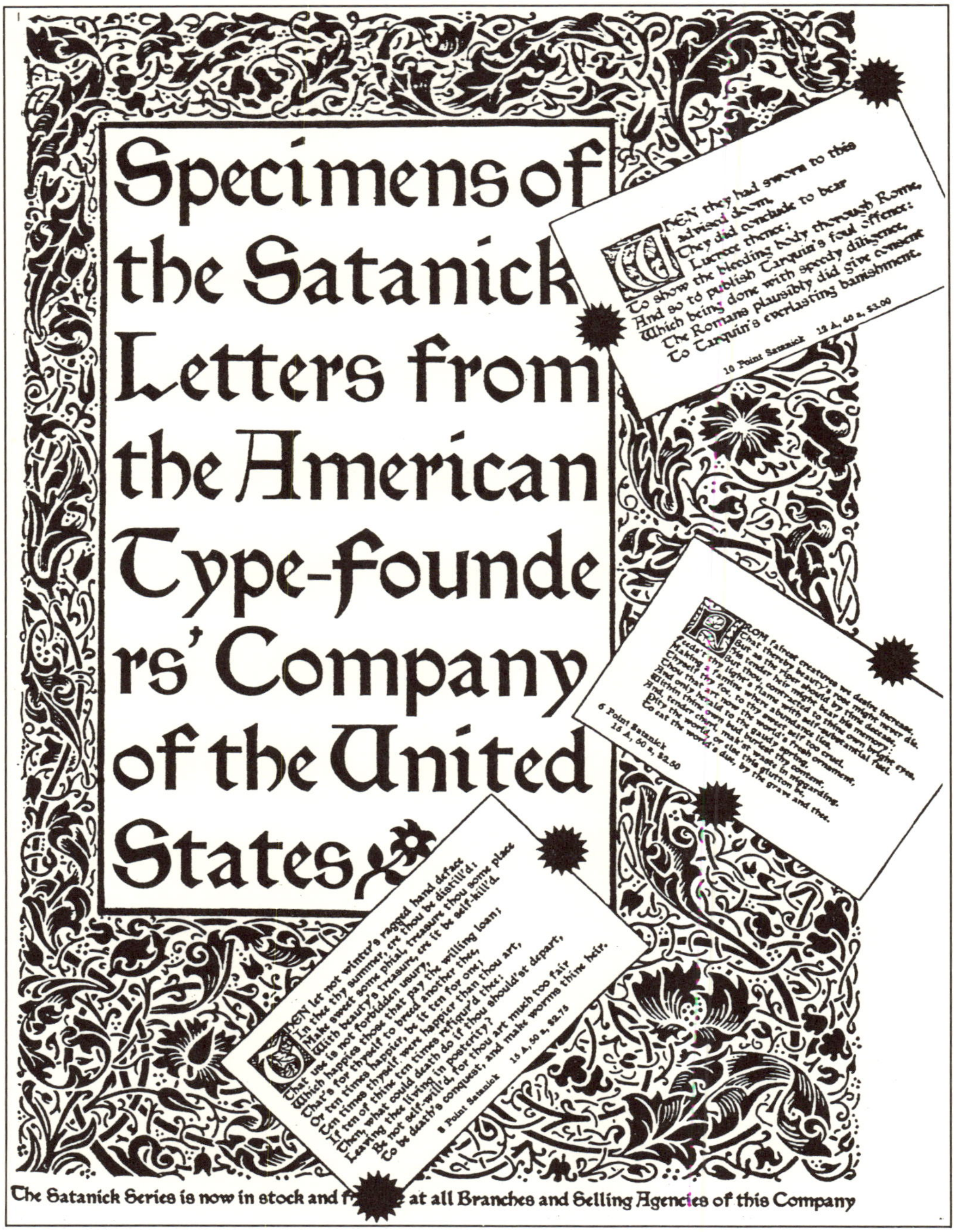

Figure 62 The Satanick type, an obvious copy of Morris's Troy type, as advertised in an American periodical, the *Inland Printer*, in 1896.

of the American Type Foundry (Jersey City, N.J.), or possibly both (the historical evidence is ambiguous), approached Morris with the proposal that they make his types and ornaments commercially available in the United States. 'Mr. Morris wrote back and told them to go to a hot place,' according to Bullen, and Phinney therefore proceeded without his permission, manufacturing initials, borders, and fourteen sizes of a typeface called Jenson Old Style which was similar to the Golden type.[88] An imitation of the Troy type named Satanick (allegedly to commemorate Morris's heated response to Phinney's letter) followed, and by mid-1895 the advertising pages of American printers' trade journals were awash with announcements of pseudo-Kelmscott types and ornaments in every conceivable shape and size. Under various names many of them became available throughout Europe.

Figure 61

Figure 62

Eventually some of these reproductions reached England, where Morris and Cockerell struggled to keep them off the market. Early in 1896, for example, Cockerell was consulting a solicitor 'about the piracy by an American firm of a great quantity of Kelmscott initials & ornaments' which were being sold in England (Cockerell diary, 15 and 20 January 1896). A more serious problem arose in December 1897 when Cockerell discovered that Sir Charles Reed and Sons, Morris's own typefounders, were advertising an American imitation under the name 'Kelmscott Old Style' in their latest specimen sheet. Cockerell, always an alarming adversary when he thought Morris's interests were in jeopardy, moved against the Reeds at once. Threatened with legal action, they agreed to withdraw the offending specimens, to announce that the name of Kelmscott could not be used by any customers who had already purchased the type, and to surrender the matrices to Cockerell. It was total capitulation. 'Reeds have climbed down & are reprinting their spec[ime]ns but I don't feel very warmly towards them,' Cockerell wrote grimly.[89]

The anxiety that Morris felt about American imitations of the Kelmscott types and books indicates more than an artistic concern that his designs might be used inappropriately. Like the Victorian industrialists that he condemned, Morris was trying to protect his market against cheap foreign imports. This point is worth emphasizing because, whenever one attempts to assess the business side of any of Morris's activities, the evidence suddenly becomes obscured by a smoke-screen of socialist piety. Jeanette L. Gilder, for instance, interviewed Morris in the last year of his life and recorded this characteristic exchange: 'I asked Mr. Morris to tell me, if it was not an impertinent question, whether it paid him pounds, shillings and pence to print such handsome books. He answered that it did; but his secretary [Cockerell], who had just entered the room, begged leave to contradict this statement. "You have not," said he, "counted your work and your time, Mr. Morris. You are thinking only of the mechanical part; but if you put in your time and your skill, at what you would be paid for them by another person, you could hardly say that your publications made money." Mr. Morris smiled but said no more on the subject. The profit did not seem to interest him very much: it was the artistic results which he accomplished that pleased him.'[90]

Cockerell's point is of course well taken: Morris's own invaluable contributions were never appraised in calculating the profits and losses of the Kelmscott Press, and we must never forget the genuinely impressive idealism that lay behind the whole of Morris's career, including the Kelmscott Press. Nevertheless – whatever Miss Gilder may have supposed – he was indeed preoccupied with the profits made by his books. This is not to say the profit motive was his main concern at the Press, which really was, as he said, a typographical adventure and therefore financially risky, but Morris never lost sight of the need to sell books and to pay his employees. This unexpected combination of hard-headed business sense and Utopian idealism has puzzled many admirers of Morris, and it has prompted others to accuse him of hypocrisy. Morris himself, as one might expect, had thought long and deeply about the problem. He argued that, until the socialist revolution came, the employer was obliged to work within the capitalist system, no matter how much he might personally despise it.[91] What Morris failed to take sufficiently into account was how tainted even the most pure craft activity might become through commercial considerations; this, after all, is what happened to the Kelmscott Press, which began as an attempt to restore aesthetic (and moral) integrity to bookmaking and ended as the publisher of expensive books purchased mainly by wealthy collectors. Morris's friend and fellow-socialist, Walter Crane, saw very clearly the extent to which the achievements of craftsmen and artists were coloured by the corrupting ethos of the market-place. 'My difficulty', he wrote to Ashbee in 1890, '. . . is how to establish any good system of working or teaching while the present economic system exists to work its destruction ultimately. Do what one will it seems one is bound to be devoured by commercialism in some form or another.'[92]

Another source of difficulty in analysing the financial state of the Kelmscott Press is that Morris, despite his business acumen, took a decidedly casual attitude towards record-keeping. When he sent Crane a cheque for his share of the profits from the second version of *The Glittering Plain*, Morris promised to 'get an account made out which I shall take care is not too much humbugged by the accountant, who (any one of them) could easily make out that there was no profit at all. But I will see that it is all fair, as I really know what the profit is effectually.'[93] There was a confusing overlap between Morris's personal bank account and that of the Press, and during Sparling's time as Secretary the financial records, like much else at the Kelmscott Press, were in disarray. 'In future we may be able to keep a closer record of expenditures than hitherto,' Cockerell reported to the Press's accountant in August 1894, shortly after replacing Sparling.[94] In addition to invoices and various other kinds of scattered records, there exists a ledger (in the Morgan Library) describing the cost of production of all the individual Kelmscott books printed between 1893 and 1896 as well as other profit-and-loss accounts and inventories; but the ledger has unfortunately not yet been subjected to careful scrutiny by scholars familiar with nineteenth-century bookkeeping conventions, and so it is not now possible to say with absolute certainty how often, and to what degree, Morris realized a profit on his books. The safest generalization one can offer at present is that he nearly always broke even (bearing in mind, again, that his services were unpaid) and that sometimes he came out slightly ahead. Selecting one title completely at random, for instance, one can readily calculate, based on the most conservative assumptions, that Morris's gross receipts from More's *Utopia*, an octavo published in an edition of 300 paper copies (thirty shillings) and eight vellum copies (ten guineas), could not have been less than £300, whereas it cost only about £255 to produce.

Morris was eager to explain to journalists that he was not a bloated capitalist: 'You have, no doubt, heard a great deal of the enormous profit I must be making on these books,' he told an interviewer in 1895, 'because of the high prices I ask for them. I assure you that, if the people who go about talking of my profits were to see my balance sheet, they would speak quite differently.'[95] The two examples he cited were *Beowulf*, which he said he sold at a loss, and the *Chaucer*, which represented Morris's largest investment and thus caused him much apprehension before it was finally published. When he feared that word about his ink troubles with the *Chaucer* might spread to the public, he wrote in great agitation to F. J. Furnivall (who had a notoriously loose tongue): 'I must beg you not to say a word about it to anybody or you will do me very serious injury.'[96] Only after the pressrun of the *Chaucer* was increased from 325 to 425 copies did Morris feel secure from financial disaster. On the other hand, Morris sold his smaller books without difficulty and, as a matter of deliberate policy, used them to finance the printing of the folios and quartos. 'I count rather on these little ones for getting cash to carry on the Chaucer, as the little books pay me much better than the big,' Morris admitted to Quaritch.[97] There is also evidence to suggest that Morris squeezed every penny he could from publishers who commissioned Kelmscott Press books: *Maud*, for example, cost him only £202 14*s.* 1*d.* to produce (according to the Press ledger), yet he charged Macmillan £400 for the paper copies alone.[98]

It is obvious that the slow, careful methods of design and production followed at the Kelmscott Press increased the cost of the books, and there are thirteen known instances of alterations being made in the Kelmscott volumes after printing had begun, usually involving the resetting and reprinting of a signature or a pair of conjugate leaves.[99] Another unusual expense at the Press was Batchelor's paper, for which he charged about five or six times the price of machine-made book paper; Walker pointed out that the cost of the paper in a single copy of the *Chaucer* was approximately 25*s.*[100] On the other hand, Morris recovered some of this expenditure by allowing a smaller trade discount than was usual, a policy which caused occasional friction between him and the booksellers. For Reeves and Turner, and Quaritch, the discount was 20 per cent; for all other booksellers, and for a few friends like Robert Proctor, it was 15 per cent.

There is an undated Kelmscott Press circular printed in the Golden type and addressed 'To the Trade only' which reads as follows: 'We send you the enclosed, and shall be glad to receive your orders. The discount to the trade will be 15 per cent., and we must ask you to understand that these terms are strictly for cash. We shall write you when the book is actually ready for delivery, and on receipt of your cheque your order will be executed.' The tone of this communication would not have been uncongenial to Dickens's Mr Bounderby. Beneath the beauty and elegance of the Kelmscott Press books lay a hard, flinty foundation of commercial shrewdness. When we are tempted to burst into rhapsodical praise for Morris's triumphant revival of medieval craftsmanship, it is useful to remember that.

SEVEN THE STORY OF THREE BOOKS

THE Kelmscott Press books, as we have seen already, come in a variety of sizes and forms: some are long (*Chaucer*, 564 folio pages), some short (*Epistola de contemptu mundi*, sixteen pages); a few are heavy with ornaments (*Godefrey of Boloyne*), while others are decorated only with a few small initials and a border on the first text page; a minority are typographically elaborate (*Atalanta in Calydon* (1894) has three typefaces on the title-page), yet many of the books are almost austerely simple in design. Despite Morris's preference for medieval texts, there is also literary diversity in the Kelmscott books, which reprint works written over a span of more than a thousand years.

For these reasons it is difficult, indeed dangerous, to offer broad generalizations about how the books of the Kelmscott Press were produced. The process by which Morris designed and printed the *Epistola*, for example, was very different from that followed in manufacturing *The Well at the World's End*, which was in production from 1892 to 1896 ('the Interminable', Morris called it). If we wish to move beyond glib truisms, therefore, we must look carefully at particular instances of Morris the printer at work and try to understand the ordinary, day-to-day problems with which he struggled at the Press. In this chapter we will examine the production history of three Kelmscott books: Morris's *Poems by the Way* (1891), Wilfrid Scawen Blunt's *The Love-Lyrics & Songs of Proteus* (1892), and *The Golden Legend* (1892). Because they are all early works of the Press (the second, third, and seventh books issued, respectively), there is an unusually large amount of documentary evidence available, and though unfortunately the Press's ledger does not cover the period when they were printed, it is possible to reconstruct financial information from other sources. These particular books also possess the merit of representing the three major types of literary texts that Morris printed: his own books, the writings of other Englishmen (mainly poets), and medieval works.

Even though it was the last of these three books to be published, *The Golden Legend* deserves to be considered first since it was the earliest book planned for the Kelmscott Press. The following notice appeared in the 'Literary Gossip' column of the *Athenaeum* on 13 September 1890: 'Students of English mediaeval literature will be gratified to learn that an exact reprint of "The Golden Legend" as originally set forth by William Caxton is in preparation, under the joint editorship of Mr. William Morris and Mr. F. S. Ellis. If it is found practicable to obtain the use of a copy of the first edition it is proposed to reproduce that text with scrupulous accuracy, supplementing it with a glossary and index. In place of the black letter, to the use of which there are manifold objections, a fount of types newly designed by Mr. Morris after the fashion of those employed by Nicholas Jenson will be substituted. . . . Some time must necessarily elapse . . . before the work can go to press, though it will be proceeded with immediately. The editors have agreed to give their labours gratuitously in consideration of Mr. Quaritch bearing all expenses of production. As the impression will be a limited one, subscribers would do well to send their names to the publisher forthwith.' This announcement, which came from a remarkably well-informed source (probably F. J. Furnivall), constitutes in fact the earliest reference in print to any of the Kelmscott Press books.

During the summer of 1890 Morris, as he was designing his Roman type, pondered the possibility of reprinting one of two Caxton books – either *The Golden Legend* (1483) or *The Recuyell of the Historyes of Troye* (1474) – which would be published by Bernard Quaritch. With this in mind, he purchased on 21 July 1890 a copy of a slightly later edition of *The Golden Legend* (Wynkyn de Worde, 1527), and, when Cockerell saw Morris on 14 August, he found that 'he thinks of reprinting [it] when his new type is all cast' (Cockerell diary). To Ellis, who had offered to edit the book, Morris wrote on 29 August: 'I did want to weigh between the Golden Legend and the Troy book for reprinting: now I have borrowed a Recueil of the Histories of Troy (the Wynkyn de Worde of course) from Quaritch, and have no doubt that the G.L. is by far the most important book of the two. . . .'[1] (Eventually Morris printed the *Troye* as well.)

What was the nature of *The Golden Legend*'s appeal to Morris? A collection of saints' lives compiled by Jacobus de Voragine (or Varagine) and originally published under the title *Legenda aurea sanctorum, sive Lombardiea historia* (Strasburg, *c.*1472–4), it struck some Victorian readers who did not share Morris's antiquarian enthusiasms as an improbable choice for the first big book of the Kelmscott Press. In his 'Memoranda, Bibliographical & Explanatory' which appeared at the end of the third volume of the Kelmscott edition, Ellis offered this justification for reprinting it: 'Among the books which serve to illustrate the religious life and mode of thought that prevailed in the middle ages, none holds a more important place than the "Legenda Aurea." Compiled and put into form in the latter half of the 13th century, the welcome with which it was received by the clergy & the lettered world is attested by the great number of manuscripts of it which are still extant, while its ready reception by the people generally is evidenced by the fact that within a short time of its appearance it was translated into the vulgar tongue of most of the nations of Europe, frequently with the hagiological preferences of the people of the country for which it was designed' (iii. 1282).

Quaritch, always the sly businessman, saw this venture as a means of snaring Morris. 'You are no doubt aware of the high standing, as a poet, of William Morris,' he remarked to his son on 5 June 1890; 'I have sold to him lately several early printed books. I also told him I should like to publish some future book of his. I am in a fair way of getting such a new book from him; – and this may lead much further afterwards.'[2] In early September 1890 letters were flying back and forth among Morris, Ellis, and Quaritch as they negotiated a contract for the book. Quaritch was surprised to discover that the only modern reprint of *The Golden Legend* was an incomplete facsimile published by the Holbein Society in 1878 (a book that Morris owned). Furnivall's Early English Text Society, however, had printed several other Caxton translations during the 1870s and 1880s, and David Nutt, a London publisher, issued a limited edition of Caxton's *Le Morte Darthur* between 1889 and 1891; the latter was, like the Kelmscott *Golden Legend*, an exact reprint, but its typography was undistinguished and two of the three volumes were devoted to learned analyses and bibliographical studies of the text. Quaritch, who did not yet fully grasp Morris's new preoccupation with book-design, inquired innocently whether he perhaps intended to make *The Golden Legend* uniform with Nutt's book.[3]

'My interest in the book is double,' wrote Quaritch to Ellis, '*first* I think every man of culture should be acquainted with the alleged Lives of Saints, and *second* I am aware of the philological value of all books printed by Caxton.' But then he offered a typical aside: 'Some of the naughty Saints stories will be relished by the numerous readers of Burton's Arabian Nights.'[4] The contract[5] (which was dated 11 September 1890) specified that Morris and Ellis were to contribute their *Figure 63*

Figure 63 The contract for *The Golden Legend*, the first book planned for the Kelmscott Press. Morris and Ellis were to be unrewarded for their labours except for some complimentary copies of the volume, but the agreement specified that Morris was 'to have absolute and sole control over choice of paper, choice of type, size of the reprint and selection of the printer'.

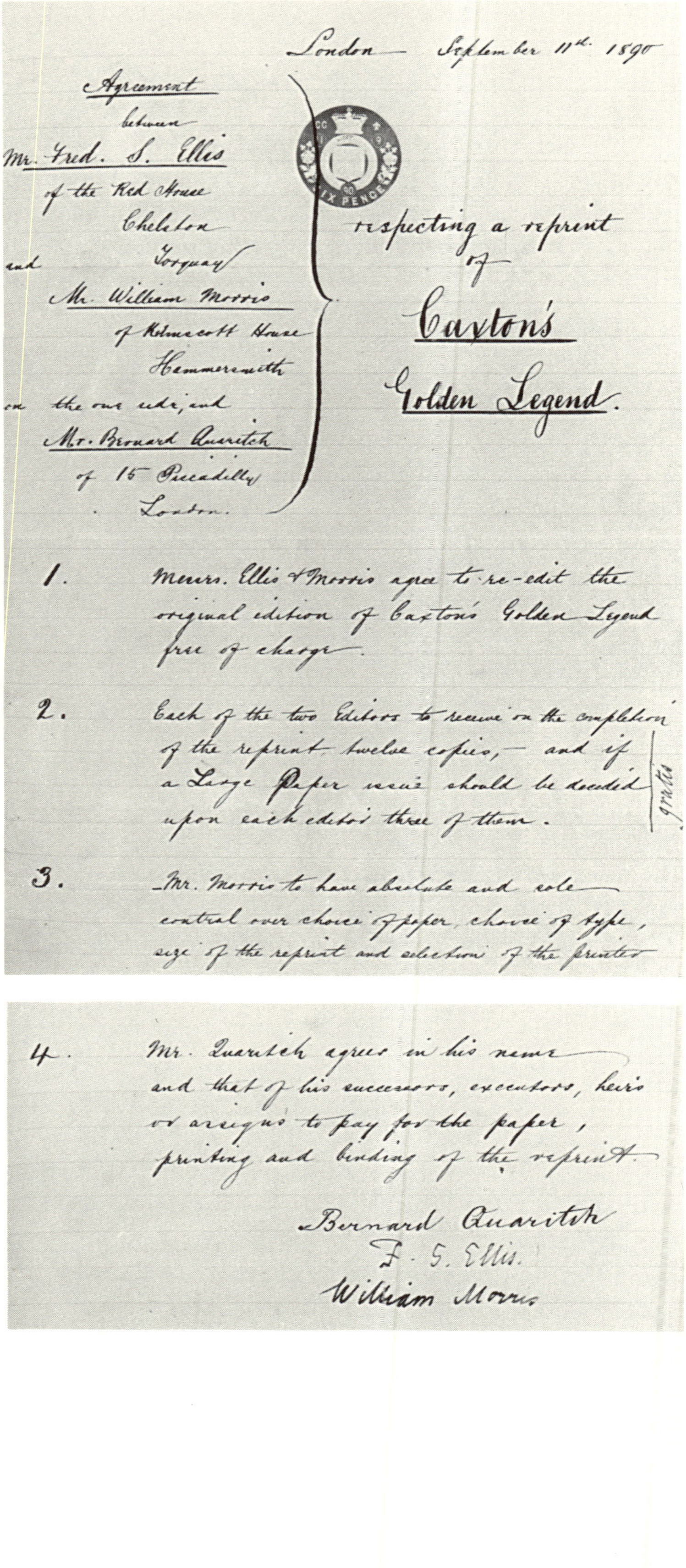

London September 11th 1890

Agreement
between
Mr. Fred. S. Ellis
of the Red House
Chelston
Torquay
and
Mr. William Morris
of Kelmscott House
Hammersmith
on the one side, and
Mr. Bernard Quaritch
of 15 Piccadilly
London.

respecting a reprint
of
Caxton's
Golden Legend.

1. Messrs. Ellis & Morris agree to re-edit the original edition of Caxton's Golden Legend free of charge.

2. Each of the two Editors to receive on the completion of the reprint twelve copies, – and if a Large Paper issue should be decided upon each editor three of them. gratis

3. Mr. Morris to have absolute and sole control over choice of paper, choice of type, size of the reprint and selection of the printer

4. Mr. Quaritch agrees in his name and that of his successors, executors, heirs or assigns to pay for the paper, printing and binding of the reprint.

Bernard Quaritch
F. S. Ellis.
William Morris

services as editors gratis, to be reimbursed only with twelve copies of the book; that Quaritch was to pay for the paper, printing, and binding of 500 copies; and that Morris was 'to have absolute and sole control over choice of paper, choice of type, size of the reprint and selection of the printer'. The odd phrasing of the last clause reminds us that Morris at this stage still had only a very indistinct notion of the future Kelmscott Press and was undecided whether to ask another printer to produce the book (with Morris's types, of course); significantly, Quaritch assumed that as publisher he would furnish the paper. By the end of 1890 it was clear that Morris would print the book himself.

In selecting the edition he would reprint, Morris displayed less hesitation – it would be Caxton's first version, not the Wynkyn de Worde edition that he owned – but this meant finding a copy that could be borrowed. The Cambridge University Library agreed to lend its Caxton *Golden Legend*, heavily insured, to Ellis, whose daughter Phyllis (later Mrs Payne) then laboriously transcribed the entire book, so that the volume itself would not have to be touched by the inky hands of Morris's printers. It was an heroic undertaking, for which she never received the public recognition she deserved, and the few surviving scraps of her manuscript confirm W. H. Bowden's testimony that not one word of it was illegible.[6] Later her father read the Kelmscott proofs directly against the book. Though the contract with Quaritch referred to Morris and Ellis as co-editors, Morris was only nominally involved in editorial decisions, thus setting the pattern for many later Kelmscott books. Ellis established the text, directing the occasional query to Morris; Morris agreed or disagreed with Ellis's proposals. That is how it always went between them, with much joking on Ellis's side and some blustering on Morris's. But in his concluding 'Memoranda' Ellis rather exaggerated Morris's participation: 'That I express myself in the first person throughout this notice, and sign it, is only because I am responsible for the statements as to matters of fact which it contains, & which I could not ask Mr. Morris to guarantee with his name, but it should be understood that he has gone carefully over the whole book, and nothing of any importance in the editing has been done without his advice and sanction' (iii. 285–6). The colophon is less equivocal, listing Ellis as the editor and Morris as the printer.

Figure 64

As usual, Ellis worried a good deal about bibliographical and textual matters while the book was in production. He found a French version of the *Legenda aurea* which he believed was Caxton's source, and corresponded enthusiastically with Francis Jenkinson, University

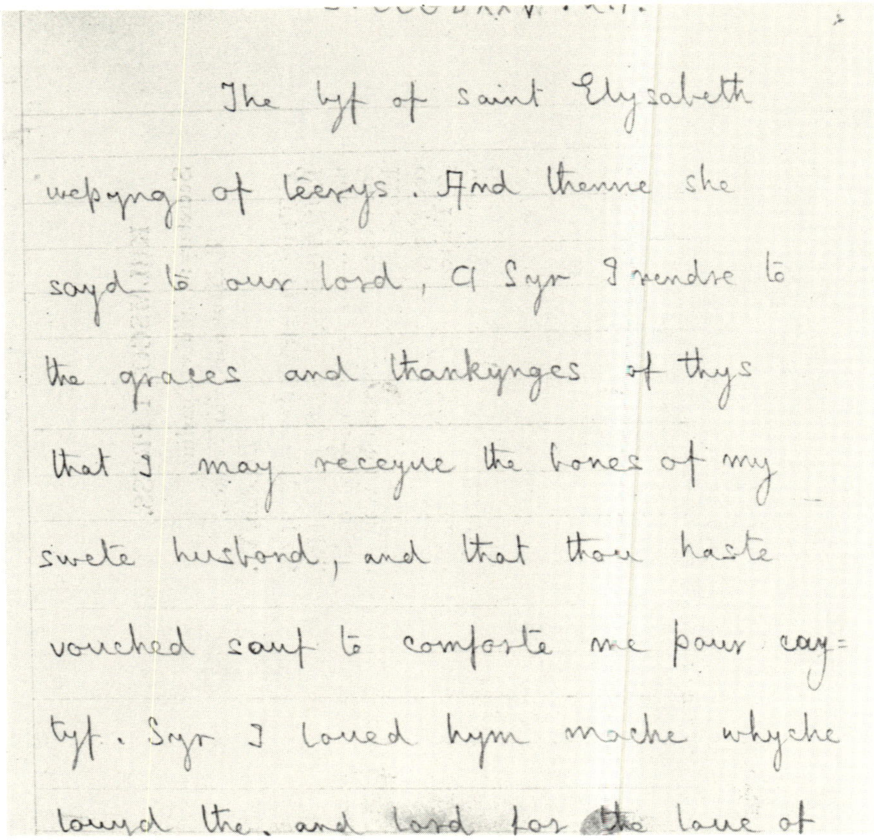

The lyf of saint Elyzabeth

wepyng of teerys. And thenne she
sayd to our lord, A Syr I rendre to
the graces and thankynges of thys
that I may receyue the bones of my
swete husbond, and that thou haste
vouched sauf to comforte me pour cay=
tyf. Syr I loued hym moche whyche
loued the, and lord for the loue of

Figure 64 The Kelmscott edition of *The Golden Legend* was set from a transcription of the book by F. S. Ellis's daughter Phyllis. This scrap of her manuscript happened to survive because a proof of one of the Press's circulars was printed on the verso.

Librarian at Cambridge, about the matter. He also struggled to persuade Morris to include a glossary and a bibliography of English editions of the book; the glossary appeared in truncated form as 'A List of Some Obsolete or Little Used Words', but Morris declined to use the bibliography prepared by Edward Gordon Duff. In September 1892 Ellis wrote to Duff, 'The difficulty is that he has not the necessary types or the small fount which it would be desirable to use. He said he would get it printed by Whittingham [i.e. the Chiswick Press] as a separate paper to be given with the Golden Legend but whether he will think of it again or not I cannot say – but seeing the many irons he has in the fire I am fearful of its getting forgotten & do not like to bother him further on the subject.'[7] The real reason for its omission, however, was, as Ellis had admitted in an earlier letter, that the 'author–printer kicks much against introductions'.[8]

The colophon of the Kelmscott Press *Golden Legend* describes it as a reprint 'in which there is no change from the original, except for the correction of errors of the press, & some few other amendments thought necessary for the understanding of the text'. Ellis's textual alterations in this and other Caxton books he edited for Morris have displeased some twentieth-century specialists, but to assess his work fairly we have to see it within the tradition of energetic amateur scholarship which flourished in Victorian England, particularly among the members of the societies founded by Furnivall. Ellis, though infinitely more scrupulous and careful in his treatment of documents than Furnivall himself was, lived at a time when the editing of historical and literary documents was not yet solely the preserve of professionals as it mainly is today. Ellis's aim in preparing the text of *The Golden Legend* for the Kelmscott Press was not only to achieve verbal accuracy but also to anticipate the needs of non-specialist readers. Although much has been written about the typography and design of Kelmscott books, historians of the Press have failed to observe that Morris and Ellis were also engaged in a pioneering attempt at intelligent popularization of literary works drawn, in some instances at least, from obscure or remote sources. Measured by that standard, Ellis's achievements as an editor are impressive.

While Ellis concentrated on the text, Morris began to think about the physical shape of *The Golden Legend*. Ellis was eager to reproduce the general appearance of Caxton's book, but Morris had misgivings about the wisdom of this, because he recognized that the Roman type he would be using was different in nature from Caxton's Gothic, and perhaps also because he had learned from *The Story of Gunnlaug* the futility of such mock-quaintness. 'Of course I should like the reprint to be of the same *form* as the original if the Roman type can do it, which I doubt, as black letter takes up less room: in any case some kind of folio it will have to be,' he reported to Ellis in September 1890. 'In one thing I think I differ from you a little, *i.e.*, about the joined letters or queer signs: since our book is to be a *reprint*, not a fac-simile, I do not think we need reproduce these: indeed I should extend the abbreviations in order to make the book more readable.'[9] By December Morris's ideas about the dissimilarities between Roman and Gothic types had been taken a step further: 'One thing may disappoint you – to wit, that we cannot make a double-column page of it, the page will not be wide enough. For my part, I don't regret it: double column seems to me chiefly fit for black letter, which prints up so close. Jenson did not print even his Pliny in double column. But it is a case of *a fortiori* in modern printing: because we have no contractions, few tied letters, and we cannot break a word with the same frankness as they could: I mean we can't put whi on one side and ch on the other. This makes the spacing difficult, and a wide page desirable.'[10]

When the first delivery of paper arrived from Batchelor early in the following year, Morris concluded that it did not have sufficient sizing and reported to Ellis (11 February 1891) that, since 'it is only a little lot (9 reams)', he intended to print 'a little edition of the Glittering Plain on it'. Meanwhile, he said, he would set up a signature of *The Golden Legend* as an experiment until more paper was available. 'But I can sympathize with my pardner's anxiety,' Morris commented; 'and accordingly send him a [trial] page of the G[littering] P[lain], of course full of defects, but on the paper and with the types. I don't know what you will think of it; but I think it precious good.'[11]

It would be gratifying to know exactly how Morris went about designing a book, but, although he wrote at length about the broad principles of book-design and left us a legacy of fifty-two books at the Kelmscott Press, there is surprisingly little documentation of his actual procedures and techniques. Unlike some designers of our century – Bruce Rogers or Jan Tschichold, say – he seems not to have made elaborate, precise drawings of the pages he envisioned. Designs of ornaments, types, and lettering for frontispieces we have in abundance from Morris's hand, but not one drawing to show the layout of a Kelmscott page. (There is, in the Bodleian Library, a rough sketch by *Figure 41* Morris of the title-page of the Reeves and Turner edition (1894) of *The*

Well at the World's End,[12] but it is not very instructive and is merely patterned after the title-page of the Reeves and Turner *Poems by the Way*.) Probably this dearth of evidence reflects the way in which Morris worked: he always distrusted excessively detailed designs by an artist, because he felt that they inhibited the freedom of the craftsman who would execute them. No doubt the instructions he gave his compositors at the Kelmscott Press were generally delivered orally, with Morris often at their elbow, and he never hesitated to demand that they produce one trial page after another until he was satisfied. Trial pages of *The Golden Legend*, which survive in particular abundance,[13] reveal that by the time they were printed (in March 1892) Morris had settled on an approximate page size, $11\frac{1}{4} \times 8\frac{1}{16}$ inches, and the length of the line of type, $5\frac{1}{8}$ inches (though he continued to experiment with the latter), but these dimensions were largely determined by the paper size. Morris was much more preoccupied with the proportions of the margins, on which he made countless small adjustments. Once the general appearance of a page was established and the margins were deemed satisfactory, Morris's compositors began setting the text, usually providing him with galley proofs, though sometimes, as in volumes of poetry, page proofs were pulled immediately. (This was probably unwise, since last-minute alterations frequently shifted lines on to following pages.) Spaces for initials were left blank, and in the proofs Morris specified which initials and borders were to be inserted by writing in their numbers from his 'bloomer book'.

The Golden Legend was by far the largest book issued by the Kelmscott Press in the first two years of its existence – more than 1,300 pages printed on a ton and a half of paper[14] – and progress was correspondingly slow. Morris had begun designing initials for it even before the first supply of type arrived in January 1891, and he hoped to ornament the book profusely: 'I think I shall make some scratch of a border to each life or section,' he wrote in April. 'I want to make it grand.'[15] Such a scheme was far beyond the resources of the Press at this early stage, but Morris was able to design four full borders (the first one drawn at Folkestone in April), corner borders, and a large number of six- and ten-line decorated initials. A book of this size strained the capabilities of Morris's small group of printers; to make matters worse, in late April a slab holding fifty pages in galleys collapsed, with the type scattering in terrible confusion on the floor.[16] By the following December, responding to the pressure of *The Golden Legend*, Morris had eight or nine men working for him in Sussex Cottage.[17]

Figure 65 A proposed design by Burne-Jones for the second illustration in *The Golden Legend*, which differs from the published wood-engraving. In 1892 Burne-Jones was still developing the distinctive style of book-illustration that came to be associated with the Kelmscott Press.

Quaritch began to express uneasiness about the slow pace of the book. On 25 July Morris wrote to him, '1st as to the extent of the book – 1500 p.p. about (this is pretty certain) 2nd as to time of completion. *probably* next June (1892) 3rd the cost: from £1200 to £1400 including doing up in boards covered with grey hand-made paper, which I think a satisfactory temporary binding. Of course if you please you might have some copies done up in a more costly way, at an increase in cost, which would mean an increase in selling price. I should add that I propose to *give* my work on designing the ornament, just as Ellis does his transcribing.' He also offered a further piece of news which must have encouraged Quaritch: 'Mr. Burne Jones has promised to design me a frontispiece for the book.'[18] Eventually Burne-Jones supplied two designs, which had to be inserted on separate leaves because the artist was so late in delivering them; and as a substitute for a frontispiece Morris designed the first of his lovely engraved titles. Burne-Jones, shown proofs of the book in September 1891, 'went crazy with pleasure'.[19] The two illustrations and the engraved title were not cut by W. H. Hooper until the summer of 1892, when the third volume was nearing completion. 'The editing of the Golden Legend has been the greatest pleasure to me,' Ellis commented that summer. 'It was with a heavy sigh of regret that I returned the last proof for press.'[20]

Figure 65

Figure 66

The colophon was dated 12 September 1892; the books were delivered to Quaritch from Leighton on about 4 October; and the official date of publication was 3 November. There were 500 copies printed on paper – but none on vellum – at a price of five guineas. On 7 October Morris sent Quaritch his receipt for the printing and binding costs (£1,350) and this valediction: 'I must needs say that, looking at the finished book, I am proud of it, and of having pushed it through so promptly. It is just 2 years ago that Ellis & I began *talking* about doing it in this very house where I am writing [Kelmscott Manor].'[21] Towards the end of the month Morris and his friends celebrated its publication at a dinner attended by Ellis and his son, Furnivall, Duff (whose bibliography never saw the light of day), Joseph Knight, Richard Garnett, William Blaikie, William Michael Rossetti, and Walker.

Figure 67

Reviewers and readers responded enthusiastically to the book. *The Times* (6 October 1892) described it as 'truly magnificent' and considered it a bargain for 'judicious bibliophiles' at its price. The *Library*, the journal of the Bibliographical Society, suggested that *The Golden Legend* marked 'a new epoch in the production of beautiful books in this country'. In some of the earlier Kelmscott Press books, the

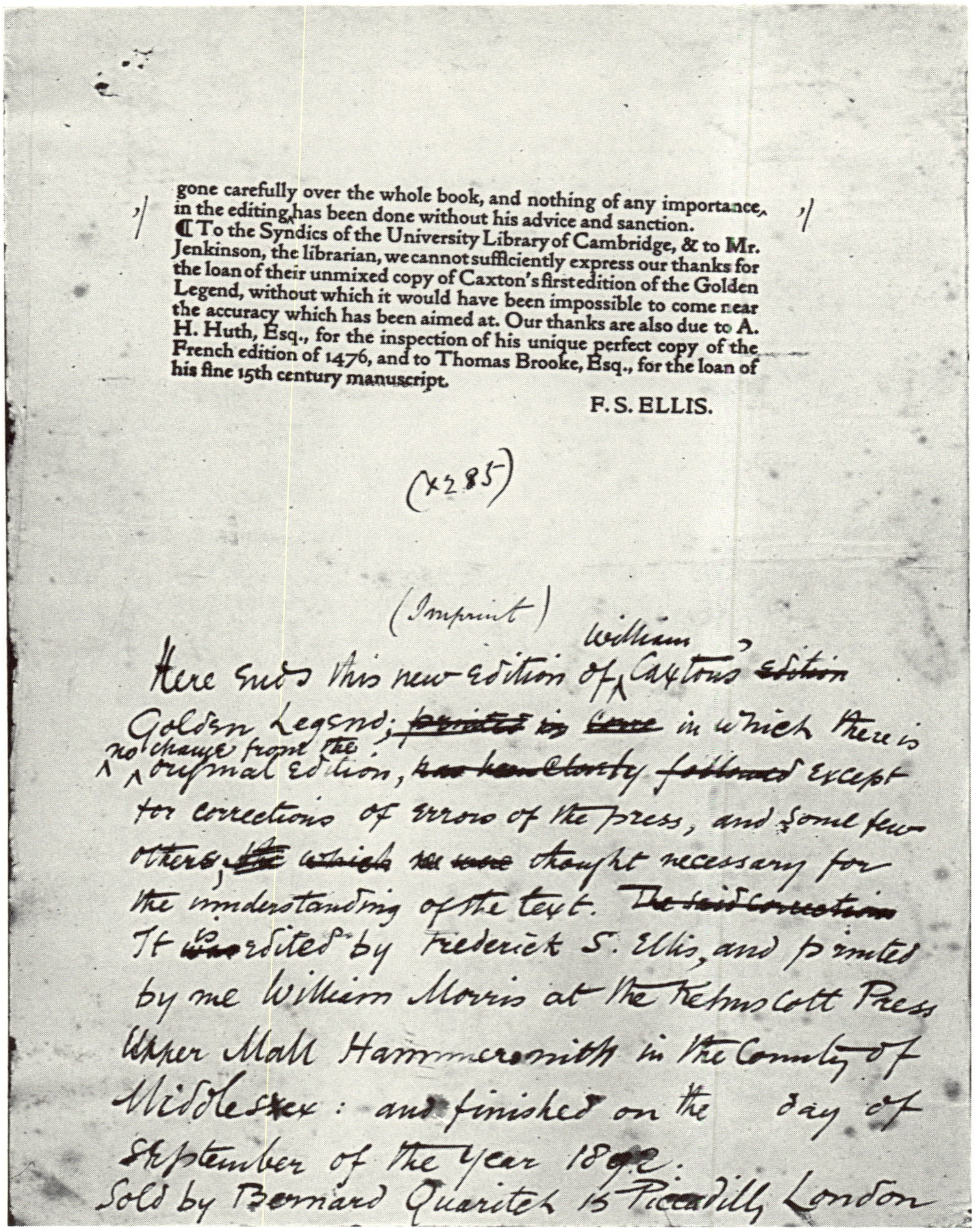

gone carefully over the whole book, and nothing of any importance, in the editing has been done without his advice and sanction.
¶ To the Syndics of the University Library of Cambridge, & to Mr. Jenkinson, the librarian, we cannot sufficiently express our thanks for the loan of their unmixed copy of Caxton's first edition of the Golden Legend, without which it would have been impossible to come near the accuracy which has been aimed at. Our thanks are also due to A. H. Huth, Esq., for the inspection of his unique perfect copy of the French edition of 1476, and to Thomas Brooke, Esq., for the loan of his fine 15th century manuscript.

F. S. ELLIS.

(x285)

(Imprint)

Here ends this new edition of William Caxton's ~~edition~~ Golden Legend; ~~printed in~~ in which there is no change from the original edition, ~~followed~~ except for corrections of errors of the press, and some few others, ~~which~~ thought necessary for the understanding of the text. ~~The said correction~~ It was edited by Frederick S. Ellis, and printed by me William Morris at the Kelmscott Press Upper Mall Hammersmith in the County of Middlesex: and finished on the day of September of the Year 1892.
Sold by Bernard Quaritch 15 Piccadilly London

Figure 66 The colophon for *The Golden Legend*, written by Morris on a proof of Ellis's 'Memoranda, Bibliographical & Explanatory', which appeared at the end of the book. The date was left blank so that it could be supplied by the printer on the day the forme for the last sheet was put on the press.

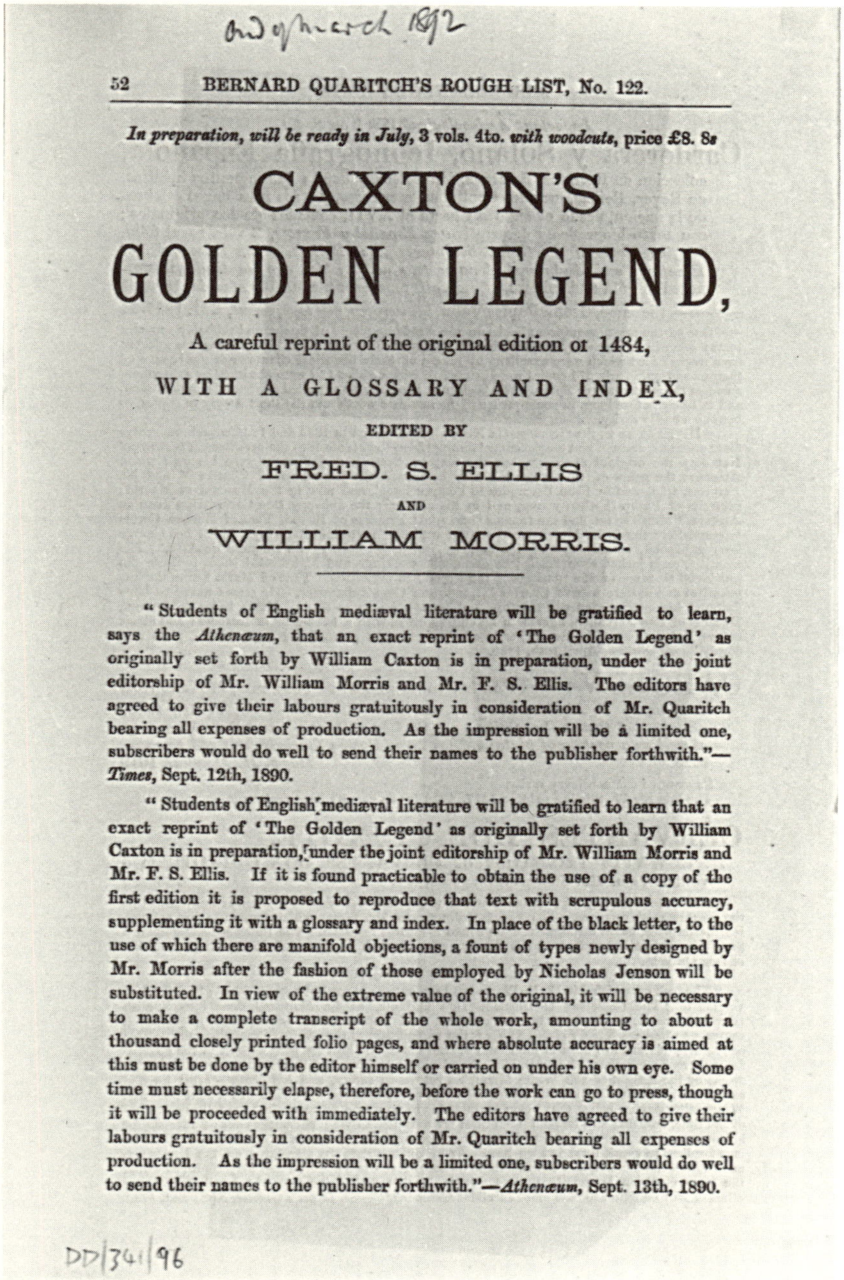

52 BERNARD QUARITCH'S ROUGH LIST, No. 122.

In preparation, will be ready in July, 3 vols. 4to. *with woodcuts,* price £8. 8s

CAXTON'S

GOLDEN LEGEND,

A careful reprint of the original edition of 1484,

WITH A GLOSSARY AND INDEX,

EDITED BY

FRED. S. ELLIS

AND

WILLIAM MORRIS.

"Students of English mediæval literature will be gratified to learn, says the *Athenæum,* that an exact reprint of 'The Golden Legend' as originally set forth by William Caxton is in preparation, under the joint editorship of Mr. William Morris and Mr. F. S. Ellis. The editors have agreed to give their labours gratuitously in consideration of Mr. Quaritch bearing all expenses of production. As the impression will be a limited one, subscribers would do well to send their names to the publisher forthwith."—*Times,* Sept. 12th, 1890.

"Students of English mediæval literature will be gratified to learn that an exact reprint of 'The Golden Legend' as originally set forth by William Caxton is in preparation, under the joint editorship of Mr. William Morris and Mr. F. S. Ellis. If it is found practicable to obtain the use of a copy of the first edition it is proposed to reproduce that text with scrupulous accuracy, supplementing it with a glossary and index. In place of the black letter, to the use of which there are manifold objections, a fount of types newly designed by Mr. Morris after the fashion of those employed by Nicholas Jenson will be substituted. In view of the extreme value of the original, it will be necessary to make a complete transcript of the whole work, amounting to about a thousand closely printed folio pages, and where absolute accuracy is aimed at this must be done by the editor himself or carried on under his own eye. Some time must necessarily elapse, therefore, before the work can go to press, though it will be proceeded with immediately. The editors have agreed to give their labours gratuitously in consideration of Mr. Quaritch bearing all expenses of production. As the impression will be a limited one, subscribers would do well to send their names to the publisher forthwith."—*Athenæum,* Sept. 13th, 1890.

Figure 67 Quaritch's trade announcement of *The Golden Legend* in 1892. Though Morris is described here as co-editor, in the book itself Ellis calls himself the sole editor but adds that Morris 'has gone carefully over the whole book, and nothing of any importance in the editing has been done without his advice and sanction'.

reviewer wrote, the initials were too large and the margins too small; but now these faults had been corrected, and he could only describe the result with the same word used by *The Times* – 'magnificent'.[22] The most lavish praise of all came from Swinburne, who declared in a letter to Ellis that *The Golden Legend* was 'the most superbly beautiful book that ever, I should think, came from any press'. Now, he observed, he would be able to 'enjoy the study of the most famous hagiology in the world under a form which would make Euclid or Adam Smith readable & Aristotle or Hallam attractive'.[23]

The production history of *Poems by the Way*, the second Kelmscott Press book, is much briefer and simpler than that of *The Golden Legend*. In appearance it is similar to its predecessor, *The Story of the Glittering Plain*: its distinguishing characteristics are a page size of about $7\frac{7}{8} \times 5\frac{5}{8}$ inches (the first Flower paper folded twice), the Golden type (the only typeface Morris then had), a full border on the first page of text, and a scattering of six- and ten-line initials originally designed for *The Golden Legend* (and thus too large). Unlike the earlier book, however, *Poems by the Way* is printed in two colours, with red ink being used for a few lines and the shoulder titles – a pattern thereafter followed by Morris. Except for the second colour, it is an unimposing little book (224 pages), one of several works Morris pushed into print during 1891 and 1892 in order to use up the smaller paper which now seemed unsatisfactory to him. For a time Morris had thought of entitling the volume *Flores atramenti*.

Morris announced his plans to Reeves and Turner in June 1892: 'The title of my new book will be Poems by the Way; the *format* the same as the Glittering Plain. It will be printed in red and black. The poems will include some recently written and some written many years ago. Some have appeared in magazines, but with the two exceptions of a little piece out of the Jason and one out of the Ogier ['Ogier the Dane', not included after all], they will none of them have been printed in any book of mine.'[24] In preparing a manuscript for his printers, Morris found it necessary to seek the assistance of Charles Fairfax Murray, who, according to May Morris, 'had preserved a number of unpublished poems of early date . . . and who also had kept a record of those that had appeared in periodicals from time to time'.[25] In order to make the book slightly longer, Morris wrote one poem, 'Goldilocks and Goldilocks', especially for it on 20 May 1891.

Printing began in July but was delayed when an error was discovered in one of the sheets, which then had to be reprinted. Blunt, after lunch at Kelmscott House on 7 August, strolled with Jane Morris over to Sussex Cottage and (as he recorded in his diary) 'saw the printing. Morris's own poems were being struck off, most beautiful they are with their rubrics. The sheet I saw being printed contained the Ballad of John a Wood.' Despite some last-minute alterations in the colophon (for the purpose of improving word-spacing), printing was completed by the following month. 'I expect the book will be all printed to-morrow,' Morris wrote to Jenny on 23 September, 'and will go to the binders on Monday. They are printing the colophon sheet to-day, which is exciting.'[26] The date of publication was 20 October 1891; the next day Cockerell, who had been watching the progress of the book with keen interest (though he was not yet employed by Morris), noted in his diary: 'J R[uskin]'s Poems, edited by [W. G.] Collingwood, published as well as Morris's "Poems by the Way" beautifully printed in red and black at the Kelmscott Press. Of course no mention of the latter in the papers. . . .' The edition consisted of 300 paper copies (two guineas) and thirteen vellum copies (approximately twelve guineas); all were bound in stiff vellum covers, a few of which were experimentally stained red, yellow, blue, and green by Morris and Company at the Merton Abbey works. The book was listed as out of print in the Kelmscott Press circular of 1 December 1893, but as late as September 1897 Leighton still had eight vellum copies in its warehouse.[27]

The name of Jane Morris does not appear frequently in this or any other history of the Kelmscott Press. She took little interest in the activities of the Press and had almost nothing to do with its daily operation. However, in the case of Wilfrid Scawen Blunt's *The Love-Lyrics & Songs of Proteus* she read and criticized both manuscript and proofs, conveyed messages between her husband and the author, and in general carried the responsibilities that were later to be borne by Cockerell in overseeing the work through the Press. The explanation for this departure from her usual behaviour is that she had been, for nearly a decade, Blunt's mistress – a fact that lends peculiar colour to the story of how the *Proteus* volume came to be printed at the Kelmscott Press.

The relationship between Blunt and Morris's wife was first revealed in a biography of Blunt by Lady Longford published in 1979 and was

Figure 68 Wilfrid Scawen Blunt as a parliamentary candidate for Camberwell in 1885. Politician, explorer, poet, and womanizer, Blunt persuaded Morris to publish his poems as the third book printed by the Kelmscott Press.

described more fully a year later by Peter Faulkner in a lecture to the William Morris Society.[28] Both of these accounts are based on Blunt's voluminous diaries in the Fitzwilliam Museum, to which scholars had earlier been denied access. Despite the massive quantity of words recorded in the diaries, Blunt strikes one as an enigmatic figure whose motives and inner life defy analysis. He was in many respects an old-fashioned Tory squire, a Member of Parliament (briefly), a traveller, a writer of some reputation, and a fearless opponent of imperialism at the zenith of the British Empire. And he was devastatingly handsome and charming. Initiated as a young man into the mysteries of sex by Catherine Walters ('Skittles'), the most celebrated Victorian courtesan, Blunt thereafter romped amorously through life, leaving behind him a trail of seduced women, outraged husbands threatening divorce suits, and numerous illegitimate offspring. These adulterous affairs, even when they produced some of his most serious and intense love poetry, often had comic overtones: when Blunt was visiting Kelmscott Manor, a jealous Morris would from time to time burst into the room where Blunt and Jane were sitting, and midnight assignations at Kelmscott were made more hazardous than usual by creaking floors. Blunt's bizarre career as an amorist inevitably calls to mind the adventures of Byron's Don Juan, and it seems wonderfully appropriate that in 1869 he married Byron's granddaughter, who mysteriously tolerated her husband's erotic adventures until they at last separated in 1906.

Figure 68

It must be emphasized that Blunt was not a cold-hearted villain who had stepped out of a Victorian melodrama. He was unusually kind to women and wrote high-minded poems about his affairs, which in time were generally transformed into solid friendships – as was to be true of his relationship with Jane Morris. Throughout his life Blunt oscillated dizzily between manic hedonism and devout Roman Catholicism. After his worst debauches he often went on retreats or tried to confess his sins to any eminent Catholic churchman he encountered: Manning, Newman, on one memorable occasion even Pope Leo XIII himself. Despite frequent introspection, especially during his more religious phases, Blunt was never able to justify or explain to himself the sexual obsessions that dominated his life. The nearest he came to doing so in his diaries is this revealing passage, in which, significantly, he falls back on a gambling metaphor: 'Friendship plain and simple with a woman is to me an impossibility. It is not that in theory I should not delight in such a friend. I have often wished for one, but in practice and in spite of all possible good will I find it freezes me. If I do not make love, I don't know what to do or say and she is bored and I am bored and insensibly the friendship and all else withers away. I lose my interest in a game where there is nothing staked at all to win or lose. It is not that I want much but there must be just something.'[29]

Blunt met Jane Morris in 1883 and her husband a year later. Exactly when she joined Blunt's stable of mistresses is not clear, nor do we know precisely when their sexual relationship subsided into a generous, mutually satisfying friendship. Blunt preserved the 145 letters Jane wrote to him between 1884 and 1913, but for the most part they are oddly formal (the two lovers never address each other by first name) and tell us only what is self-evident: that she felt great affection for Blunt and was flattered by his attentions.[30] Blunt's diaries – our other chief source of information – are slightly more helpful and contain accounts of his 'love-making' (a phrase that he uses in both the Victorian and modern senses) in the Morris household. It is obvious that Blunt was attracted to Jane partly because she was one of the legendary beauties of the nineteenth century, partly because she had earlier been the mistress of his hero Rossetti, and partly because she was the wife of a man he admired very deeply. Seducing the wife of Morris may seem a peculiar way to honour him, but Blunt's mind always moved in a mysterious way.

Faulkner has suggested that the romantic entanglement reached its peak in 1889; certainly Blunt's references to Jane in his diaries during the 1890s are elegaic in tone, speaking of both her beauty and their love in the past tense. When Morris died in 1896, Blunt called at Hammersmith to pay his respects (5 October). The coffin lay in Morris's bedroom on the ground floor; upstairs he found Jane reclining on the sofa as usual, not – he emphasized – wearing black, and he kissed her 'which I had not done for long and comforted her as well as I could'. Jane's words, as recorded by Blunt, were chilling: 'I am not unhappy . . . though it is a terrible thing, for I have been with him since I first knew everything. I was 18 when I married – but I never loved him.' Blunt, always resourceful, seized the occasion to persuade Jane to join him in Egypt that winter – something he had been urging her to do for years, since he found the Levant a highly satisfactory backdrop for amorous encounters.

After the drama of that scene at Kelmscott House, the expedition to Egypt, was an anticlimax, for Jane brought along May, whom Blunt found awkward and intimidating. On 29 November Blunt made this unhappy, self-mocking entry in his diary: 'Mrs. Morris's visit has been rather a disappointment – her daughter May is an obstinately silent woman[31] and Judith [Blunt's daughter] is bored by her and I fear they are likely to be a little bored by us. Neither of them can ride, even donkeys, though Mrs. Morris has made an attempt and life without riding here is impossible. One cannot walk on account of the heavy sand, so they sit indoors all day. I go to see them twice a day but with May it is impossible to keep up any conversation and I am now at my wit's end how to amuse them for I cannot make love to either of them and what else is there to be done.' Jane and May stayed on until early April.

When we turn to the production history of *The Love-Lyrics & Songs of Proteus*, we are constantly aware of this strange triangular relationship in the background. In November 1890, before the Kelmscott Press was officially in operation, Blunt convinced Morris to print a volume of his poems, and Jane's comment on the decision (in a letter to Blunt dated 4 December 1890) captures perfectly Morris's mood of euphoria in the early days of the Press: 'My husband wrote to you consenting to print the poems with his new type. he does not begin till after Xmas – when he will first print his "News from Nowhere". if it succeeds I believe he contemplates bringing forth quartos and folios in large numbers.' Another letter (29 December) also reflects Morris's naïvety about the costs of the kind of printing he proposed to do at the Kelmscott Press: 'About the book, my husband says he thinks the cost of printing will be very little above the ordinary cost. he has not started yet, but will soon. he has taken a little place where he can set up the press.'

Morris's letter of agreement to Blunt (19 November 1890) was brisk and businesslike: 'I have put off writing to have more news first of my type-founding adventure. I think now that I shall be ready to set to work in about 3 months perhaps earlier. I shall of course be very pleased to print your book, if I may have my own way about the get up – which, I take it is what you want.' There are in fact only two known instances in which Morris succumbed to outside pressure in designing the Kelmscott books: Charles Fairfax Murray claimed that he was responsible for the format of the *Epistola de contemptu mundi*,[32] and Blunt, who had found the rubrics in *Poems by the Way* extremely attractive, persuaded Morris, against his better judgement, to print the initials of the *Love-Lyrics* in red. Morris's restrained comment, when he saw the first sheet of Blunt's book, was that 'it looks very gay &

Figure 69

SONG. YOU HAVE LET THE BEAUTY OF THE DAY GO OVER.

Song. You have let the beauty of the day go over.

YOU have let the beauty of the day go over,
You have let the glory of the noon go by.
Clouds from the West have gathered close and cover
All but a remnant now of our proud sky.

2.

Dumbly the rain beats on our darkened faces.
Hushed are the woods. Alas, for us no bird
Shall sing to-day of pleasure in green places,
No touch shall thrill, no soul of leaves be stirred.

3.

Why did we wait? What faith was ours in fortune?
What was our pride that fate should kneel to us?
Oh, we were fools. Love loves not to importune,
And he is silent here in this sad house.

4.

Alas, dear love, the day for us is ended,
The pleasure of green fields, of streams, of skies;
One hour remains, one only of joy blended
With coming night; ah! seize it ere it flies.

61

Figure 69 At the urging of Wilfrid Scawen Blunt, Morris printed the initials of Blunt's *The Love-Lyrics & Songs of Proteus* (1892) in red, but Morris was not happy with the experiment and never repeated it.

pretty with its red letters, but I think I prefer mine [i.e. *Poems by the Way*] in style of printing'.[33] It is not surprising that Blunt held strong views on fine printing, for it was a subject that had long interested him; several of his books were handsomely produced at the Chiswick Press, and for one of them, *In vinculis* (1889), Jane had designed the shamrock border which appeared on its front cover.

Morris announced to Blunt on 23 February 1891 that his type had now arrived and he could begin printing Blunt's poems. A few days later (28 February) Jane offered this report to her lover: 'The printing [of *The Story of the Glittering Plain*] is now fairly started, though the work goes very slowly with our two men, soon we shall take two more and get on much faster, your book I believe will be the third produced by the "Kelmscott Press".' Blunt's book – the partial manuscript of which he delivered to Morris in January – was to consist mainly of poems previously published in *Sonnets and Songs by Proteus* (1875) and *The Love Sonnets of Proteus* (1880), 'with [as the Kelmscott title-page phrased it] many sonnets omitted from the earlier editions'. The expressive pseudonym 'Proteus' had also been employed by Blunt in an exchange of letters about religious belief published in 1878, and in the preface to the first edition of *The Love Sonnets* he remarked, 'The author of these sonnets, styling himself Proteus, acknowledges thereby a natural mood of change.'

However, in February 1891 Blunt completed an additional sequence of thirty-one sonnets entitled 'Natalia's Resurrection' which he submitted to Morris for inclusion in the Kelmscott volume. It is a peculiar group of poems, filled to overflowing with surrealistic allegory, exotic imagery, and supernatural events. Young Adrian falls in love with a Roman matron (the setting is ambiguous but probably medieval), who rejects him because she feels an obligation to bear a child for her decidedly unpleasant husband. The baby is stillborn, Natalia dies in childbirth, and Adrian falls into deep mourning. In a dream Natalia comes to him declaring her love, for she says (and here the logic of the sequence becomes rather tenuous) that she is now free of the legal restraints of marriage; strangely, even her husband, who is also present in the dream, agrees. As Natalia and Adrian are about to marry, he awakens to discover that her funeral procession is at that moment passing by on the road. That evening Adrian goes to Natalia's grave and opens it, whereupon Natalia rises up, very much alive, and they dash off to get married. 'Natalia's Resurrection' is not an impressive poem (though on other occasions both Yeats and Pound expressed admiration for Blunt's verse), but it requires no great psychological insight to understand why it appealed so strongly to Jane Morris. Writing to Blunt on 13 April 1891, she commented, 'Yesterday I finished reading the 2nd part of your M.S. poems. I like the "Natalia's Resurrection" so very much. I should like to see that placed first in the book. What do you say?' A comment on the sonnets by Blunt's wife Anne in her diary offers a different perspective: 'Poets' theories of "love" ignore honourable feelings.'[34] Her husband, following Jane's advice, agreed to put 'Natalia's Resurrection' at the beginning of the book, and on 21 April, from a ship bound to Naples, he posted the completed manuscript to Morris.

Printing of the *Love-Lyrics* went much more slowly than Morris had anticipated: he was still naïvely optimistic about the amount of work his few printers could do on a hand-press, the rubrication required double presswork, and *The Golden Legend* was in production at the same time. Typesetting did not begin until the summer of 1891, and proofs of Blunt's volume reached him in the country by early September, but his pleasure in seeing his words set in the beauty of the Golden type was considerably diminished by an unexpected letter from one of his former lovers.

Nearly all of Blunt's love poetry reflects his relationships with particular women – *Esther*, for example, offers a fictionalized account of his prolonged, unhappy passion for Skittles – and the evidence in Lady Longford's biography suggests that Blunt's ladies were usually flattered to find their amours celebrated in graceful verse, with of course no real names or incriminating circumstances mentioned. During 1890, however, Blunt had carried on an affair with Margaret Talbot, the wife of the Military Secretary at the British Embassy in Paris, that ended abruptly when Mrs Talbot confessed everything to her husband; uncharacteristically, there were bitter recriminations on both sides.

During the summer of 1891 Mrs Talbot learned about 'Natalia's Resurrection' and concluded, no doubt rightly, that it contained indirect references to her. Like Natalia, she was married to an older man of considerable social standing but was childless; she too had turned away a lover because of loyalty to her husband. Whatever message she now sent to Blunt was sufficiently peremptory to force him into action at once. On 28 September, Blunt, his wife, and daughter spent the day in London: Blunt went to his club while the two women shopped. That evening, when Blunt did not return home,

Anne surmised correctly that he had decamped to Kelmscott Manor, where he had an urgent conference with Morris. Blunt made the following entry in his diary on 30 September (though part of the explanation he offers here appears to be a later interpolation): 'I have been away for two nights to Kelmscott to see Morris about the printing and to wish Mrs. Morris good bye for the winter. It was very perfect weather, and we did our gudgeon fishing and walks as usual, and I made a little love to Mrs. Morris, poor woman, for the quite last time. . . . We have arranged that Natalia is to be suppressed and instead of it that the old Proteus Sonnets shall be reprinted as the second part of the Kelmscott volume with certain new sonnets and some of those omitted from the earlier edition. The suppression of Natalia was decided on in consequence of a wild letter from Margaret who imagined that the poem described the relations between herself and her husband and me. And, although the poem had really nothing to do with her, she was so urgent that I could not refuse. It put Morris to great inconvenience, for Natalia was already in print, and I was obliged to explain the circumstances to him, when he at once and most good naturedly consented to the change.'

In a volume of proofs of the *Love-Lyrics* Cockerell added one of his characteristic notes indicating that it contained a portion of the cancelled 'Natalia's Resurrection' – 'with which', Cockerell wrote, 'the K.P. volumes was to have opened – but a lady, so WSB told me, pretended that the poem was about herself, and, to please her, he removed it'.[35] Though Cockerell is usually a reliable witness, we must recall that he always displayed a fierce loyalty to his friends and turned a blind eye to their faults: asked once whether he was aware of Blunt's infidelities, Cockerell replied, 'I realized that there was usually a lady in the offing. But Blunt always behaved like a gentleman: he never told a dirty story.'[36] The proofs preserved by Cockerell show that eighteen pages of type for the *Love-Lyrics* had been set in vain, and, despite his complacent note, we can assume that Morris and Blunt did not make this decision lightly or merely on the basis of an unfounded accusation.

How much did Jane Morris know about the 'Natalia's Resurrection' episode? Since she proof-read Blunt's book (something that she did for no other Kelmscott Press volume), she must have learned what had happened. Yet she never mentions it in her letters; what does emerge in her correspondence with Blunt during 1891 is her great admiration for a group of twelve poems entitled 'A Woman's Sonnets' in the *Love-Lyrics*: 'the "Woman's Sonnets" I liked immensely', she wrote to

Blunt on 11 December, 'and I think them some of the best, though I don't agree with all the sentiments therein contained.' The sequence is obviously different from others in the volume, for it describes the breakup of a love-affair from a woman's point of view. What Blunt apparently never told Jane – or the readers of the Kelmscott Press edition – was that the sonnets were in fact written by Lady Gregory, another of Blunt's former mistresses, about the change in their relationship. Lady Gregory also read the proofs.

The Love-Lyrics & Songs of Proteus – a 264-page small quarto – was finally issued on 27 February 1892 in an edition of 300 paper copies (two guineas), and it sold out very quickly. On 26 April, having received an inscribed copy from Blunt, Jane sent him her final evaluation of it: 'I think it is beautiful in all ways. I spent much time alone with it yesterday, and read again and again with the greatest pleasure my favourite poems. I admire more than ever the "Women's Sonnets" series, they flow so sweetly and smoothly, no hesitation anywhere. . . . I had no copy of your book. We bought two and gave one away, then we sent for another, but all had been sold within the fortnight of publication.'

Jane's husband, meanwhile, was supplying financial reports to Blunt from time to time. Blunt was in effect his own publisher, paying Morris for the printing and providing a fee of ten shillings to Reeves and Turner for each copy sold. On 12 March 1892 Morris was able to offer him good news: 'The 220 copies offered for sale have all been sold, and Reeves has handed me £200 on account for you[.] . . . I am charging you £150 for the printing which was a good deal more than I thought it would cost: but the book finally was much longer than I had estimated and the red printing, there being a great deal of it nearly doubled the cost of the press-work. . . . I am glad that you will make something out of the book; for when Reeves pays up the whole there will be a good balance in your favour, without reckoning the copies you keep. So I think we may congratulate ourselves on a success.' A more detailed account, sent by Morris to Blunt on 30 April, indicates that the total cost of production was £195 and that Blunt's net profit was £142.

In the second letter Morris rejoiced that the book had been well received by reviewers: 'The only discordant note on the subject was by [Joseph] Pennell of the Star. However as he is a cur by profession, one should not be annoyed at his making his livelihood by the only art he understands.' But there were also mutterings of discontent from Ellis, who took a dim view of Blunt's poetic pretensions: 'The fact is Morris is so good natured that he does not like to say "no" but is printing this book surely against the grain.'[37] The great unanswered, perhaps unanswerable, question (which lies beyond the scope of this history of the Kelmscott Press but is irresistibly fascinating nevertheless) is whether 'good natured' Morris was aware, as he designed and printed *The Love-Lyrics & Songs of Proteus*, that he was making a book for his wife's lover.

What general conclusions, if any, can we draw from the story of these three richly diverse books? *The Golden Legend* illustrates Morris's weakness for the large, sumptuous volume that he was not yet really capable of producing; impressive as *The Golden Legend* may be, it is not the book that Morris envisioned or the sort of book that he was later able to create at the Kelmscott Press. *Poems by the Way*, a much less grandiose undertaking, shows the slow evolution of Morris's typographical style while the Kelmscott books were still in an experimental stage. *The Love-Lyrics & Songs of Proteus* gives off echoes of very odd psychological resonances from the Morris marriage. If the three books have anything at all in common, it is a certain note of improvisation (sometimes urgent, sometimes amusing) in the history of their production; if we grasp this, we will perhaps avoid the temptation of treating the Kelmscott Press, with reverential solemnity, as merely a portentous episode in nineteenth-century cultural history.

EIGHT ❧ THE KELMSCOTT CHAUCER

LONG before it came off the press, the Kelmscott *Chaucer* was famous. One of the earliest references to the book (in the *Athenaeum*, 22 October 1892) struck the note of admiration bordering on adoration that recurs in nearly all future allusions to it: 'Lovers of Chaucer on both sides of the Atlantic will rejoice to hear that [Edward Burne-Jones] has made very great progress with a series of designs, fifty or sixty in all, which are to be cut in wood under his own superintendence, and intended to illustrate the "Canterbury Tales" and the other poems of Chaucer. These designs promise to be charmingly graceful and beautiful in execution. Indeed, we are confident that since, precisely a hundred years ago, Flaxman ... designed for the first Countess Spencer the famous "Illustrations to the Tragedies of Æschylus," poet and artist have never been better fitted to each other. The text to which these designs are adapted will be collated with the best manuscripts and carefully edited by Mr. W. Morris and Mr. F. S. Ellis. The typography will be worthy of the occasion, and the volume a stately quarto [*sic*].' In 1892 *The Works of Geoffrey Chaucer* was still in embryo, but as its birth neared there was widespread gossip about the astonishing folio in production at the Kelmscott Press. A New York periodical, the *Book Buyer* (18 November 1895), published a facsimile of the first page of *The Canterbury Tales* and declared that the *Chaucer* 'will be without question the grandest book of the century'; in 1895 some sheets of the *Chaucer* were shown at the Arts and Crafts Exhibition in Dublin, prompting the *Studio* (January 1896) to describe the book, not yet published, as holding 'the place of honour among the other productions of this press'.

Figure 70 Morris's associates did not hesitate to bestow superlatives upon it. 'It's the finest book ever printed and it's more than likely there'll never be such another,' Burne-Jones said. Ellis argued that 'for typography, ornament and illustrations combined', the *Chaucer* was 'the grandest book that has issued from the press since the invention of typography'.[1] Many outside the Morris circle have repeated this extravagant language: a reviewer in the *Library* (1896), for example, described it as 'the finest book ever printed', and the *Athenaeum* (3 October 1896) called the *Chaucer* 'the crowning achievement of the Kelmscott Press'.

The Kelmscott Chaucer

Even in the second half of the twentieth century there is still a tendency to abandon all rhetorical restraint when discussing the Kelmscott *Chaucer*, as evidenced by this passage from an American library exhibition catalogue of 1976: 'This is one of the two greatest ikons of the bibliographic faith, the other being the Gutenberg Bible. As the cross of the Christian community proclaims by its very presence the gospel of Christ, so the Kelmscott Chaucer on vellum stands surrogate for all that was important in the book world before it was completed in 1896. . . . While there have been many litanies of praise heaped upon William Morris, and upon the Kelmscott Chaucer on vellum in particular, none of them have been adequate. Not until one has had the opportunity to go through a copy does the full majesty of the Kelmscott accomplishment appear. . . . The ultimate limitations of words are reached when one tries to describe the Kelmscott Chaucer on vellum.'[2] And so on for several more breathless paragraphs.

Naturally such ecstatic approbation invites a cynical response, and there have always been some readers willing to argue that the Kelmscott *Chaucer* is a ponderous, overdecorated non-book which threatens the eyesight and physical safety of anyone foolhardy enough to hold it on his lap (a vellum copy in the pigskin binding weighs more than thirteen pounds). We have seen that the *Chaucer* is far from being a typical Kelmscott volume, but it is certainly the most celebrated and frequently photographed, and it represents, in a sense, Morris's tendencies in book-design and ornamentation carried to their logical conclusion; thus, rightly or wrongly, the *Chaucer* has become a litmus test which measures one's response to Morris's work as a printer.

That the most ambitious, richly ornamented book of the Kelmscott Press proved to be an edition of Chaucer came as no surprise to any of Morris's admirers, for he and Burne-Jones had counted Chaucer as one of their great discoveries at Oxford, and throughout his life Morris's poetry was constantly being compared by critics with that of his medieval predecessor.[3] Though Morris himself protested occasionally against the injustice of the analogy, he had only himself to blame for his identification in the public mind with Chaucer: he had, after all, addressed Chaucer as 'my Master' in both Book XVII of *The Life and Death of Jason* and the 'Envoi' to *The Earthly Paradise*, and the latter poem was plainly patterned after *The Canterbury Tales*. Still, Morris remained slightly uneasy about the comparison, and it is noteworthy

Two fyres on the auter gan she beete,
And dide hir thynges, as men may biholde
In Stace of Thebes, and thise bookes olde.
Whan kyndled was the fyr, with pitous cheere,
Unto Dyane she spak, as ye may heere.
O CHASTE goddesse of the wodes grene,
To whom bothe hevene and erthe & see is sene,
Queene of the regne of Pluto derk and lowe,
Goddesse of maydens, that myn herte hast knowe
ful many a yeer, and woost what I desire,
As keepe me fro thy vengeaunce and thyn ire,
That Attheon aboughte cruelly.
Chaste goddesse, wel wostow that I
Desire to ben a mayden al my lyf,
Ne nevere wol I be no love, ne wyf.
I am, thow woost, yet of thy compaignye
A mayde, and love huntynge and venerye,
And for to walken in the wodes wilde,
And noght to ben a wyf and be with childe;
Noght wol I knowe the compaignye of man.
Now helpe me, lady, sith ye may and kan,
For tho thre formes that thou hast in thee.
And Palamon, that hast swich love to me,
And eek Arcite, that loveth me so soore,
This grace I preye thee withoute moore,
As sende love and pees bitwixe hem two,
And fro me turne awey hir hertes so,
That al hire hoote love and hir desir,
And al hir bisy torment and hir fir,
Be queynt, or turned in another place.
And if so be thou wolt do me no grace,
Or if my destynee be shapen so
That I shal nedes have oon of hem two,
As sende me hym that moost desireth me.
Bihoold, goddesse of clene chastitee,
The bittre teeres that on my chekes falle.
Syn thou art mayde, and kepere of us alle,
My maydenhede thou kepe and wel conserve,
And whil I lyve a mayde, I wol thee serve.
THE fires brenne upon the auter cleere
Whil Emelye was thus in hir preyere;
But sodeynly she saugh a sighte queynte,
For right anon, oon of the fyres queynte
And quyked agayn, and after that, anon
That oother fyr was queynt, and al agon,
And as it queynte it made a whistelynge,
As doon thise wete brondes in hir brennynge;
And at the brondes ende out ran anoon
As it were blody dropes many oon;

Figure 70 Though many of the earlier Kelmscott Press books had been illustrated,

for which so soore agast was Emelye,
That she was wel ny mad, and gan to crye,
for she ne wiste what it signyfied;
But oonly for the feere thus hath she cried,
And weepe, that it was pitee for to heere.
And therwithal Dyane gan appeere,
With bowe in honde, right as an hunteresse,
And seyde, Doghter, stynt thyn hevynesse.
Among the goddes hye it is affermed,
And by eterne word writt and confermed,
Thou shalt ben wedded unto oon of tho
That han for thee so muchel care and wo;
But unto which of hem I may nat telle.
farwel, for I ne may no lenger dwelle.
The fires whiche that on myn auter brenne
Shulle thee declaren, er that thou go henne,
Thyn aventure of love, as in this caas.
AND with that word the arwes in the caas
Of the goddesse clateren faste & rynge,
And forth she wente, and made a vanysshynge;
for which this Emelye astoned was,
And seyde, What amounteth this, allas!
I putte me in thy proteccioun,
Dyane, and in thy disposicioun.
And hoom she goth anon the nexte weye.
This is theffect, ther is namoore to seye.
THE nexte houre of Mars folwynge this,
Arcite unto the temple walked is
Of fierse Mars, to doon his sacrifise,
With alle the rytes of his payen wyse.
With pitous herte and heigh devocioun,
Right thus to Mars he seyde his orisoun:
O STRONGE god, that in the regnes colde
Of Trace honoured art and lord yholde,
And hast in every regne and every lond
Of armes al the brydel in thyn hond,
And hem fortunest as thee lyst devyse,
Accepte of me my pitous sacrifise.
If so be that my youthe may deserve,
And that my myght be worthy for to serve
Thy godhede, that I may been oon of thyne,
Thanne preye I thee to rewe upon my pyne.
for thilke peyne, and thilke hoote fir,
In which thou whilom brendest for desir,
Whan that thou usedeste the beautee
Of faire, yonge, fresshe Venus free,
And haddest hire in armes at thy wille,
Although thee ones on a tyme mysfille,

only in the *Chaucer* (1896) were the pictures (by Burne-Jones) and text fully integrated.

that he declined to write an introduction to Chaucer for T. H. Ward's *The English Poets* in 1878. Not only did he feel ill-equipped to supply such an introduction, but Morris also objected to Ward's plan to print merely a small selection from Chaucer's works.[4]

Behind the Kelmscott *Chaucer* lies a long lineage of Chaucer manuscripts and printed editions, many of them illustrated, with which Morris, as a collector and antiquarian, was well acquainted.[5] The practice of wedding picture with Chaucer's text extends back to the earliest manuscripts, the most celebrated of which is the fifteenth-century Ellesmere manuscript of *The Canterbury Tales* (now in the Huntington Library) containing twenty-three portraits of the individual pilgrims, each mounted and placed in the margin near the decorated initial at the beginning of his tale. The tales themselves are not illustrated; this convention, which may seem rather odd to us today, was observed by all illustrators of Chaucer until the late-eighteenth century, and only in the Kelmscott edition is there at last a serious attempt to represent the action of the pilgrim's stories beyond a few simple tableaux. Burne-Jones's departure from this iconographic pattern is so complete, in fact, that he shows us none of the pilgrims except Chaucer himself (and he is not mounted), nor does he draw the pilgrims in procession or at table in the Tabard Inn, two other conventions of Chaucer's early illustrators.

The portrait of Chaucer in the Ellesmere manuscript is one of only two presumed to be reliable likenesses; the other is known as the Hoccleve portrait, also found in an early manuscript. A third likeness of Chaucer which has captured the imagination of later generations is an early fifteenth-century painting of Chaucer reading his poems to the royal court – a theme that recurs in a mid-Victorian Pre-Raphaelite representation by Ford Madox Brown. It is remarkable that, even though these images of Chaucer had been frequently engraved and were familiar to Victorian readers, Burne-Jones scarcely alludes to them in his illustrations for the Kelmscott *Chaucer*. His portrait of Chaucer does show a general indebtedness to both the Hoccleve and Ellesmere likenesses; there are broad similarities in pose and costume, the outstretched hand, even the forked beard. Yet Burne-Jones's Chaucer is more gaunt and solitary, more inward-looking (he is
Figure 71 reading a book), and more visionary. One notices that by the end of *The Canterbury Tales* he has aged perceptibly, and by the last page of the volume he looks quite decrepit.

The first printed edition of *The Canterbury Tales* was issued by

And loveth him, the which that right for love
Upon a cros, our soules for to beye,
First starf, and roos, and sit in hevene above;
For he nil falsen no wight, dar I seye,
That wol his herte al hoolly on him leye.
And sin he best to love is, and most meke,
What nedeth feyned loves for to seke?

Lo here, of Payens corsed olde rytes,
Lo here, what alle hir goddes may availle;
Lo here, these wrecched worldes appetytes;
Lo here, the fyn and guerdon for travaille
Of Jove, Appollo, of Mars of swich rascaille!
Lo here, the forme of olde clerkes speche
In poetrye, if ye hir bokes seche.

O moral Gower, this book I directe
To thee, and to the philosophical Strode,
To vouchen sauf, ther nede is, to corecte,
Of your benignitees and zeles gode.
And to that sothfast Crist, that starf on
rode,
With al myn herte of mercy ever I preye;
And to the Lord right thus I speke and seye:

Thou oon, and two, and three, eterne on lyve,
That regnest ay in three and two and oon,
Uncircumscript, and al mayst circumscryve,
Us from visible and invisible foon
Defende; and to thy mercy, everichoon,
So make us, Jesus, for thy grace digne,
For love of mayde and moder thyn benigne!
Amen.
Explicit Liber Troili et Criseydis.

Figure 71 The last page of the *Chaucer* displays Morris's printer's mark and a wood-engraving of an old, enfeebled Chaucer.

Caxton in 1478, followed by a second edition with illustrations and an improved text in 1484. The crude woodcuts in the latter, the fourth illustrated book published in English, followed the pattern of the Ellesmere manuscript except that each mounted pilgrim appears twice: once in the General Prologue, and once at the beginning of his tale. Caxton also added a charming picture of the Canterbury pilgrims seated at an enormous round table presided over by Harry Baily, landlord of the Tabard Inn. Another fifteenth-century edition of the *Tales* (Richard Pynson, 1491 or 1492) contains a smaller and less striking representation of the same scene, and it reappears – in a very Augustan style, with Harry Baily looking like Dr Johnson in a spirited discussion at a coffee house – in John Bell's edition of Chaucer (1782). Caxton's other Chaucerian woodcuts were also imitated and adapted by later printers for several centuries.

Figure 72

Figure 72 Caxton's second edition of the *Canterbury Tales* (1484) includes this woodcut of Chaucer's pilgrims seated around a table at the Tabard Inn. Burne-Jones ignored all such precedents when he created his illustrations for the Kelmscott *Chaucer*.

The Kelmscott Chaucer

John Urry's edition (1721) brings together, for the first time, all the mounted pilgrims and displays them at the moment of their departure from the Tabard. This edition appears to have been a financial and critical failure: readers disliked Urry's textual emendations and – as a symptom of the growing nostalgia for all things medieval – they regretted the abandonment of Gothic for Roman type. (One of the sponsors of the book, Christ Church, Oxford, was able to dispose of its remaining copies only by forcing every undergraduate to buy one when he entered the college.) In Bell's edition there is also an engraving by Thomas Stothard of a small group of pilgrims on horseback, and early in the nineteenth century both Stothard and Blake produced large-scale engravings – not intended as book illustrations – of Chaucer's pilgrims leaving the Tabard.

What is most noticeable about all the eighteenth-century editions of Chaucer is how abruptly they depart from the medieval spirit of the earlier blackletter versions. To read *The Canterbury Tales* in, for instance, Urry's Roman type, with neo-classical initials and ornaments and an extraordinarily anachronistic Squire, is a very different experience from that of studying a typical sixteenth-century page in Gothic type, illustrated with a woodcut of a pilgrim who is lineally descended from Caxton and the manuscripts before Caxton. Bell's fourteen-volume edition of Chaucer has been praised for the excellence of its typography by Stanley Morison, and it deserves our attention also as the first attempt to illustrate the stories of *The Canterbury Tales*; but however admirable Bell's printing may be, one must recognize that it was precisely this eighteenth-century rendering of Chaucer – made over into the image of an Augustan gentleman of impeccable manners – that Morris attempted to obliterate at the Kelmscott Press. Morris's aim was to recapture, without direct copying, the simplicity and vitality of fifteenth- and sixteenth-century editions of Chaucer that were closer in time and spirit to Chaucer's own age. Morris believed that only by peeling off the Renaissance and neo-classical layers of cultural interpretation could he recover something like the Chaucer of the Middle Ages, and this involved both a careful restoration of Chaucer's text (including his spelling) and a return to a more medieval style of typography and ornamentation.

The Kelmscott *Chaucer* is in fact the splendid culmination of the Victorian rediscovery of Chaucer and of early English literature in general. The person who was largely responsible for this revival was Morris's friend F. J. Furnivall, who founded, among others, the Early

English Text Society (1864), the Ballad Society (1868), the New Shakspere Society (1873), and the Wyclif Society (1882). Furnivall created the Chaucer Society in 1868 in order (according to its prospectus) 'to do honour to Chaucer, and to let the lovers and students of him see how far the best unprinted manuscripts of his works differ from the printed texts'. The Chaucer Society was never a large organization – Furnivall frequently complained of its meagre membership in the annual reports, and in 1900 Ellis observed sarcastically that by then only sixteen private individuals belonged, 'about one in every two million and a quarter of the population of the British Isles'[6] – and it was disbanded shortly after Furnivall's death in 1910. Nevertheless, during the four decades of its existence the Society laid the foundations of modern Chaucer scholarship by publishing early manuscript versions of Chaucer's works and other primary materials, mostly edited by Furnivall himself, a man of unflagging energy. For one of these publications, a Chaucerian manuscript in the Cambridge University Library, W. H. Hooper, who was to engrave Burne-Jones's designs at the Kelmscott Press, made facsimile engravings of the original illustrations.[7] The most important achievement of the Society, issued in instalments over several years from 1868 onwards, was Furnivall's parallel-text edition of the six Chaucer manuscripts judged to be the most trustworthy. Like Morris, Furnivall believed passionately in the importance of digging down through the centuries of accumulated debris in order to find Chaucer's actual words rather than the glosses and conjectural emendations of later editors; like Morris, too, he took a relatively uncomplicated view of Chaucer, emphasizing what he called 'the most genial and humourful healthy-souled man that England had ever seen'.[8] Morris, of course, never remoulded Chaucer into a Muscular Christian as Furnivall did, but both men tended to ignore Chaucer the subtle Italianate poet, sometimes introspective and melancholy, seeing him instead as a cheerful realist whose value to modern readers lay chiefly in his portrayal of medieval life and manners.

During the 1860s the University Presses of both Oxford and Cambridge were laying plans for scholarly editions of Chaucer, and, although Furnivall hoped that the former might print his six-text version, Oxford decided instead to publish a single text based on a collation of all the authoritative manuscripts, to be edited by the Revd Walter W. Skeat, Professor of Anglo-Saxon at Cambridge. Skeat was by no means an ideal editor: following the example of Furnivall (who was in many respects his mentor),[9] Skeat constructed a text based on intuitive, eclectic principles and drawn exclusively from Furnivall's six-text edition. Both Ellis and Burne-Jones commented on the donnish humourlessness of Skeat's notes. 'Every student of Chaucer must be grateful for the labour and learning bestowed on his works by Professor Skeat, but to be a parson without any sense of humour is not an advantage for an editor of Chaucer,' Ellis remarked wryly. Burne-Jones said to Sebastian Evans, 'Skeat is a bore, I must say. He never edits a poem without saying it's dull and unimportant and long-winded. How does he know and what business is that of his – if it's unimportant what does he edit it for? He is himself always long-winded.'[10]

Yet Skeat's monumental six-volume edition of Chaucer – published under Oxford's Clarendon Press imprint in 1894, with a supplementary volume issued in 1897 – seemed a godsend to Ellis and Morris, for it was incomparably superior to all previous editions and relieved Ellis of the heavy burden of constructing his own text for the Kelmscott version. Writing to the University Librarian in April 1892, Ellis declared, 'I shall also take the opportunity of calling on Mr. Skeat if he is in Cambridge to hear what he says as to the text of the projected Kelmscott Press folio Chaucer.'[11] Skeat shared Ellis's enthusiasm for the planned Kelmscott edition, but he could not by his own authority permit Ellis and Morris to reprint the text he was preparing for the Clarendon Press, and, not surprisingly, the Delegates of the Press displayed reluctance to allow Morris to produce an edition of Chaucer that would be potentially in competition with their own. The Delegates first considered the request in November 1892, and their initial response was negative: 'With regard to the Kelmscott *Chaucer*,' C. E. Doble wrote to Skeat, 'they do not think that you would gain anything by permitting Mr. Morris to reprint your text, and they are not disposed to recommend you to sanction the use which it is proposed to make of it. Would Mr. Pollard's text be good enough for Mr. Morris' purpose? It is difficult to believe that the adoption of your text by Mr. Morris would increase the sale of your books by a single copy.'[12]

Skeat suggested to Morris, as an alternative, that he might reprint Caxton's first edition of Chaucer, and meanwhile Ellis continued to study various Chaucerian texts; for *The Canterbury Tales* he intended to adopt the version found in the Ellesmere manuscript, but the other poems presented more complex editorial problems, and neither he nor Morris was willing to abandon their hope of using Skeat's text. At the meeting of the Delegates of the Press on 26 May 1894, therefore, the request was renewed, again without success. The minutes record the following action: 'Letter read from the Rev. Profr. W. W. Skeat proposing that the Delegates should grant permission to Mr. Wm. Morris to reprint the Oxford Text of "Chaucer" amongst his Kelmscott Editions on condition of a contribution being made by him to the proposed English Lectureship Funds at Cambridge Resolved that the Delegates think that it would be injurious to the Clarendon Press edition to accede to Mr. Morris's proposal.'

Morris refused to admit defeat. Aided by the intervention of Dr Arthur Evans (Keeper of the Ashmolean Museum, Oxford), he submitted the request yet again, but this time with a fuller statement of his intentions: 'May I ask you to be so kind as to lay before the Delegates the request to allow me to make use of Professor Skeat's Chaucer under the following circumstances,' he wrote to the Secretary of the Clarendon Press on 9 August 1894. 'I am publishing a Chaucer, which is being printed at the Kelmscott Press, & will be ornamented with picture[s] designed by Sir Edward Burne-Jones, and borders &c by myself. This book I hope to make a specially beautiful one as to typography & decoration, and I naturally wish to make the text as good as possible. I ask therefore to be allowed to avail myself of the corrections of errors which Professor Skeat's learning & acumen are making public, though we (Mr F. S. Ellis, the editor, & myself) by no means intend to produce a literal reprint of his text. We should be pleased to make full acknowledgment of our obligations to Professor Skeat in the book itself.

'I should add that it is impossible that this book could come into competition with the text now being published at the University Press, as only 325 copies will be issued, at a high price (£20), & without notes or commentary. It is intended to be essentially a work of art.'[13]

Morris explained in a second letter that he was willing to wait until the October meeting of the Delegates for his proposal to be considered. 'I do not think there is any hurry', he wrote, 'for I have only just begun printing, and it will be some weeks before I am far on in the Canterbury Tales, over which I take it no question could arise as we have the advantage of good MS.S. (the Ellesmere is our foundation) and the emendations are few & slight. It would [be] chiefly in the poems where there is no 15th century M.S. such as the House of Fame where we should need help.'[14] Finally, on 19 October, the Delegates gave Morris permission 'to reproduce in an illustrated edition de luxe of the Chaucer to be printed at the Kelmscott Press a limited number of readings from Prof. Skeat's Text of "Chaucer" as printed in the Library Edition, provided due acknowledgement of the sanction of the Editor and the Delegates be made'.

Such an acknowledgement eventually appeared in the colophon of the Kelmscott *Chaucer*: 'The hearty thanks of the Editor and Printer are due to the Reverend Professor Skeat for kindly allowing the use of his emendations to the Ellesmere MS. of the Canterbury Tales, and also of his emended texts of Chaucer's other writings. The like thanks also the Editor and Printer give to the Delegates of the Oxford University Press for allowing them to avail themselves of Professor Skeat's permission.' But Morris's claim that he and Ellis did not intend 'to produce a literal reprint of [Skeat's] text' was disingenuous, for their marked copy of the Skeat edition (which is owned today by the Beinecke Library, Yale University) discloses that they very rarely strayed from his textual readings. Except for *The Canterbury Tales*, the Kelmscott *Chaucer* is, for all practical purposes, a reprint of Skeat's edition, but shorn of the pedantic notes that irritated Ellis and Burne-Jones. 'And I want particularly to draw your exceptionally sweet attention to the fact that there is no preface to Chaucer, and no introduction, and no essay on his position as a poet, and no notes, and *no* glossary,' Burne-Jones announced to one of his female correspondents when the book was published. 'So that all is prepared for you to thoroughly enjoy him – unvexed by the impertinence of the intruder.'[15] The thwarted intruder in this case was not Ellis but a solemn clerical professor in Cambridge.

It was not merely the search for a suitable text that kept the *Chaucer* in production for so long; all aspects of the book – paper, ink, ornaments, illustrations, binding – presented formidable technical problems, if only because everything was on a much larger scale than for any other Kelmscott Press volume. Even *The Golden Legend* and *The Earthly Paradise* were minor undertakings by comparison.

An edition of Chaucer had been on Morris's mind from the earliest period of the Kelmscott Press's history. 'He thinks of printing a Chaucer from a blackletter fount which he hopes to design,' Cockerell recorded in his diary on 11 June 1891. In the spirit of optimism which characterized the Press's announcements in those days, Morris intended to advertise the *Chaucer* in the first list distributed in May 1892, but at the last moment he deleted this line from the proof:

The Kelmscott Chaucer

'THE POEMS OF GEOFFREY CHAUCER. Black Letter. Large 4to.'[16] In a list issued the following December, however, the book was at last announced: 'Chaucer's Works. With about 60 designs by E. Burne-Jones.' The reason for the postponement is that Morris had reluctantly reached the conclusion, early in 1892, that the Troy type was too large for the projected *Chaucer*. When the type was first delivered from Sir Charles Reed and Son in January, Morris immediately had trial passages from Chaucer set up, since he then hoped the Troy would be satisfactory for the *Chaucer*, but Ellis, in a letter written in March, explained Morris's dilemma: 'I enclose you a specimen of the new type in which as you see Caxton's "Recuyel" is to be & Chaucer is to follow but in a size smaller type but same pattern – and in double col[umn]s. To print Chaucer in this size w[oul]d be delightful but I fear buyers would turn resentful[?] at the size which the book would necessarily make.'[17] Morris's own bemused observation on the Troy type was that 'it is so big that it is no use thinking of printing the Chaucer in double columns with it unless the book were to be as big as Eggestein's Gratian's Decretum [1471]' – an elephantine volume weighing nearly thirty pounds.[18]

Morris was never happy with the reduced Chaucer type, which was in his hands by the summer of 1892: 'I was driven by the necessities of the Chaucer (a double-columned book) to get a smaller Gothic type of Pica size,' he explained apologetically in his *Note* on the Kelmscott Press.[19] Blunt, visiting Kelmscott Manor on 5 July 1892, found himself in the midst of 'a great discussion as to whether it was to be printed in single or double columns'. Double column it was, as Morris had foreseen from the beginning, and during July trial settings were struck off in the Chaucer type, at least one of which was completely unleaded.[20] A new trial page established the basic design of the book: a folio (on the Perch paper) with two columns of sixty-three lines each. A note by Morris on one of the few surviving proofs of the *Chaucer* indicates that the book was to some extent modelled after a specimen page, already set up, of the Froissart volume which was never completed at the Kelmscott Press.[21] With a trial page in hand, Morris was able to work out a remarkably accurate estimate of the book's length (544 pages, just a few pages short of the size it proved to be); the same document, on which he listed the proposed sequence of Chaucer's works, suggests that he may have originally intended to print them in their order of composition.[22] But the arrangement that he eventually adopted was, as he said, 'purely architectural'.[23]

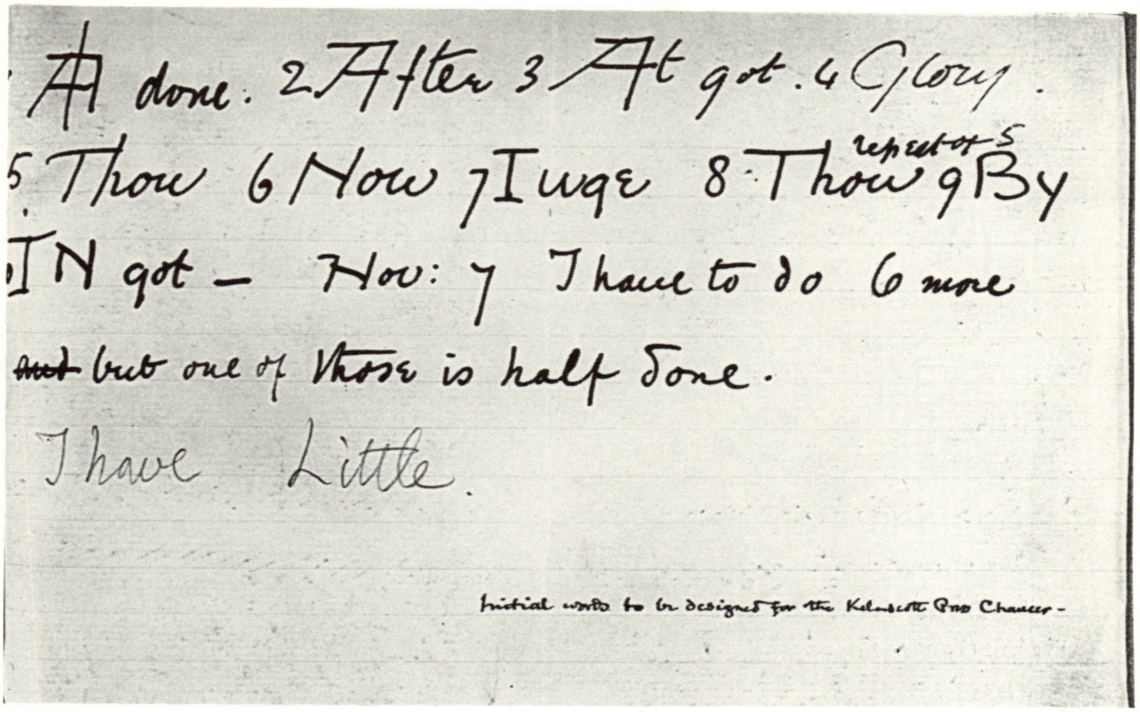
A done. 2 After 3 At got. 4 Glory.
5 Thow 6 Now 7 Iuge 8 Thow repeat of 5 9 By
IN got — Now: 7 I have to do 6 more
~~and~~ but one of those is half done.
I have Little.
Initial words to be designed for the Kelmscott Press Chaucer -

Figure 73 A list in Morris's hand of ornamented initials to be designed for the *Chaucer*.

The Kelmscott Chaucer

Morris began designing the borders for the *Chaucer* on 1 February 1893, and for the first text page of *Canterbury Tales* – which is overwhelming in its richness of effect – Hooper engraved the border, the frame for the illustrations, the large initial word ('Whan'), and the hand-lettered title on a single block of wood. During the next three years Morris's prolific brush produced one ornament after another for the book, all on the same massive scale: according to Cockerell's concise summary, he designed 'a woodcut title, fourteen large borders, eighteen different frames round the illustrations, and twenty-six large initial words'[24] – a list that does not include the scores of decorated initials, since many of them had been used already in earlier books. It was a staggering accomplishment, and it took its toll on Morris's health. By the spring of 1896 Burne-Jones was terrified that Morris might die before he could complete the engraved title-page: 'the Chaucer title page is done at last,' Burne-Jones commented on 28 March. 'So that's all safe now, and there's nothing but the printing' (Rooke diary).

Figure 73

Figure 74

Printing of the first sheet of the *Chaucer* started on 8 August 1894: 'BEGAN PRINTING THE CHAUCER,' Cockerell wrote in his diary in uncharacteristic capitals. Morris also recognized the momentousness of the occasion: 'we printed the 1st side of the Chaucer on vellum; successfully, I think, after some trouble,' he reported to Jenny.[25] Towards the end of 1894 Morris acquired another Albion press – this one very large and reinforced with steel bands in order to withstand the additional pressure of printing his large borders – and rented Sunderland Cottage, 21 Upper Mall, to hasten the progress of the book. 'Morris is very busy,' Charles Fairfax Murray wrote in November. 'He is just setting up a new press to print the Chaucer faster, which he calculated would [take] him otherwise 2 years to print.'[26] During the spring and summer of 1895 the book fell several months further behind schedule when yellow stains began to appear on some of the sheets already printed (see Chapter Four). Burne-Jones was, as always, dilatory in supplying his designs for the illustrations, and the difficulties that arose in printing the pictures and ornaments were legion. Hooper inadvertently caused a last-minute delay when he made the engraved title too small, leaving an excessively broad white gap between it and the surrounding border, and Morris – with his strength rapidly failing – had to extend the design by about a fifth of an inch.[27] The colophon was dated 8 May 1896, and on 20 May T. M. Rooke recorded in his diary Burne-Jones's relieved response: 'Mr

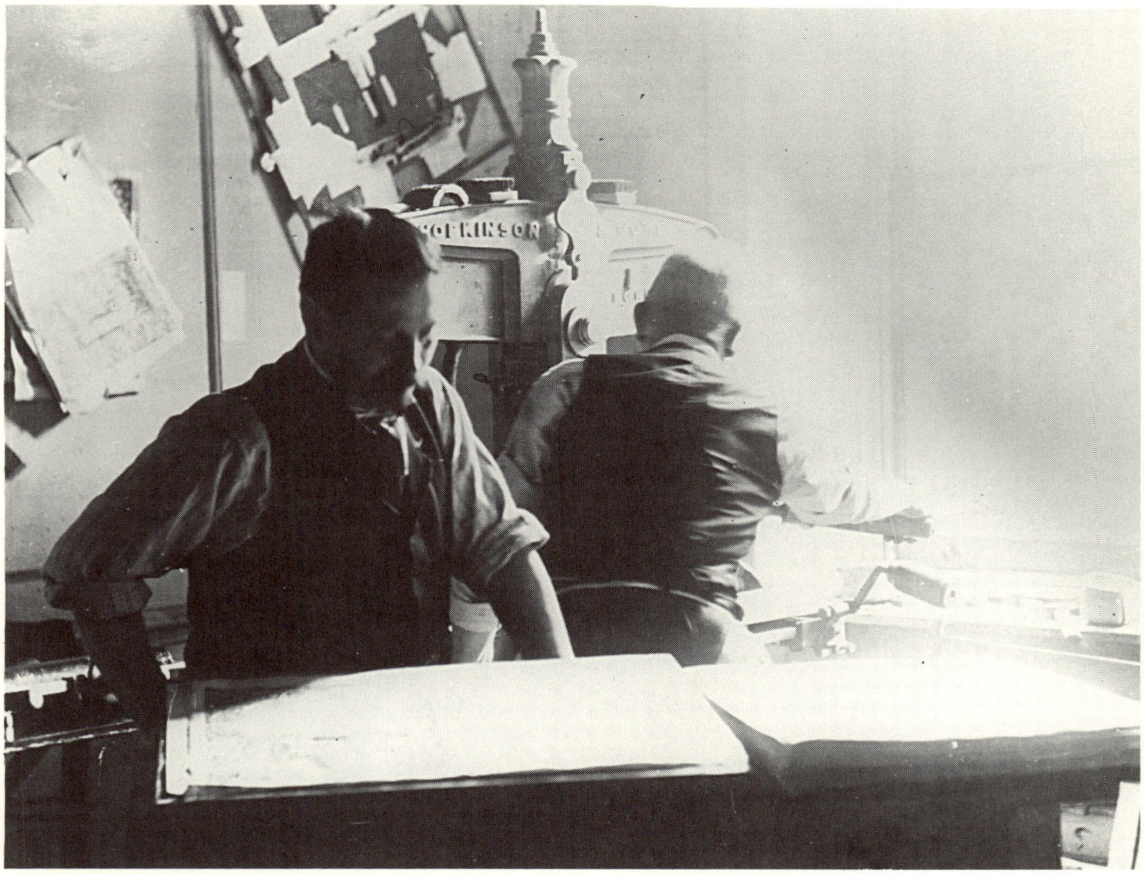

Figure 74 Two of Morris's pressmen, William H. Bowden and W. Collins, at work on the *Chaucer*.

Morris didn't look well yesterday, but the printing of the Chaucer's every bit of it done and in a month it'll be all dry and tightly bound. They daren't bind it heavily as yet or the print would come off.'

Most of the copies of the *Chaucer* were issued in the standard Kelmscott quarter-linen binding with blue paper-covered boards, but *Figure 75* for forty-eight of the books Morris designed a pigskin binding – patterned after a binding by Ulrich Schreier of Salzburg (1478) that he owned[28] – which was executed at the Doves Bindery. (Cobden-Sanderson also produced a binding for two copies of the *Chaucer* tooled from a more delicate design of his own.) A circular distributed by the Kelmscott Press in February 1896 announced that Morris would design four *Chaucer* bindings in all: (1) a full pigskin from the Doves Bindery, (2) a half pigskin from the Doves Bindery, (3) a full pigskin from J. and J. Leighton, and (4) a half pigskin from Leighton. Although Cockerell believed that Morris was able to carry out only the first of these schemes, there is a dummy copy of the *Chaucer* in Columbia University's Butler Library that seems to match the description of the third binding announced in the circular – i.e. a full pigskin executed by Leighton and evidently designed by Morris. A note in the book indicates that it is a 'specimen copy of the Kelmscott Chaucer 1896, made up from odd sheets [mostly pages 39–42, which were obviously cancelled] and bound by J. & J. Leighton, to the order of William Morris; carried out in thorough 15th century style oak boards, covered with pigskin and elaborately ornamented in blind; the tools specially cut after the original designs as found on the Apocalypse Block Book and the Richel Bible, c. 1475 . . . '. The title on the spine is stamped in gilt. It is likely that this less expensive pigskin binding was never produced by Leighton either because Morris was dissatisfied with it or because the dummy was completed when he was too ill to make a decision.[29]

The typography, ornaments, and binding of the Kelmscott *Chaucer* are indeed sumptuous – the book is 'essentially a work of art', as Morris predicted – but a special source of its appeal for many readers has always been the eighty-seven wood-engraved illustrations by Edward Burne-Jones. In the earliest announcement of the *Chaucer* (December 1892) it was said that there would be 'about 60 designs', but that number crept upwards inexorably as time went by. In a little trial *Figure 76* dummy of eleven leaves stitched together, on which Burne-Jones sketched out his preliminary plans for the pictures, is this note in the artist's hand: 'There are 48 planned here but I may add to the Knights

Figure 75 The special pigskin binding for forty-eight copies of the Kelmscott *Chaucer* was designed by Morris and executed at the Doves Bindery.

Figure 76 Burne-Jones made some of his preliminary sketches for the *Chaucer* illustrations in a dummy book consisting of eleven leaves stitched together, of which the first page is shown here. In the upper right corner he wrote this note to himself: 'There are 48 [illustrations] planned here but I may add to the Knights tale & the early part – there may be 60 in all if I like.' By the time the book was published in 1896, there were in fact eighty-seven wood-engravings.

tale & the early part – there may be 60 in all if I like.'[30] Burne-Jones wrote on another sheet, '72 in all and he [presumably Morris] won't have more'.[31] Emery Walker claimed, in a later lecture on Morris, that the number of illustrations continued to grow as long as the volume was in production: 'When the book was printed no one, not even Burne-Jones, knew how many pictures there were in it. He had gone on putting in as many as he liked, knowing that the more he drew the better Morris would be pleased. If the artist liked the story particularly, then he did more pictures. As the volume neared its end the number of illustrations increased; when they were counted after the first two volumes were bound it was found that a copy contained 87.'[32] Was Walker exaggerating slightly? He did have a weakness for colourful anecdotes, and it seems improbable that, even at the Kelmscott Press, where improvisation ruled supreme, no one had counted the illustrations until the book was bound.

But Walker was undoubtedly correct in his assertion that Burne-Jones illustrated those stories most abundantly which appealed to his own taste. It was Chaucer the poet of courtly love and sophisticated teller of French and Italianate tales who most interested Burne-Jones (not the sturdy medieval bard that Morris found in *The Canterbury Tales*), and it was accordingly the stories of the Knight, the Clerk of Oxenford, and the Franklin which attracted him; the tales of bawdy realism he passed over in silence. 'I have abstained from decorating certain of the Canterbury Tales,' he wrote to Swinburne. '. . . Morris has been urgent with me that I should by no means exclude these stories from our scheme of adornment – especially he had hopes of my treatment of the Miller's Tale, but he ever had more robust and daring parts than I could assume.'[33] When his studio assistant congratulated him for ignoring Chaucer's 'base stories', Burne-Jones replied, 'I'd like to pretend Chaucer didn't do them. Besides, pictures to them would have spoiled the book. You don't want funny pictures either. Pictures are too good to be funny. Literature's good enough for that.'[34] In private Burne-Jones did have a taste for ribald comedy: his notebooks and letters are filled with lively caricatures, mainly of emaciated, terrified men threatened by grotesquely obese women. But his public art was reserved for high seriousness, which means that the *Chaucer* is populated with the usual mysterious human figures – willowy and graceful, intense but expressionless – that are so familiar from his paintings. It is certainly not true, as one recent commentator has claimed, that Burne-Jones 'ruined' the Kelmscott *Chaucer* by this partisan approach to illustrating it,[35] yet it would be fair to say that his wood-engravings, by ignoring Chaucer's humour and irony, convey a very incomplete interpretation of the poetry.

Because the older English woodcut illustrations of Chaucer held no interest for him, Burne-Jones was compelled to turn to other visual sources. We are told that he and Morris regularly pored over illuminated manuscripts at both the Bodleian Library and the British Museum, and an exhaustive study of the manuscripts he could have seen there would probably turn up a few miniatures from which he borrowed or adapted some details in his *Chaucer* illustrations. There is (to choose a more remote example) a miniature in a manuscript of Boccaccio's *Teseida* which is remarkably similar to the wood-engraving which appears on page 9 of the *Chaucer* to illustrate the Knight's Tale: it is not clear whether or how he saw this particular manuscript (which is now in the national library of Austria), but probably the miniature represents an iconographic tradition which Burne-Jones encountered in other manuscripts as well.[36]

Figure 77 He was also able to draw upon his own earlier work. In 1857 Burne-Jones painted scenes from the Prioress's tale on a wardrobe as a wedding gift to the Morrises, of which the left panel is almost identical to the illustration on page 60 of the Kelmscott *Chaucer*. (Another version of this scene appears in the last work Burne-Jones completed, a water-colour exhibited at the New Gallery in 1898.) On the right panel of the wardrobe Chaucer himself is portrayed, looking very much like the Ellesmere miniature and not yet as gaunt and melancholy as he was to become in 1896. Chaucer also reappears in a series of cartoons Burne-Jones did for Ruskin in 1864 to illustrate *The Legend of Goode Wimmen*; Ruskin intended that the girls at Winnington Hall should embroider a set of hangings under Georgiana Burne-Jones's supervision, but the scheme came to nought, and a water-colour of 1871 – on which the Kelmscott illustration on page 416 is based – was the only tangible result of the aborted project.

Burne-Jones insisted that he was trying to see Chaucer's poetry afresh, not merely through the eyes of earlier illustrators, and this often led to a peculiar sort of literalism in his interpretation of the text: the example often noticed by readers of the *Chaucer* is the extraordinary image of the House 'made of twigges' as a large wicker basket in *The Hous of Fame*. Such an approach was not always easy for the illustrator. 'I wish Chaucer would, once & for all, make up his unrivalled & precious mind whether he is talking of a picture or a statue – I do wish

Figure 77 Several of Burne-Jones's designs for the *Chaucer* were based on earlier works of his, such as this scene from the Prioress's tale in *The Canterbury Tales* which he painted on a wardrobe for the Morrises in 1857.

it,' wrote Burne-Jones, '– for in this book I am putting myself wholly aside, & trying to see things as he saw them – not once have I invaded his kingdom with one hostile thought –'[37] Burne-Jones has also left a revealing, albeit disjointed and impressionistic, account of the thoughts that passed through his mind while he was designing the first of the wood-engravings for *The Romaunt of the Rose*: like many of his epistles to female correspondents, it is a thinly disguised love-letter. 'You come today & be with me while I draw in the little garden room,' he wrote to Helen Gaskell (whose face is recognizable in some of the Kelmscott *Chaucer* illustrations[38]) in July 1894. '. . . It's the Romance of the Rose I am beginning – all new I shall make it & nothing that I have done before shall be here – all shall be better[.]

'I shall only picture the little part that Chaucer wrote – hope Morris will only print the Chaucer part[39] up to Narcissus at the well – a fragment it is – a kind of Pilgrim's progress – only Love is the goal & not heaven – at least not Bunyan's heaven – dont want his heaven I dont. . . .

'now come & stand by me while I draw the first scene – it will take two days' time – dont leave me – we shall be in Chaucers bedroom – but very discreetly – There I've sketched it – put him in bed little books by him – a curtain – & a window in his bed – & I've drawn his spirit going out at a door – they do – spirits do go out & travel far – all this is true – the spirit of him shall not be drawn faintly, but stronger than the sleeping figure – & thats true – we are getting to the truth at last. but be sure not to leave me or all the dream will go.'[40]

Figures 78–9

The illustration described by Burne-Jones in this letter is typical of many in the *Chaucer*: the space is claustrophobically enclosed by the walls of the tiny bedroom, the surface of which is broken only by a tiny window and an arched doorway. Both Chaucer's gown and the hangings of the bed give Burne-Jones an opportunity to display his well-known virtuosity in drawing the sinuous folds of fabrics. Duncan Robinson, who has done an extended analysis of the *Chaucer* illustrations, finds a recurring pattern of a 'flat back of the room, often recessed, with a gently curved vault and pierced further with either a window or a door'; in the outdoor scenes, he adds, there is frequently a 'three-layer composition, using a wall or battlements to divide the picture plane horizontally'.[41] The *English Illustrated Magazine* (of which Emery Walker was an assistant editor) had published in September 1884 a series of five wood-engravings designed by H. Ryland to illustrate an article by Alfred Ainger entitled 'The Women of Chaucer',

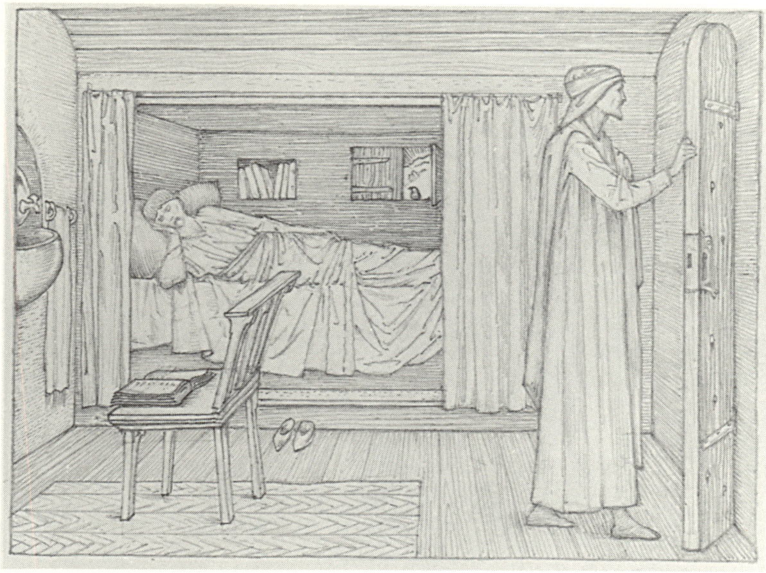

Figure 78 Burne-Jones's design for the wood-engraving which appears on page 241 of the Kelmscott *Chaucer*. In a letter to Helen Gaskell, Burne-Jones remarked of the illustration: 'I've drawn his [Chaucer's] spirit going out at a door - they do - spirits do go out & travel far - all this is true - the spirit of him shall not be drawn faintly, but stronger than the sleeping figure . . .'

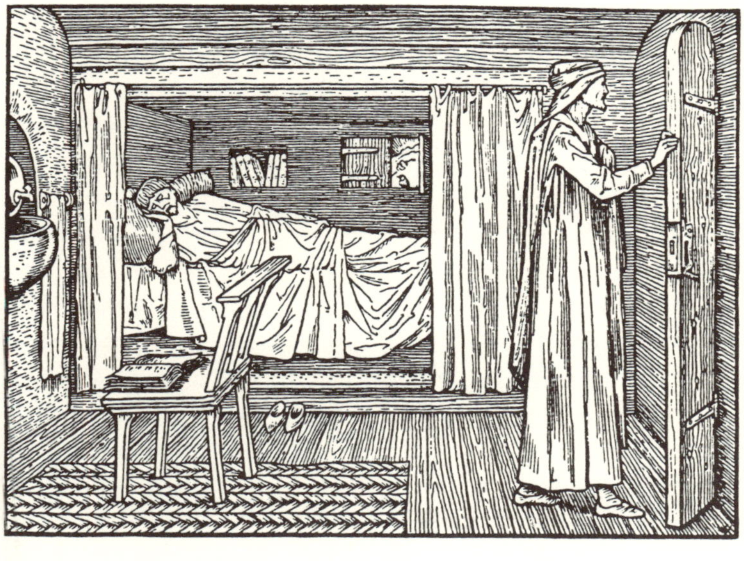

Figure 79 The final wood-engraving based on the design above.

and they are eerily similar to Burne-Jones's later drawings of the same scenes; but the most plausible explanation of this correspondence is that Ryland was merely imitating stylistic conventions that Burne-Jones had already established in his paintings, which had been frequently exhibited.

Burne-Jones began the *Chaucer* illustrations in late 1892 or very early 1893, usually working on the designs on Sunday when Morris came to his studio to talk and read aloud. As the task drew to a close, he was able to look back on it with satisfaction. 'And I have been calculating that . . . the time I have given to the Chaucer work in the past two years and a half is exactly to an hour the time I should have spent in visits from Saturday till Monday at "houses," if I had been amiable and sociable – for I haven't let it invade the week's work, but have designed only on Sunday with very little exception –. . . . I have fourteen more designs to make and then my share is done, and I may permit myself the luxury of one smile. I have always been happy over it: it has never tired me but refreshed me always.'[42]

Yet this mellow retrospective mood fails to convey the genuine anxiety both he and Morris felt at earlier stages in the book's history. Many of the illustrations had to be redone several times. Burne-Jones said their rule was 'that in illustrating a story we must always choose the essential incidents',[43] but their radically differing conceptions of Chaucer's poetry must have led to some very lively discussions between him and Morris; moreover, the elaborate procedure by which Robert Catterson-Smith 'translated' Burne-Jones's delicate pencil strokes into thick, bold lines suitable for engraving (see Chapter Five) was time-consuming and expensive. There were moments when it seemed that the book would never be done. Burne-Jones completed the last designs at the end of December 1895 and wrote one of his delightfully whimsical letters about them: 'I have just finished my Chaucer work and in May I hope the book will see the light. I hope sincerely it will be all the age does not want – I have omitted nothing I could think of to obstruct the onward march of the world. The designs are carved in wood . . . the lines as thick as I could get them. I have done all I can to impede progress – you will always bear me witness that I have not faltered – and that having put my hand to the plough I invariably look back.'[44]

The Kelmscott Chaucer

Burne-Jones received £500 for his *Chaucer* designs; the full cost of the book was £7,217 11*s.*, an enormous investment (extending over four years) that was a source of worry to Morris, who occasionally used the word 'panic' in his diary to describe his response to the multiple problems encountered in its production. Because of the unexpected delays and expenses, he announced to his customers on 14 November 1894 that 'the number printed will be 425 instead of 325' and that they would be allowed to cancel their original orders if they wished to do so. There is no evidence that any customer did. Jane Morris wrote to Blunt that 'at first it seemed likely to be a failure financially, then an extra 100 were printed, which has made all the difference between loss and profit'.[45]

This increase in the size of the edition, however, produced further disagreements with Bernard Quaritch, who was still sulking because he had not been allowed to publish the *Chaucer*. The terms under which Morris sold the book to dealers were a 15 per cent discount on the paper copies, with an additional 5 per cent allowed for cash payment within a week of delivery, and a flat 15 per cent discount on the vellum copies (cash only). Morris had discovered that Quaritch, who was given a further discount, was offering the *Chaucer* to the trade at a lower price than he himself was, and he was therefore reluctant to sell him any of the additional hundred copies. After protracted, unpleasant negotiations Morris agreed to supply fifty of the copies to Quaritch on condition that he not undersell the Kelmscott Press itself. As Morris had remarked dryly to his wife a few months earlier, 'They are printing the Chaucer very well. Item it is all sold except 3 vellum copies & people are quarrelling over the privilege of buying it.'[46]

Figure 80

The 425 paper copies of the *Chaucer* sold for £20 each, though that price was rising among the dealers before it was published; the thirteen vellum copies went for 120 guineas each. The pigskin binding cost an additional £13. Despite the fact that the entire edition had been sold before the date of publication, the Kelmscott Press ledger reveals that the book made no profit for Morris. As of 3 October 1896 (slightly more than three months after publication) the *Chaucer* revenues amounted to £7,005 16*s.* 10*d* – less than production costs – though an additional £123 14*s.* dribbled in later from dealers who had taken the book on credit. Nevertheless, the *Chaucer* was not a financial success: it had been subsidized, as Morris pointed out on several occasions, by the smaller Kelmscott Press books which had a healthy profit margin.

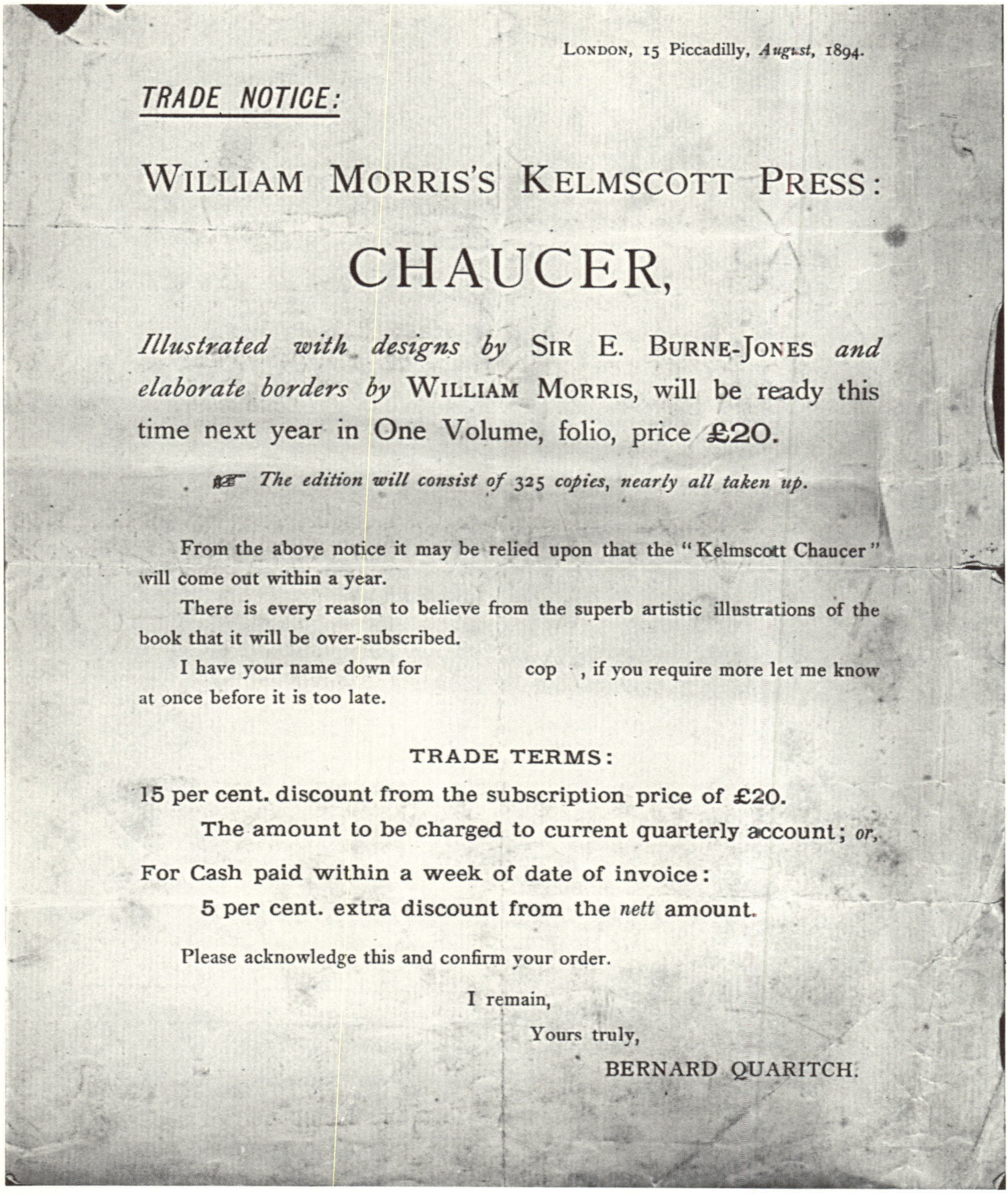

LONDON, 15 Piccadilly, *August*, 1894.

TRADE NOTICE:

WILLIAM MORRIS'S KELMSCOTT PRESS:

CHAUCER,

Illustrated with designs by SIR E. BURNE-JONES *and elaborate borders by* WILLIAM MORRIS, will be ready this time next year in One Volume, folio, price £20.

☞ *The edition will consist of 325 copies, nearly all taken up.*

From the above notice it may be relied upon that the "Kelmscott Chaucer" will come out within a year.

There is every reason to believe from the superb artistic illustrations of the book that it will be over-subscribed.

I have your name down for cop , if you require more let me know at once before it is too late.

TRADE TERMS:

15 per cent. discount from the subscription price of £20.

The amount to be charged to current quarterly account; *or*,

For Cash paid within a week of date of invoice:

5 per cent. extra discount from the *nett* amount.

Please acknowledge this and confirm your order.

I remain,

Yours truly,

BERNARD QUARITCH.

Figure 80 The discounts offered by Bernard Quaritch were so generous that Morris feared some customers would be tempted to buy the *Chaucer* from Quaritch rather than from the Kelmscott Press itself.

The Kelmscott Chaucer

Even by the leisurely standards of the Kelmscott Press, the *Chaucer* was a long time in the making, and the abundant references to the book in contemporary letters and diaries make it possible for us to follow its progress with some precision. Trial pages were pulled in July 1892, and during the same month Morris began to discuss the illustrations with Burne-Jones. Morris designed the first border in February 1893. On 8 August 1894 the first sheet was printed (on vellum). Cobden-Sanderson discussed the binding with Morris in October 1894. On 21 December 1894 another Albion press arrived and was installed in Sunderland Cottage exclusively for printing the *Chaucer*. By April 1895 all the designs for *The Canterbury Tales* had been redrawn by Catterson-Smith, and in the following month the printing of the *Tales* was completed. Between May and August, however, there was a hiatus while Morris tried to solve the mystery of the yellow stains on the sheets already printed. Burne-Jones drew the last illustration in December 1895; at about the same time Morris began to design the pigskin binding,[47] and in February 1896 he started the engraved title-page, finishing it on 27 March. At the end of May the printing of the *Chaucer* was done, and Morris celebrated by giving presents to his pressmen.

Aside from the magnitude and complexity of the book, a major source of delay (at least during the final year of its production) was Morris's failing health. His friends wondered whether he would live to finish it. T. M. Rooke has left a moving account in his diary of a visit Morris paid to Burne-Jones's home on 2 March 1896: Morris appeared at the door 'very ghostlike, feeble and old looking', and Burne-Jones declared that the *Chaucer* was 'the finest book ever printed – if W.M. had done nothing else it would be enough'. Morris responded, 'I must own I'm complacent about it – must try though not to be too conceited. And there's one thing that's not to be forgotten, that you backed me up well in it, old chap. If you'd been at all slack over it and hadn't been as much excited about it as I was, we should never have got through with it.'

The first two bound copies of the *Chaucer* were delivered to Morris and Burne-Jones on 2 June, the latter passing his copy on to his daughter the following day as a birthday gift. On 24 June Douglas Cockerell took a train to Folkestone, where Morris had gone on the advice of his doctor, and handed him a copy in the pigskin binding. 'Douglas came with first bound Chaucer about 1. very satisfactory,' Morris wrote in his diary. When he gave May her copy, she afterwards

Figure 81 In this characteristically light-hearted caricature by Burne-Jones, Chaucer blesses Morris and Burne-Jones (who look like awkward schoolboys) for their work on the Kelmscott *Chaucer*.

recalled, 'the look of profound melancholy that (perhaps unconsciously) he turned on me in smiling tenderly seemed like a glimpse into a very far country. What it quite meant I don't think I realized at the time, though it moved me deeply: all was made clear not long afterwards . . . it was the look of an intensely lonely man – never to be forgotten while memory serves.'[48] The *Chaucer* was Morris's last great achievement: he died the following October.

Figure 81

The Works of Geoffrey Chaucer represents the culmination of Morris's lifelong search for the Ideal Book; it is the kind of book that he and Burne-Jones dreamed of creating when they were undergraduates and that they attempted unsuccessfully to produce during the 1860s and 1870s: 'it should have come out when Morris and I were at Oxford in our first term,' Burne-Jones wrote, 'and we should have lost our senses with bliss.'[49] Instead it came at the end of Morris's life, when he was able to pour into this single volume all his bibliophilic passion and his unmatched skill as a designer of ornaments; it is significant that even those readers who find Morris's pages too heavy or congested for their tastes feel compelled, in the end, to pay tribute to the *Chaucer* as one of the great books of the world. 'To me it is the most beautiful of all printed books,' Yeats declared;[50] it is a sentiment that many, sometimes grudgingly, have echoed through the years.

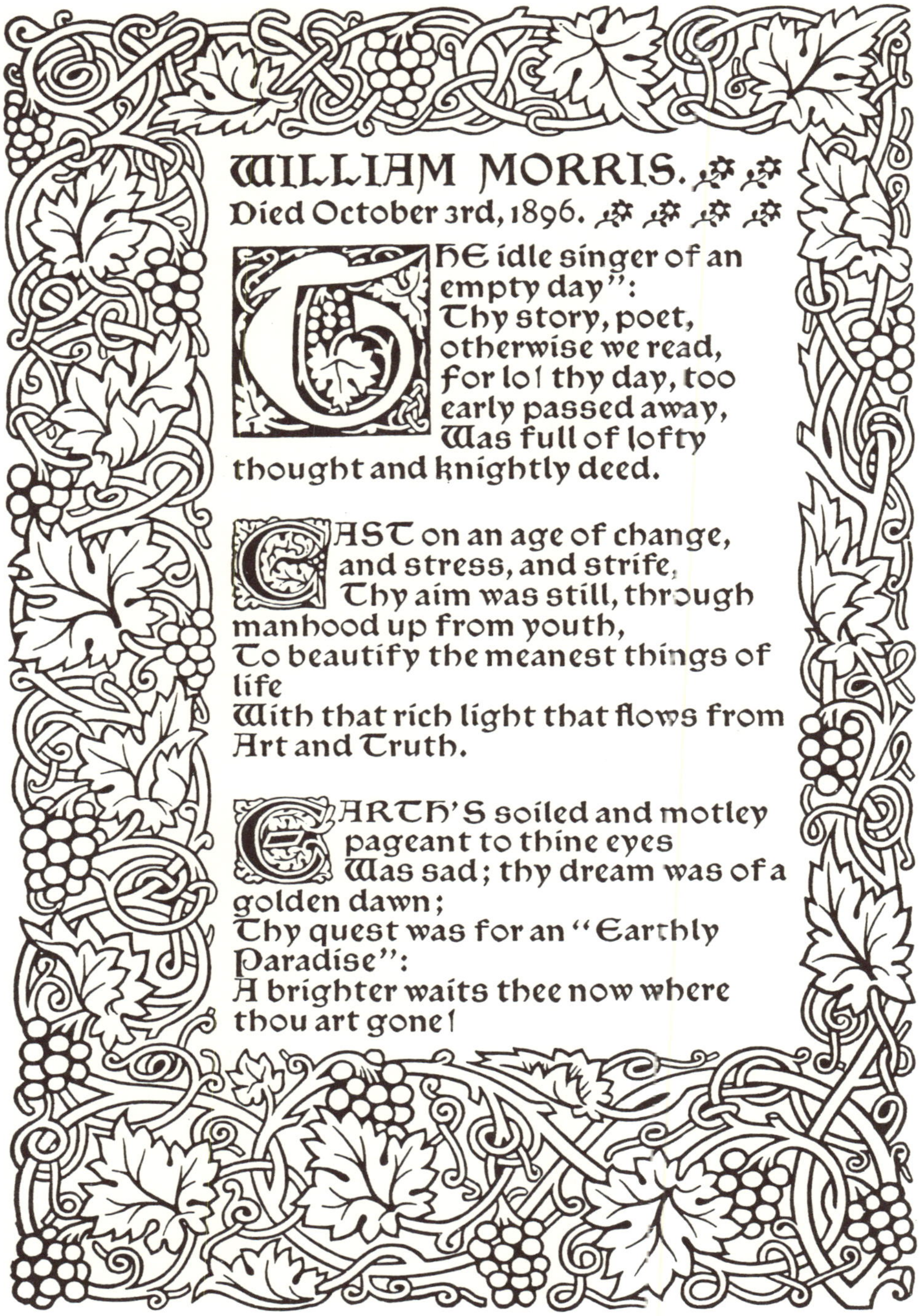

WILLIAM MORRIS.
Died October 3rd, 1896.

THE idle singer of an
empty day":
Thy story, poet,
otherwise we read,
for lo! thy day, too
early passed away,
Was full of lofty
thought and knightly deed.

CAST on an age of change,
and stress, and strife,
Thy aim was still, through
manhood up from youth,
To beautify the meanest things of
life
With that rich light that flows from
Art and Truth.

EARTH'S soiled and motley
pageant to thine eyes
Was sad; thy dream was of a
golden dawn;
Thy quest was for an "Earthly
Paradise":
A brighter waits thee now where
thou art gone!

Figure 82 A poem composed and printed by Morris's printers shortly after his death. Morris might well have had misgivings about the typography.

NINE ❧ CLOSING THE PRESS

IF William Morris had not died in 1896, how many more books might the Kelmscott Press have published? We can never know the answer to that question; but it is evident that Morris had numerous schemes in mind at the time of his death, the most important of which were, according to Ellis, an edition of Shakespeare's plays, the English Bible, and Froissart's *Chronicles.*[1] Morris on his deathbed made one last, feeble effort to prolong the life of the Press, as recorded in Cockerell's diary on 30 August 1896: 'He asked whether I should be prepared to carry on the Press after his death, with Walker, & I said that I was in favour of its ceasing – as otherwise it would fizzle out by degrees, & the books already issued would suffer by inferior ones following them. He said he thought I was perhaps right.' Nevertheless, there was much unfinished business which Cockerell and Walker had to supervise after Morris was gone. The remaining five volumes of *The Earthly Paradise* were issued, and ten more titles (as well as the Froissart trial pages) were published during the last phase of the Press's history. Not until the spring of 1898 – more than a year and a half after Morris's funeral – was Cockerell able to bring the Kelmscott Press to a definitive close. For the most part the books produced during this period had been planned or even begun while Morris was still alive, but it is enlightening to discover that fully a fifth of the Kelmscott volumes were produced without his direct supervision by associates and employees who had learned to emulate his principles of typography.

Morris showed signs of deteriorating health in 1895. Although one medical diagnosis suggested that he merely suffered from the disease of 'being William Morris, and working 18 hours a day',[2] it was obvious to his family and friends that – as Burne-Jones put it – his 'enormous vitality [was] diminishing'.[3] By February 1896 the symptoms were alarming, and Morris was seeing doctors regularly, though Cockerell still reassured himself that all would be well in the end. 'W.M. very unwell & weak,' he wrote in his diary on 20 February. '. . . Surely it is influenza.' Throughout the spring of 1896 there were further signs of decline, but of Morris's intimate circle only Burne-Jones and Blunt seemed to recognize the gravity of his illness. Blunt's diary describes a visit on 14 May to Hammersmith, 'where I found my poor Morris looking very ill and aged, doddling feebly in front of his house. We went in together however and he brightened up and told me of his maladies in a cheerful and not too desponding way. . . .' When a fortnight later the Morrises visited Blunt at his country house, he observed in his diary (29 May) that Morris's health had worsened: he was now 'very feeble & aged. I fear from the luck of things that it is some form of consumption and that he will not recover, but his spirits are fairly good and he talks at times as brilliantly as ever. . . . his illness does not make him gloomy, only it troubles him in his work.' (Blunt's amateur diagnosis of consumption was correct.) These developments plunged Burne-Jones into misery: 'now I am frightened – & can think of little else – steadily he seems to be consuming away – without pain – but growing weaker & weaker and I can't think about anything else,' he declared in a letter to Blunt on 6 June.[4]

A few days later, on his doctor's recommendation, the Morrises left London for Folkestone in search of the beneficial effects of sea air. In July Morris, again on professional advice, took a voyage to Norway, but when he returned the following month to Hammersmith, his health displayed no sign of improvement. He now withdrew to his bedroom at Kelmscott House, which he never thereafter left for more than a few hours, though during the last month of his life there was a steady flow of visitors – the Burne-Joneses, Ellis, Mary De Morgan, Charles Fairfax Murray, Blunt, and others – to the room. On 9 September 1896 Morris made a new will, naming his wife, Ellis, and Cockerell as executors. When Quaritch called at Kelmscott House on 17 September, he discovered Morris 'worn out and feeble', emaciated, and staring out of the window at the river. 'I fear a catastrophe is impending,' Quaritch wrote later the same day.[5] To raise Morris's spirits, Burne-Jones came on Sundays and sat in the bedroom drawing illustrations for the Kelmscott Press edition of *Sigurd the Volsung*, and Ellis, elsewhere in the house, transcribed traditional Border ballads for another Kelmscott book that Morris hoped to publish. Morris himself, at intervals, struggled with the design for an initial word ('Empty') that was to appear in *The Water of the Wondrous Isles* (1897). 'He sat up & was going to design "Empty" . . . but only just managed to black in the spaces, & just rough in the word,' Cockerell wrote in his diary on 4 September. And two days later: 'W.M. rather tired, but decidedly no worse. He had done a little more to "Empty".' But Morris never completed the design, and after his death Catterson-Smith had to put the finishing touches on it.

Closing the Press

On 28 September Blunt witnessed Morris entering the dining-room at Kelmscott House 'like a man risen from the grave'.[6] The following day Morris was in better spirits and insisted upon being taken outdoors in a bath chair and then walking about within the house; a short time later Cockerell, returning from an errand, found him upstairs with blood streaming from his mouth. Morris died on the morning of 3 October. The effect on his friends was devastating. Burne-Jones fell on his knees next to the body and burst into tears; Cockerell retreated to Morris's study opposite the bedroom and wept in private; Crane paid tribute to him by placing a sonnet and a wreath next to the Kelmscott *Chaucer* that went on display the following Monday at the Arts and Crafts Exhibition. *Figure 82*

But it is entirely characteristic of Cockerell that he did not indulge his personal grief for more than a few minutes. When his sobbing abated, as he testified many years later, 'I set to work on all there was to be done.'[7] Jane Morris had little experience in handling money, Ellis was in poor health himself and lived in Torquay, and therefore almost the entire responsibility for managing the Morris family's financial affairs fell on Cockerell's shoulders. During the following months he engaged in ceaseless, almost frenzied activity. A buyer had to be found for Kelmscott House; arrangements were made for Morris's library to be sold; Jane Morris and her daughters required constant attention and sympathetic understanding. Above all, Cockerell had to avert the immediate collapse of the Kelmscott Press by pushing through to completion the books already in production and by devising a rational scheme for gradually closing down the Press.

To the discerning eye there were soon signs of Morris's departure from the Kelmscott Press. The largest book then passing through the press was *The Earthly Paradise*, three volumes of which had appeared before Morris's death. In the fourth volume the colophon (dated 25 November 1896) was altered so that 'Printed by William Morris at the Kelmscott Press' became 'Printed by the Trustees of the late William Morris at the Kelmscott Press'.[8] An announcement, issued on 12 November, of *The Shepheardes Calender* contained this unsettling final sentence: 'The folio editions of Froissart & of Sigurd the Volsung, described in the June list, have been abandoned.' Even in the books that were completed, the absence of Morris's guiding hand was evident, particularly in the lack of new borders, initials, and engraved title-pages. (The *Athenaeum*, on 27 November 1897, was sufficiently alert to notice these deficiencies in its review of *The Water of the Wondrous* *Figure 83*

THESE too with little patience seemed to hear
That story end with shame and grief and fear;
A little thing the man had had to do,
They said, if longing burned within him so.
But at their words the older men must bow
Their heads, and, smiling, somewhat thoughtful grow,
Remembering well how fear in days gone by
Had dealt with them, and poisoned wretchedly
Good days, good deeds, and longings for all good:
Yet on the evil times they would not brood,
But sighing, strove to raise the weight of years,
And no more memory of their hopes and fears
They nourished, but such gentle thoughts as fed
The pensiveness which that sweet season bred.

End of Vol. III.

Printed by William Morris at the Kelmscott Press,
and finished on the 24th day of August, 1896.

169

They were not yet of every joy bereft,
For long ago was past the agony
Midst which they found that they indeed must die;
And now well-nigh as much their pain was past
As though death's veil already had been cast
Over their heads; so, midst some little mirth,
They watched the dark night hide the gloomy earth.

End of Vol. IV.

Printed by the Trustees of the late William Morris at
the Kelmscott Press, and finished on the 25th day of
November, 1896.

Figure 83 The death of Morris in 1896 while *The Earthly Paradise* was in production at the Kelmscott Press required a last-minute alteration in the colophon of the fourth volume (below); the colophon of the third volume is shown above it.

Isles.) All the books published after Morris's death were of necessity much more lightly ornamented. *Love is Enough*, the penultimate book printed at the Kelmscott Press, is an interesting example of how Cockerell was compelled to solve book-design difficulties on his own, since no work had been done on it before October 1896; Cockerell confessed that 'it offered many difficult typographical problems on account of the varying length of the lines'.[9] It must be said, in fairness to Cockerell, that these last books of the Kelmscott Press display no dramatic falling off in quality, but, with the exception of the Froissart specimen pages, they represent a reversion to the simpler style of ornamentation characteristic of the very earliest Kelmscott books when Morris had not yet designed very many borders and initials.

There were countless books that Morris had thought of printing – the Bible (Wycliffe's version), Bunyan's *Pilgrim's Progress*, FitzGerald's translation of *The Rubáiyát of Omar Khayyám*, Malory's *Morte Darthur*, *Piers Plowman*, to name only a few – and Cockerell recognized that there was now no hope of ever doing them at the Kelmscott Press. Two projected folios were abandoned, however, with greater reluctance. In 1893 Furnivall had persuaded Morris to undertake an edition of Shakespeare's plays, with either P. A. Daniel or Furnivall as editor and Longmans as publisher. Morris printed at least two trial pages[10] in the Golden type: the design (two columns, stage directions and speakers' names in red, a six-line initial at the beginning of each scene) is simple to the point of austerity, and it is not known whether he intended to include illustrations. In any event, the project never went any farther than this, because work on the *Chaucer* absorbed Morris's attention entirely. *Figure 84*

The other aborted folio was Froissart's *Chronicles*, which was in a more advanced state of production than the Shakespeare. Morris enthusiastically announced his plans to a journalist in 1895: 'My reprint is a full folio and will take up two volumes. I also intend to publish it in four parts . . . no book that I could do would give me half the pleasure I am getting from the *Froissart*. I am simply revelling in it. It's such a noble and glorious work, and every page as it leaves the press delights me more than I can say. I am taking great pains with it, and doing all I can to realise what I have long wished. I am printing it in my Chaucer type.'[11] Despite Morris's assertion, no sheets of the book were actually printed, but by November 1896 some thirty-six pages were in type, and Cockerell hit upon the ingenious solution of publishing a two-leaf vellum specimen (as well as a larger specimen privately *Figure 85*

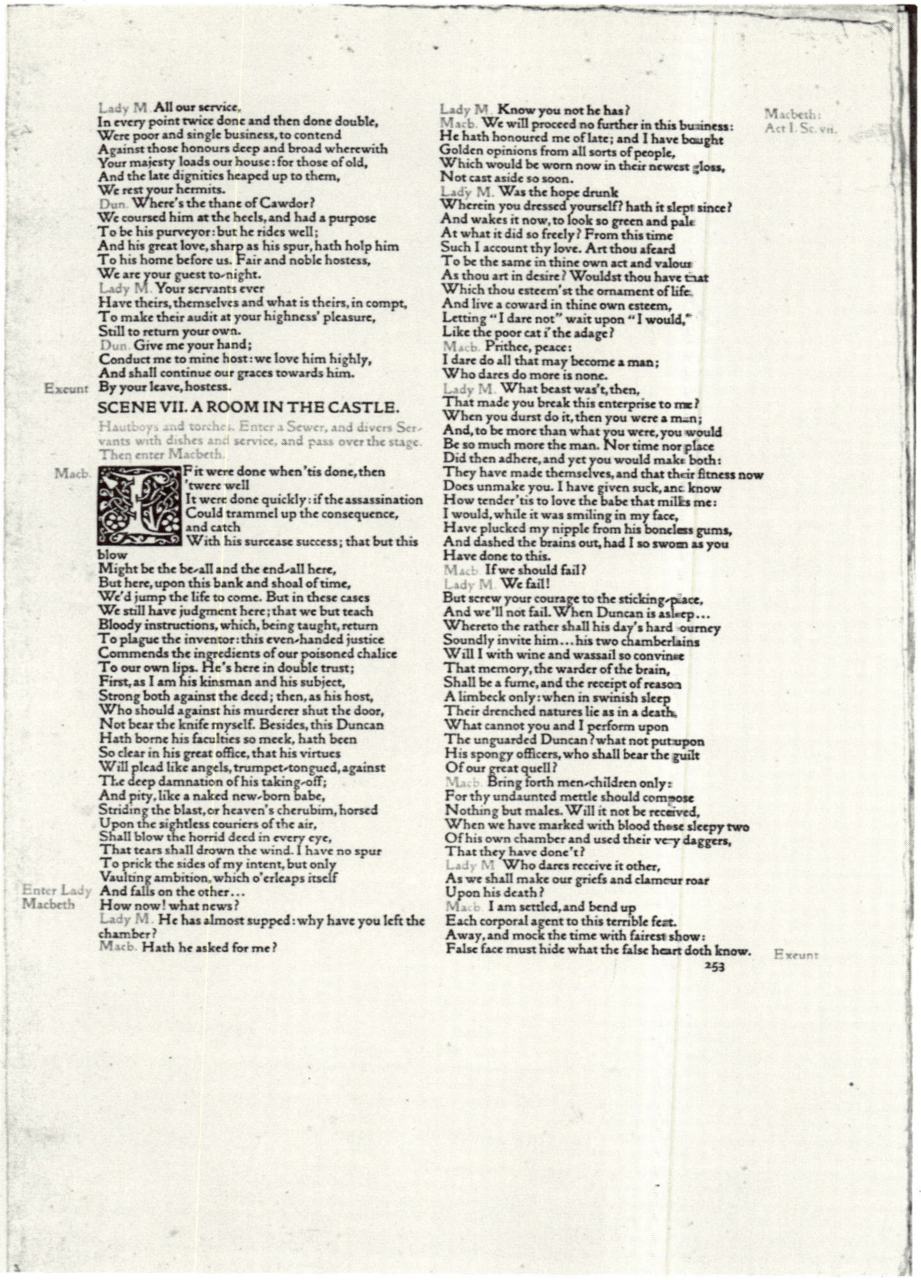

Lady M. All our service,
In every point twice done and then done double,
Were poor and single business, to contend
Against those honours deep and broad wherewith
Your majesty loads our house: for those of old,
And the late dignities heaped up to them,
We rest your hermits.
Dun. Where's the thane of Cawdor?
We coursed him at the heels, and had a purpose
To be his purveyor: but he rides well;
And his great love, sharp as his spur, hath holp him
To his home before us. Fair and noble hostess,
We are your guest to-night.
Lady M. Your servants ever
Have theirs, themselves and what is theirs, in compt,
To make their audit at your highness' pleasure,
Still to return your own.
Dun. Give me your hand;
Conduct me to mine host: we love him highly,
And shall continue our graces towards him.
Exeunt By your leave, hostess.

SCENE VII. A ROOM IN THE CASTLE.

Hautboys and torches. Enter a Sewer, and divers Servants with dishes and service, and pass over the stage. Then enter Macbeth.

Macb. If it were done when 'tis done, then 'twere well
It were done quickly: if the assassination
Could trammel up the consequence, and catch
With his surcease success; that but this blow
Might be the be-all and the end-all here,
But here, upon this bank and shoal of time,
We'd jump the life to come. But in these cases
We still have judgment here; that we but teach
Bloody instructions, which, being taught, return
To plague the inventor: this even-handed justice
Commends the ingredients of our poisoned chalice
To our own lips. He's here in double trust;
First, as I am his kinsman and his subject,
Strong both against the deed; then, as his host,
Who should against his murderer shut the door,
Not bear the knife myself. Besides, this Duncan
Hath borne his faculties so meek, hath been
So clear in his great office, that his virtues
Will plead like angels, trumpet-tongued, against
The deep damnation of his taking-off;
And pity, like a naked new-born babe,
Striding the blast, or heaven's cherubim, horsed
Upon the sightless couriers of the air,
Shall blow the horrid deed in every eye,
That tears shall drown the wind. I have no spur
To prick the sides of my intent, but only
Vaulting ambition, which o'erleaps itself
Enter Lady Macbeth And falls on the other...
How now! what news?
Lady M. He has almost supped: why have you left the chamber?
Macb. Hath he asked for me?
Lady M. Know you not he has?
Macb. We will proceed no further in this business:
He hath honoured me of late; and I have bought
Golden opinions from all sorts of people,
Which would be worn now in their newest gloss,
Not cast aside so soon.
Lady M. Was the hope drunk
Wherein you dressed yourself? hath it slept since?
And wakes it now, to look so green and pale
At what it did so freely? From this time
Such I account thy love. Art thou afeard
To be the same in thine own act and valour
As thou art in desire? Wouldst thou have that
Which thou esteem'st the ornament of life,
And live a coward in thine own esteem,
Letting "I dare not" wait upon "I would,"
Like the poor cat i' the adage?
Macb. Prithee, peace:
I dare do all that may become a man;
Who dares do more is none.
Lady M. What beast was't, then,
That made you break this enterprise to me?
When you durst do it, then you were a man;
And, to be more than what you were, you would
Be so much more the man. Nor time nor place
Did then adhere, and yet you would make both:
They have made themselves, and that their fitness now
Does unmake you. I have given suck, and know
How tender 'tis to love the babe that milks me:
I would, while it was smiling in my face,
Have plucked my nipple from his boneless gums,
And dashed the brains out, had I so sworn as you
Have done to this.
Macb. If we should fail?
Lady M. We fail!
But screw your courage to the sticking-place,
And we'll not fail. When Duncan is asleep...
Whereto the rather shall his day's hard journey
Soundly invite him... his two chamberlains
Will I with wine and wassail so convince
That memory, the warder of the brain,
Shall be a fume, and the receipt of reason
A limbeck only: when in swinish sleep
Their drenched natures lie as in a death,
What cannot you and I perform upon
The unguarded Duncan? what not put upon
His spongy officers, who shall bear the guilt
Of our great quell?
Macb. Bring forth men-children only:
For thy undaunted mettle should compose
Nothing but males. Will it not be received,
When we have marked with blood those sleepy two
Of his own chamber and used their very daggers,
That they have done't?
Lady M. Who dares receive it other,
As we shall make our griefs and clamour roar
Upon his death?
Macb. I am settled, and bend up
Each corporal agent to this terrible feat.
Away, and mock the time with fairest show:
False face must hide what the false heart doth know. Exeunt

Macbeth: Act I. Sc. vii.

253

Figure 84 Trial page (from the first act of *Macbeth*) of a projected edition of Shakespeare that never materialized at the Kelmscott Press. The design – two columns of the Golden type with sparing use of ornamented initials and red ink – is rather austere, and it is not known whether Morris intended to use illustrations.

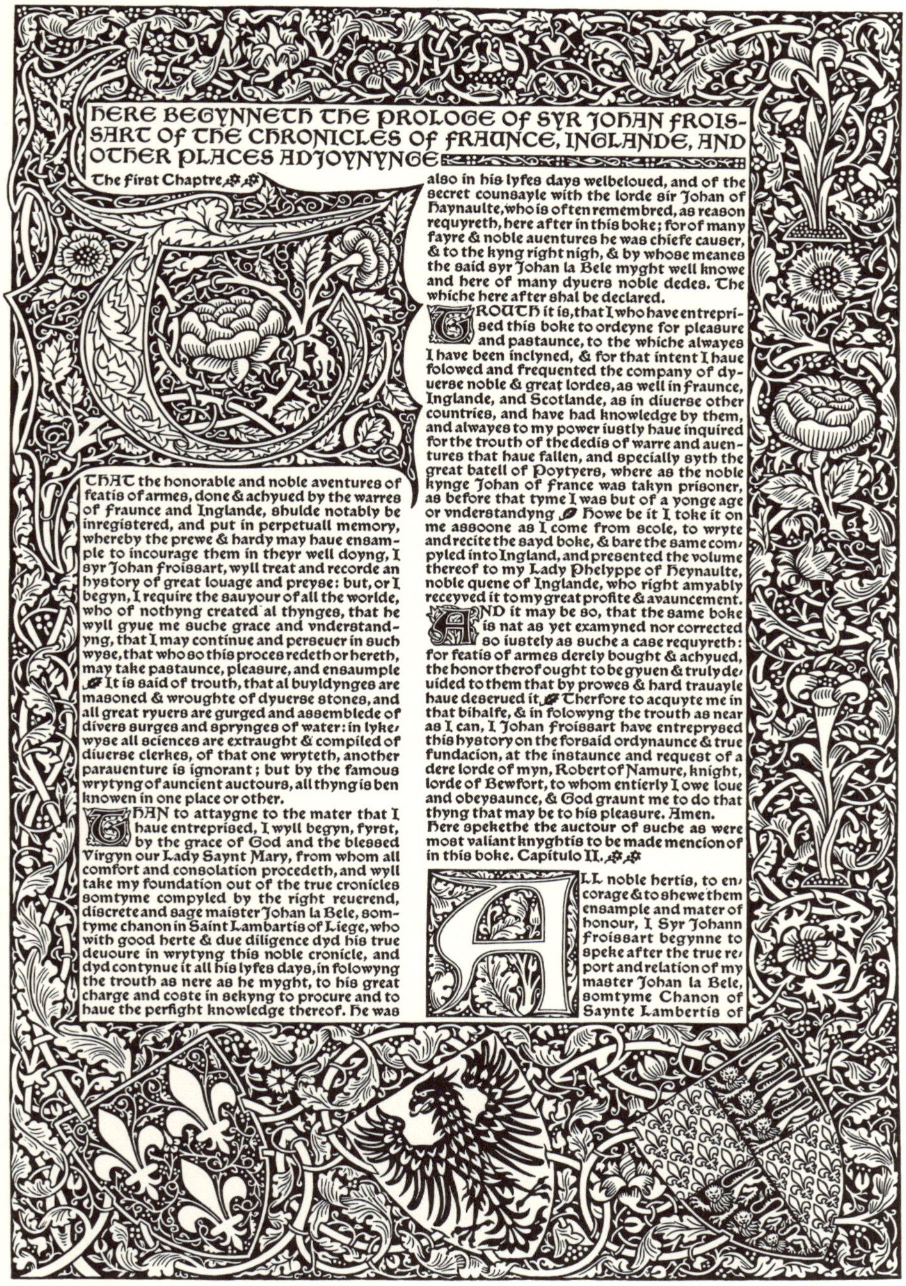

HERE BEGYNNETH THE PROLOGE OF SYR JOHAN FROISSART OF THE CHRONICLES OF FRAUNCE, INGLANDE, AND OTHER PLACES ADJOYNYNGE

The first Chaptre

THAT the honorable and noble aventures of featis of armes, done & achyued by the warres of fraunce and Inglande, shulde notably be inregistered, and put in perpetuall memory, whereby the prewe & hardy may haue ensample to incourage them in theyr well doyng, I syr Johan froissart, wyll treat and recorde an hystory of great louage and preyse: but, or I begyn, I require the sauyour of all the worlde, who of nothyng created al thynges, that he wyll gyue me suche grace and vnderstandyng, that I may continue and perseuer in such wyse, that who so this proces redeth or hereth, may take pastaunce, pleasure, and ensaumple. It is said of trouth, that al buyldynges are masoned & wroughte of dyuerse stones, and all great ryuers are gurged and assemblede of divers surges and sprynges of water: in lykewyse all sciences are extraught & compiled of diuerse clerkes, of that one wryteth, another parauenture is ignorant; but by the famous wrytyng of auncient auctours, all thyng is ben knowen in one place or other.

THAN to attaygne to the mater that I haue entreprised, I wyll begyn, fyrst, by the grace of God and the blessed Virgyn our Lady Saynt Mary, from whom all comfort and consolation procedeth, and wyll take my foundation out of the true cronicles somtyme compyled by the right reuerend, discrete and sage maister Johan la Bele, somtyme chanon in Saint Lambartis of Liege, who with good herte & due diligence dyd his true deuoure in wrytyng this noble cronicle, and dyd contynue it all his lyfes days, in folowyng the trouth as nere as he myght, to his great charge and coste in sekyng to procure and to haue the perfight knowledge thereof. He was also in his lyfes days welbeloued, and of the secret counsayle with the lorde sir Johan of Haynaulte, who is often remembred, as reason requyreth, here after in this boke; for of many fayre & noble auentures he was chiefe causer, & to the kyng right nigh, & by whose meanes the said syr Johan la Bele myght well knowe and here of many dyuers noble dedes. The whiche here after shal be declared.

TROUTH it is, that I, who have entreprised this boke to ordeyne for pleasure and pastaunce, to the whiche alwayes I have been inclyned, & for that intent I haue folowed and frequented the company of dyuerse noble & great lordes, as well in fraunce, Inglande, and Scotlande, as in diuerse other countries, and have had knowledge by them, and alwayes to my power iustly haue inquired for the trouth of the dedis of warre and auentures that haue fallen, and specially syth the great batell of Poytyers, where as the noble kynge Johan of france was takyn prisoner, as before that tyme I was but of a yonge age or vnderstandyng. Howe be it I toke it on me assoone as I come from scole, to wryte and recite the sayd boke, & bare the same compyled into Ingland, and presented the volume thereof to my Lady Phelyppe of Heynaulte, noble quene of Inglande, who right amyably receyved it to my great profite & avauncement.

AND it may be so, that the same boke is nat as yet examyned nor corrected so iustely as suche a case requyreth: for featis of armes derely bought & achyued, the honor therof ought to be gyuen & truly deuided to them that by prowes & hard trauayle haue deserued it. Therfore to acquyte me in that bihalfe, & in folowyng the trouth as near as I can, I Johan froissart have entreprysed this hystory on the forsaid ordynaunce & true fundacion, at the instaunce and request of a dere lorde of myn, Robert of Namure, knight, lorde of Bewfort, to whom entierly I owe loue and obeysaunce, & God graunt me to do that thyng that may be to his pleasure. Amen.

Here spekethe the auctour of suche as were most valiant knyghtis to be made mencion of in this boke. Capitulo II.

ALL noble hertis, to encorage & to shewe them ensample and mater of honour, I Syr Johann froissart begynne to speke after the true report and relation of my master Johan la Bele, somtyme Chanon of Saynte Lambertis of

Figure 85 One of the great unfinished undertakings of the Kelmscott Press was Froissart's *Chronicles*, which was envisioned as a lavishly ornamented folio that would rival the *Chaucer* in splendour. After Morris's death, only a few specimen pages were issued.

printed on paper as a gift to some friends of Morris and the Press). The Froissart specimen reveals a stunning new departure in Morrisian ornamentation: though the two-column text is arranged in the same pattern as that of the *Chaucer*, the borders, incorporating coats of arms, are extraordinarily vigorous and spiky in appearance, an initial word is connected to one of the borders, and the initials extend into the margins. (Morris also designed some similar initials, with vines and tendrils trailing below, for the unfinished *Sigurd*.[12]) Such innovations will startle us only if we have mistakenly assumed that Morris's principles of book-design and ornamentation were static and wholly incapable of development. The Froissart specimen pages leave no doubt that Morris's comparatively early death robbed the world of a work which would have rivalled the *Chaucer* as one of the finest printed books of the post-medieval era. Swinburne's astute comment was, 'Why did not Morris take up & finish Froissart earlier – instead of the less appropriate work (Shelley, for one instance) on which the Kelmscott Press was occasionally wasted.'[13]

Of the final group of books published at the Press, the one that caused Cockerell the greatest anxiety was Morris's *The Story of Sigurd the Volsung*, which was originally intended to be a folio adorned by some twenty-five Burne-Jones illustrations. The book had been planned as early as 1891, but Burne-Jones was not able to turn his attention to it until January 1896 after he completed the *Chaucer* designs. 'To pacify and please Morris I will do what I can to adorn the Volsung Tale,' he wrote to Lady Leighton at about this time.[14] But to his assistant he confessed, 'I shan't like doing it so much as Chaucer. The scenery for one thing isn't quite a painter's scenery – it's more a poet's.'[15] By May, Burne-Jones was still working fitfully on the *Sigurd* designs and still complaining that he found the assignment uncongenial. 'It's quite wonderful how distasteful the task of making drawings of it is to me,' he said. '. . . Terrible the tragedy of it is, and the fine things in it are so much fitter for literature than for my art, and I've no turn for the dramatic.'[16] As Morris lay dying at Kelmscott House during September, Burne-Jones kept his promise by bringing the designs to his friend's bedroom each Sunday, thus extending to the last moment the intense intellectual and artistic relationship that had existed between them since their Oxford days. Burne-Jones was plainly relieved when the folio *Sigurd* was abandoned, but on the evening of 26 January 1897 Cockerell convinced him to furnish two illustrations for a scaled-down quarto version of the book, which was

Figure 86

Figure 86 Designs by Morris for initials in *The Story of Sigurd the Volsung*. Because of Morris's death, the ornamentation of the book had to be drastically curtailed when it was published in 1898. The style of these unfinished initials, whose tails extend down into the margin, is similar to that of the initials Morris created for the *Froissart*, another aborted project.

eventually published in February 1898. It was his last piece of work for the Kelmscott Press, and both he and Cockerell were conscious that it was the end of an epoch in Burne-Jones's career. Cockerell had lunch with him at the Grange, his home in South Kensington, on 7 October 1897, and made this terse record in his diary: 'EB-J alone & very nice. He was working on the picture of Gudrun for the Sigurd which bids fair to be as fine as anything he has done for the Kelmscott Press – a fitting conclusion.'

There were in fact two more Burne-Jones illustrations in *Love is Enough*, published just a few days before the Kelmscott Press officially closed in March 1898, but they had been designed earlier. As the colophon explained, 'The picture on the opposite page was not designed for this edition of Love is Enough, but for an edition projected about twenty-five years ago, which was never carried out.' Similarly, the frontispiece of the last Kelmscott book, *A Note by William Morris on his Aims in Founding the Kelmscott Press*, was one of the wood-engravings designed by Burne-Jones and cut by Morris in the 1860s for the unfinished *Earthly Paradise*. And so, in these final two volumes, there were touching reminders of Morris's and Burne-Jones's lifelong preoccupation with the Ideal Book, and things came full circle at last. Cockerell, who struck some observers as a coolly efficient technician-manager, was in reality a man of deep emotion, especially in his loyalty to Morris, and he wound up the affairs of the Kelmscott Press with evocative flourishes.

During 1897 and early 1898 Cockerell was also kept busy sorting through the abundant records and papers of the Press. The most interesting proofs and other documents he kept, with Jane Morris's permission, for his own collection; ever methodical, he even made use of cancelled cheques by clipping Morris's signatures and pasting them in his Kelmscott books. Proofs and trial pages were also distributed to Cockerell's friends, particularly Robert Proctor, the brilliant young incunabulist and Morris-admirer at the British Museum. 'You are not unique in your taste for first proofs &c – but I will see what I can find for you later on,' Cockerell had written to him in November 1896,[17] and during the following year he supplied Proctor with a large quantity of such material. Proctor attempted to make use of it in a Morris bibliography, never published, which he was compiling shortly before his untimely death in 1903.[18] Of the vast quantities of

Kelmscott proofs and related documents that survive today in both public and private collections, it is safe to generalize that nearly all of them have passed through the hands of Cockerell or Proctor at one time.

The gradual closing of the Kelmscott Press was punctuated by one reminder after another of Morris's death. On 24 March 1897, for example, Cockerell, Walker, and Ellis met together again at Kelmscott House to celebrate Morris's birthday, with Ellis reading aloud part of *The Lovers of Gudrun* before they went to bed. Between then and August Cockerell found himself burdened with the melancholy task of supervising the removal of Morris's belongings from Kelmscott House, and on 18 August he sold Morris's vellum set of the Kelmscott Press books (for £650) to Henry Gamman, an old friend from the Coal Exchange. The same morning, by coincidence, he disposed of the last set printed on paper. In July he approached Sir Edward Maunde Thompson, director of the British Museum, about depositing the woodblocks of the Kelmscott Press ornaments and illustrations in the Museum, a transaction that took place in December. 'F. S. Ellis came in the aft[ernoo]n & discussed the future of the K.P. type & ornaments,' Cockerell recorded in his diary on 12 November 1897. 'I declined to consent to the ornaments being used after the close of the Press.' The remaining stock of Batchelor paper was transferred to the Chiswick Press, as were the founts of type.

To Jane Morris he offered this summary in December of the state of the Press: 'The winding up is very anxious work – . . . There are 5 books still to come out[:] Sigurd, Love is Enough, Sundering Flood, German Woodcuts, & the Note on the Press. The first three are waiting for pictures that are nearly ready & the 4th is at the binders. I shall have to see these books launched before I can take up other matters of a heavy kind.'[19] Because of his added responsibilities – Cockerell was in effect directing the Kelmscott Press after Morris's death, with advice and occasional assistance from Walker and Ellis – Jane Morris paid him £100 beyond his usual salary in March 1898, extending to him also 'my heartiest thanks for having brought the Kelmscott Press to such a successful end, it's career has been a marvel since the beginning, but the completeness of the success was certainly owing to your very skilful steering towards the end'.[20]

Figure 87

The last two announcements issued by the Press had a note of finality: that of 22 November 1897 declared that the Kelmscott Press 'will close early in the New Year', and that of 25 February offered for

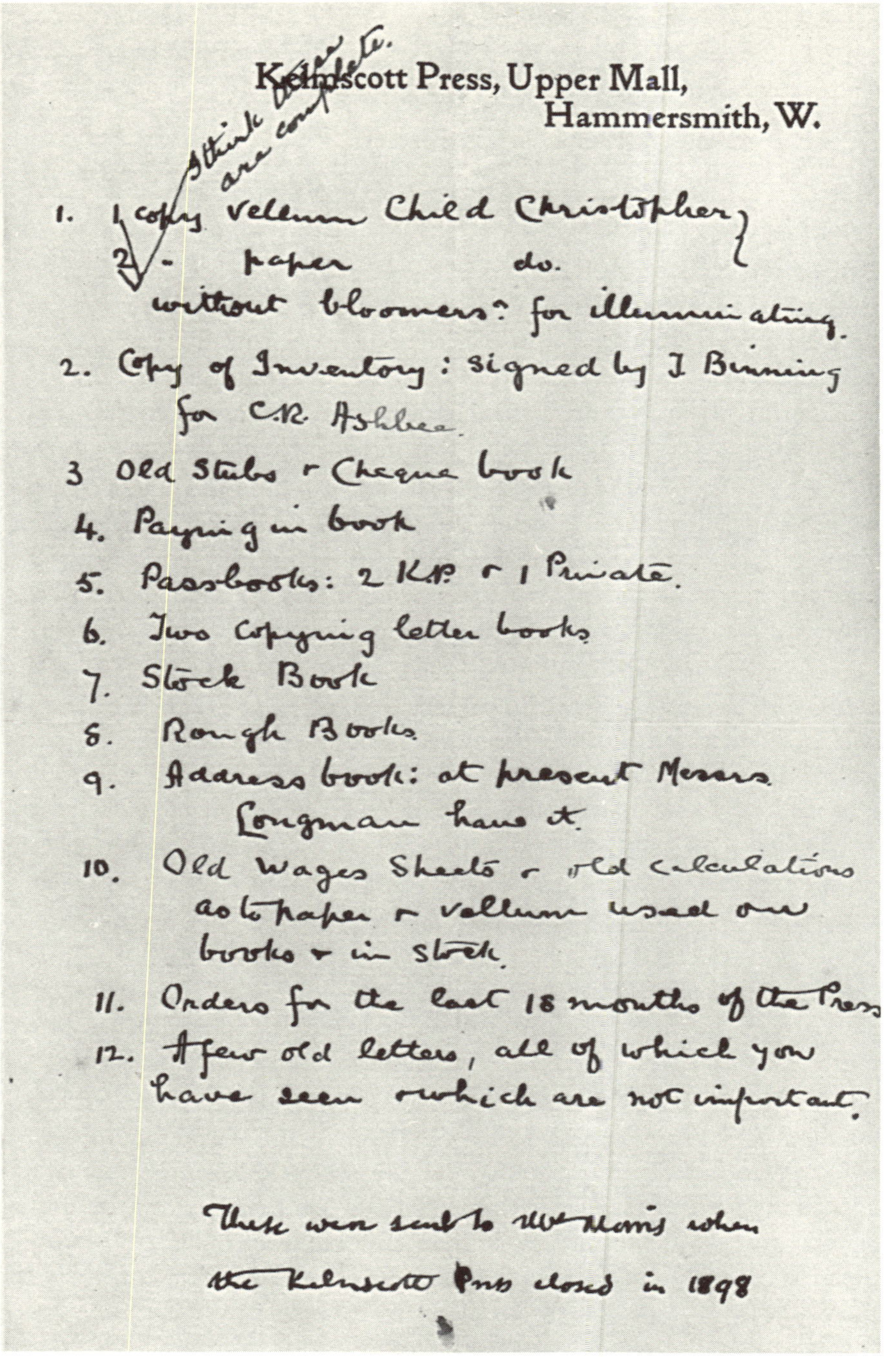

Kelmscott Press, Upper Mall,
Hammersmith, W.

Strike these are complete.

1. 1 copy vellum Child Christopher
2 - paper do.
without bloomers? for illuminating.

2. Copy of Inventory: signed by J Binning for C.R. Ashbee.

3 Old Stubs & Cheque book

4. Paying in book

5. Passbooks: 2 K.P. & 1 Private.

6. Two copying letter books

7. Stock Book

8. Rough Books

9. Address book: at present Messrs. Longman have it.

10. Old Wages Sheets & old calculations as to paper & vellum used on books & in stock.

11. Orders for the last 18 months of the Press

12. A few old letters, all of which you have seen & which are not important.

These were sent to Mrs Morris when the Kelmscott Press closed in 1898

Figure 87 An inventory of Kelmscott Press records sent by Sydney Cockerell to Jane Morris in 1898.

sale all the books still in stock and added that 'three hand-presses and the remaining type of the Kelmscott Press (exclusive of the type and ornaments), will be sold by private treaty'. On 24 December Cockerell 'spent the day in moving from Sunderland Cottage to 14 Upper Mall [Sussex Cottage], thus closing another chapter, the last but one, in the history of the Press'. Those were the words he entered in his diary that Christmas Eve; earlier in the day, as the workmen waited to carry off his desk, Cockerell wrote a final letter to Jenny Morris from his office in Sunderland Cottage, regretting that the fog prevented a last nostalgic view across the Thames from his window. Builders at that moment were already beginning to turn the cottage into a granary.[21]

The last few sheets were printed in Sussex Cottage during the early months of 1898. Following two springlike days, Cockerell reported to Jenny on 7 January: 'Our pressman, there is only one left now, says that we need a frost for the ink rollers, which get soft & difficult to print with in damp weather. I hope it will come, as we have the second Sigurd picture to do next week, & the map for the Sundering Flood – and perhaps the first picture for Love is Enough, which Mr Hooper is now engraving – all of these will need his best skill –'[22] The sole surviving pressman was Stephen Mowlem, who on 19 March printed the last copies of the last signature of the last Kelmscott Press book, *A Note by William Morris*. The final sheet, appropriately, was given to Walker; the penultimate sheet, signed by Mowlem and W.H. Bowden, the only other printer still on the premises, went into Cockerell's collection. The official publication date of the book was 24 March. The time had come, as Cockerell remarked to a correspondent the previous month, 'to write EXPLICIT over the door of the Kelmscott Press'.[23] On 31 March Cockerell sent a benedictory letter to J. and J. Leighton (thanking them 'for all your care & attention to the binding of the Kelmscott Press during the seven years of its existence'[24]), said goodbye to Jane Morris, had lunch with Georgiana Burne-Jones at the Grange, and sailed the next day to Siena, leaving behind Mrs Peddie to deal with the Press correspondence until he returned in early May.

The demise of the Kelmscott Press did not go unnoticed by the book trade or the press. 'There has been a great rush on the German woodcuts & they were all sold out last week,' Cockerell observed on 30 November 1897. '. . . A good many people have been left out in the cold, as orders keep coming in! I suppose it is the announcement that the Press is closing that has made people so eager.'[25] Frank

Hollings, the London dealer, trumpeted to his customers in December that 'there is no doubt that the dissolution of the "Kelmscott Press", . . . and the strictly limited number printed of each work, will cause them to rise in value by leaps and bounds', and he added with becoming modesty, 'I have a remarkable stock of these desirable volumes'.[26] The *Academy*, in a long article entitled 'The End of the Kelmscott Press' (12 December 1896), offered a balanced appraisal of Morris's achievement as a printer, concluding that 'on one thing, at any rate, we are all agreed, that for better or for worse, the folio edition of Chaucer which appeared this summer forms a great landmark in the history of printing, and were sufficiently monumental in itself, had he produced no other book, to render the names of the Kelmscott Press and of William Morris memorable for all times'.

Yet, when Cockerell walked out of the door of Sussex Cottage on the last day of March 1898, he did not succeed in putting the problems of the Kelmscott Press behind him. He had agreed to sell the presses and other equipment to C. R. Ashbee's Essex House Press (which will figure again in the next chapter), and several of Morris's printers, including Binning, decided to join forces with Ashbee as well. Eventually Hooper, Catterson-Smith, and Douglas Cockerell also contributed their services to the Essex House Press; Ashbee's decorated initials were freely adapted from Morris's, and in one of his books he even invoked the name of the Kelmscott Press in the colophon. Morris's old friends and associates were outraged by what they considered to be a shameless exploitation of a minor connection. To Walker it seemed futile to try to imitate the Kelmscott Press. 'The main-spring is broken,' he remarked with his usual moderation to Ashbee, 'and when the watch stops there is nothing further to be done.'[27] Others used less temperate language. 'The Press was brought to a definite end, and no one has the right to pose as successor to William Morris,' W. H. Bowden declared in a letter to a newspaper.[28] Cockerell breathed threats of lawsuits, but Ellis counselled restraint. 'I certainly do not think Ashbee worth taking any heed of,' he wrote to Cockerell in June 1899. 'It is only the inmates or candidates for Earlswood [an Asylum for Idiots in Sussex], who are likely to confuse Ashbee with Morris, & I don't think they are worth troubling about. . . . Ashbee's initial difficulty I take to be entire ignorance of [printing] whether from the literary or artistic point of view.'[29] Ellis was slightly embarrassed about the uproar, because he had earlier agreed to edit Shakespeare's poems for Ashbee (just as he had done for Morris), and he added jovially in his next letter, 'Perhaps you will learn in his next circular that he has bought me as a freehold. But I still think him quite unworthy of powder and shot.'[30]

Cockerell devoted a good deal of time and energy during the closing years of the century to matters of this sort. In general he was much more disturbed than Ellis and Walker were by supposed threats to the reputation of Morris or the Kelmscott Press. When a copy of the *Chaucer* in unbound sheets appeared in a London auction room in the summer of 1899, Cockerell became agitated because he was convinced that it must have been stolen from the pressroom; a lengthy correspondence between him and Ellis ensued, with Ellis again advocating a calm response. (The mystery was never solved, though another set of *Chaucer* sheets was offered for sale by Sotheby's the following year; in both instances May Morris's friend Robert Steele and Sparling appear to have been implicated.)

Walker favoured a proposal to print Morris's collected works in the Golden type at the Chiswick Press, but Cockerell vetoed it, and only five slender volumes of lectures by Morris were printed thus. The question of the ultimate disposition of the Kelmscott types continued to vex Cockerell for several decades. When May Morris was revising her will in 1919, Cockerell suggested that the types 'might perhaps go to Oxford University, if Walker has nothing better to propose'.[31] Seven years later he wrote to her that, 'with Mr Hornby's approval [St John Hornby was now one of the trustees of the Morris estate], I am transferring the Kelmscott Press type from the Chiswick Press (which has now passed into hands that we know not of) to the Cambridge University Press, which is now doing very good work. Apart from the fact that I am in Cambridge for a while the Press here is a permanent concern, whereas the Chiswick Press was recently in difficulties – so I hope you will also think it a wise change.'[32] Cockerell recommended to May in 1933 that she give the types to the Cambridge University Press, but it was not until 1940 that they became the property of the Press by purchase.[33] Morris's founts have recently been deposited in the Cambridge University Library.

Today the flotsam and jetsam of the Kelmscott Press are scattered round the world. The types are in Cambridge; the woodblocks are in the British Museum; the presses are in Hammersmith, New York, and California; Morris's 'bloomer book' is in Oxford; the address book and the ledger of the Press are in New York; a smaller account book is in Japan; and huge mounds of paper – all those precious scraps of proofs and trial pages so assiduously collected by Cockerell – have floated away to several continents. Confronted with these sad reminders of the impermanence of all human endeavours, we may perhaps find some consolation in the thought that the more durable monument to the memory of William Morris and the Kelmscott Press is, as the *Academy* suggested, the fifty-two wonderful books, particularly the *Chaucer*, printed in Hammersmith between 1891 and 1898.

TEN ❧ THE LEGACY OF MORRIS AND HIS PRESS

THERE must have been moments when Morris, deafened by applause for his brilliant and famous adventure in printing, wondered privately whether the Kelmscott Press measured up to his own standards of achievement. 'I aim at the idea and the ideal, and never get beyond the "collector",' Cobden-Sanderson lamented, and Morris might well have echoed the complaint.[1] In 1900 C. R. Ashbee, visiting the Caxton Club, a collectors' organization in Chicago, heard the members speak of 'Morrises' as though they were particularly lucrative shares: one man expressed his enthusiasm for the rising value of the Kelmscott books by declaring that 'no trust that ever was formed has paid such a rate of interest'.[2] In *News from Nowhere* Morris envisioned a socialist Utopia in which all men and women would have the opportunity to create beautiful objects, and the fruits of their labour would be generously distributed according to need; but, as Morris acknowledged in his evocative, bitter-sweet last chapter, such a world exists now only in our dreams and visions, and for the moment the ugly reality of 'dingy Hammersmith' (or greedy Chicago) prevails. The Kelmscott Press books, intended to symbolize a protest against the ethos of Victorian industrial capitalism, became themselves, in all their opulent splendour, an example of conspicuous consumption. The irony of this improbable transmutation was not lost on Morris.

If Morris found it difficult to evaluate the success of his Press, it is not surprising that others, from 1891 onwards, have offered remarkably contradictory appraisals of the Kelmscott books. Probably the majority of bibliophiles have always been willing to accept the proposition trenchantly expressed by Emery Walker in 1930, that 'Morris showed how a printed book might be on its own plane a work of art'; but the *Daily Chronicle* was undoubtedly also speaking for a vocal minority when it declared in 1896, 'Truly it is awful to contemplate the terrible effect that the Morris doctrine of decorative illustration has had upon the world.'[3] What is puzzling is that evidence to support either of these views lies abundantly at hand: the immediate consequence of the Kelmscott Press was to produce a horde of imitators who lacked Morris's skill and sensitivity as a book-designer, yet the ultimate impact of the Press has been so powerful and pervasive that it is not easy even now, a century later, to assess. The issue is further clouded by Morris's insistence that his typographical adventure was linked with a root-and-branch rejection of modern life. If we demand to know, therefore, what Morris's experiment in bookmaking 'means' to us today, we are asking a difficult question for which there are no unambiguous answers.

The most ardent imitator of the Kelmscott Press typography in England was C. R. Ashbee, who had established the Guild of Handicraft in London's East End in 1888.[4] Ashbee saw himself as merely putting Morris's principles into practice, though with greater emphasis on the workshop rather than on the individual craftsman, but he received little encouragement from Morris, who suspected rightly that Ashbee wished to reform, not overthrow, the capitalist system. Laurence Hodson, a wealthy young businessman with an interest in Pre-Raphaelitism and the Arts and Crafts movement, visited the Kelmscott Press in December 1897, and early in the following year he provided funds that made it possible for Ashbee to buy the Press's equipment. Ashbee was overjoyed, since the transaction offered him a symbolic claim to discipleship that had been denied while Morris was still alive: 'the great event of the week had been the purchase of the Morris Press by the Guild of Handicraft,' he wrote at the end of March 1898. 'Great jubilation! . . . So next Monday there will be vans of presses and gages, and chairs and tables and paper and machinery – lock, stock and barrel moving East from Hammersmith. We are thinking of a musical procession. We engage Morris's printers and carry on the great tradition.'[5]

To Cockerell's annoyance, Ashbee emphasized the Kelmscott connection by adopting the same ink, paper, and vellum as Morris had used at the Kelmscott Press; illustrations were to take the form of wood-engravings, with, in some instances, illuminations supplied by Florence Kate Kingsford (who later became Mrs Cockerell). Since Morris's types did not constitute part of his purchase, Ashbee printed the earliest books of the Essex House Press in Caslon, but he was eager to follow Morris's example by creating his own typefaces. Ashbee's designs were known as the Endeavour type (named after the first book in which it appeared, *An Endeavour towards the Teaching of John Ruskin and William Morris*, published in 1901) and the King's Prayer Book type, a larger, modified version of the Endeavour. 'In the design of these founts I sought to follow upon the lines laid down by Morris,' Ashbee wrote. 'Development appeared possible in the direction of combined letters, and my bias was towards a black (in preference to a grey) page and the introduction of a rather gothic and perhaps uncial feeling into the roman character.'[6] Ashbee's types are in fact ugly and eccentric, and, if one wished to argue the case that Morris's influence on subsequent type-designs was in the main pernicious, the Endeavour and King's Prayer Book types would probably constitute Exhibit A.

Figure 88

The Essex House Press printed eighty-three books over a period of eleven years, and often they rise above the limitations of Ashbee's types and wood-engraved initials (the latter more art nouveau than Morrisian): the illustrations, some of them by Ashbee himself, are undeniably attractive, and even at their weakest the books now have a certain period charm, not unlike that of an awkward Victorian building which has mellowed with age. In 1902 the Essex House Press, along with the Guild's other craft workshops, moved to Chipping Campden in the Cotswolds, and there, for six years, Ashbee and his men strenuously searched for the simple life far from the Modern Babylon. They sang newly revised texts of old folk-songs (one stanza of 'Joan's Ale' celebrated a compositor who joins the 'jovial crew' and declares, 'The printer's worth his pay | Especially if he come Kelmscott way!'), explored the countryside together, shared their profits, and attempted to revive a somewhat sentimentalized version of the medieval guild. It was a noble experiment, but, as Morris had predicted, the guild ideal was difficult to sustain in a society dominated by a hostile economic system. The Essex House Press found itself increasingly in competition with machine-press books patterned after the Kelmscott Press style, and the remoteness of Chipping Campden interfered seriously with efforts to market the Press's products. Ashbee later insisted that all the English private presses were victims of an unfounded optimism: 'What we failed to understand was (*a*) that our public was limited; (*b*) that the competitive system was so devised that the various presses which should have been co-operating for the maintenance of standards were at times forced into competition; and (*c*) that ultimately "the Trade", seeing there was money in it, would set up sham presses to catch that money.'[7]

As early as the summer of 1904 Ashbee's printers were put on half-time, yet he was reluctant to close the Press because of its potent symbolism suggesting that Morris's mantle had fallen upon his

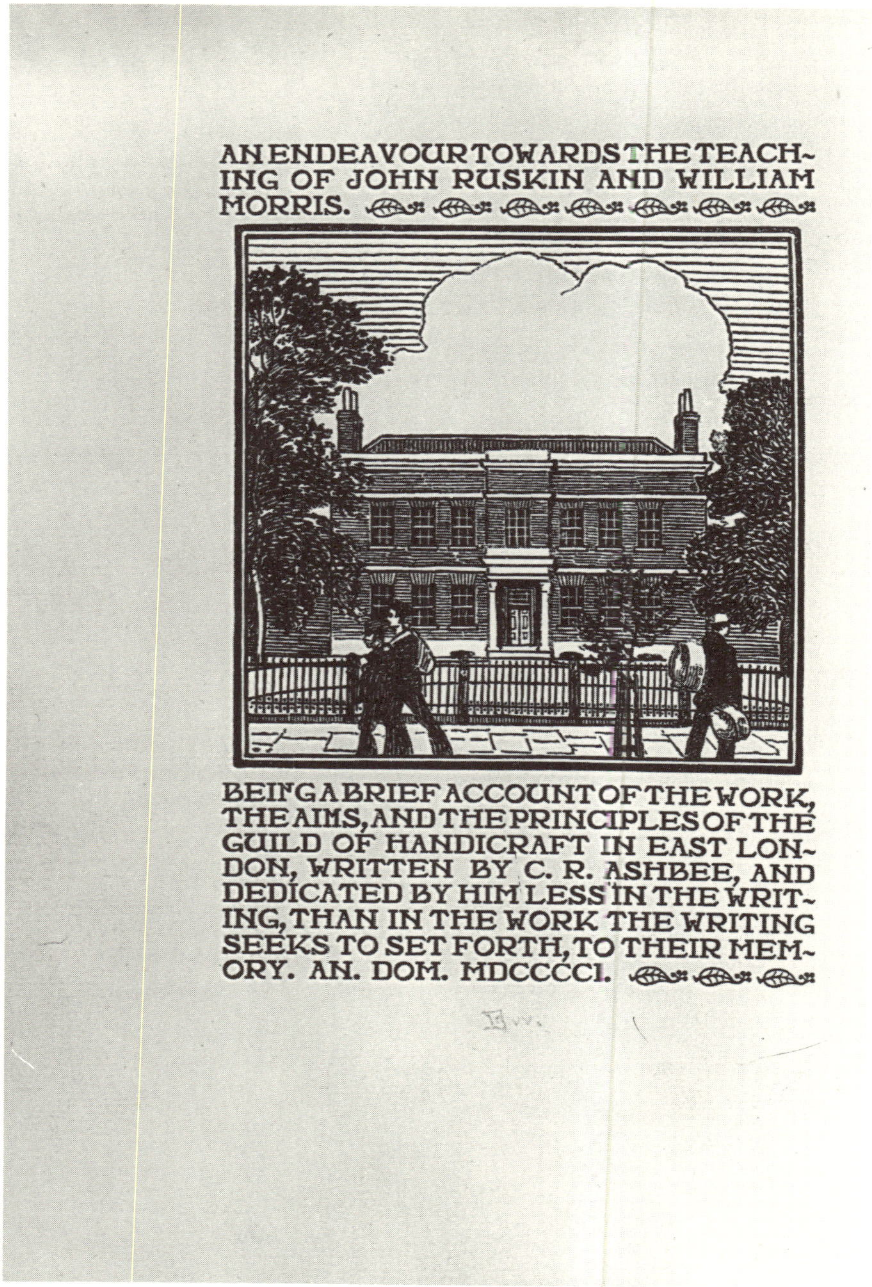
AN ENDEAVOUR TOWARDS THE TEACHING OF JOHN RUSKIN AND WILLIAM MORRIS.

BEING A BRIEF ACCOUNT OF THE WORK, THE AIMS, AND THE PRINCIPLES OF THE GUILD OF HANDICRAFT IN EAST LONDON, WRITTEN BY C. R. ASHBEE, AND DEDICATED BY HIM LESS IN THE WRITING, THAN IN THE WORK THE WRITING SEEKS TO SET FORTH, TO THEIR MEMORY. AN. DOM. MDCCCCI.

Figure 88 When the Kelmscott Press closed in the spring of 1898, much of its equipment (and some of its staff) was acquired by C. R. Ashbee, who believed that his Essex House Press would carry on Morris's work. Ashbee, despite good intentions, lacked Morris's gifts as a type-designer.

shoulders. In August 1906, however, Ashbee and Binning, the stubborn socialist compositor who had worked under Morris at the Kelmscott Press, bore the sad responsibility of stock-taking before Morris's presses and other equipment were sold: 'old Binning and I entered them up and chalked them down and set out value on them – so much for the working price and so much for the sentiment, except when we came to sizing up the Press on which the Chaucer was printed and we raised that a full £5 above its market value.'[8]

If the Essex House Press books represent a debased rendition of the Arts and Crafts typographical style, there were other English private presses at the turn of the century which displayed the influence of Morris more positively, and three of them deserve separate consideration; the Doves Press, the Ashendene Press, and the Vale Press. The Doves Press, founded in 1900 by Cobden-Sanderson and Emery Walker, was the most distinguished successor of the Kelmscott Press and was patterned after the Kelmscott model very closely indeed. The Batchelor paper and vellum bindings were virtually identical; the Doves type was based on the same photographs of pages printed by Jenson and Rubeus that Morris had used in designing the Golden type;[9] and a Kelmscott page stripped of its ornaments does not differ fundamentally from a Doves page. Given the close relationship of both its proprietors to Morris, the similarities are of course hardly unexpected. One of the pressmen at the Doves Press had even served an apprenticeship at the Kelmscott Press, and J. H. Mason records that a portrait of Morris was pinned over the fireplace in the Doves composing room.[10]

Figure 89

But it is misleading to emphasize too strongly the parallels between Hammersmith's two celebrated private presses. Cobden-Sanderson did not share Morris's bias in favour of medieval typography and ornament; his own preferences (and Emery Walker's preferences too, for that matter) led him to sixteenth-century French and Italian books as exemplars, and the volumes he produced at the Doves Press were severely simple in design, unadorned by ornament or illustration. The only form of decoration that Cobden-Sanderson would permit himself was the occasional calligraphic initial, the most-photographed example of which is Edward Johnston's beautiful 'I' on the opening page of the Doves Bible (1903–5). 'Candidly, I do not think that William Morris himself is a great printer,' Cobden-Sanderson wrote to Cockerell in 1909: 'that his work is remarkable and in itself constitutes an epoch, is, again, another matter. William Morris came to

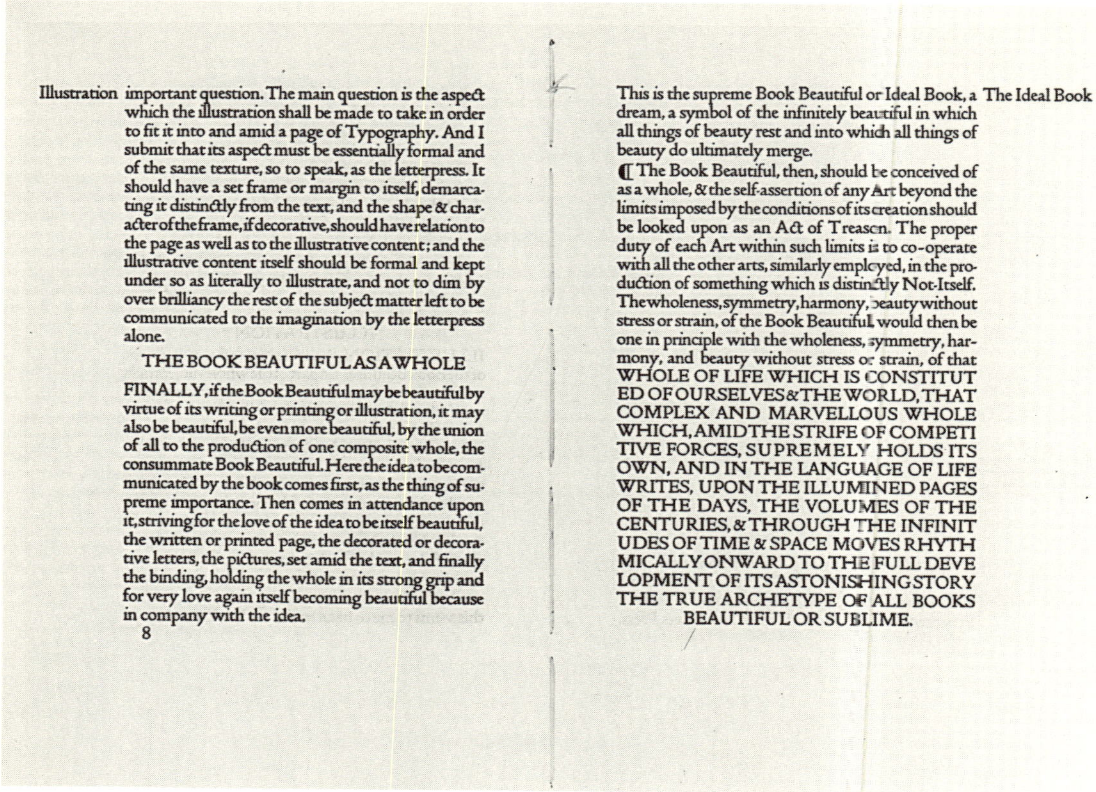

Illustration important question. The main question is the aspect which the illustration shall be made to take in order to fit it into and amid a page of Typography. And I submit that its aspect must be essentially formal and of the same texture, so to speak, as the letterpress. It should have a set frame or margin to itself, demarcating it distinctly from the text, and the shape & character of the frame, if decorative, should have relation to the page as well as to the illustrative content; and the illustrative content itself should be formal and kept under so as literally to illustrate, and not to dim by over brilliancy the rest of the subject matter left to be communicated to the imagination by the letterpress alone.

THE BOOK BEAUTIFUL AS A WHOLE

FINALLY, if the Book Beautiful may be beautiful by virtue of its writing or printing or illustration, it may also be beautiful, be even more beautiful, by the union of all to the production of one composite whole, the consummate Book Beautiful. Here the idea to be communicated by the book comes first, as the thing of supreme importance. Then comes in attendance upon it, striving for the love of the idea to be itself beautiful, the written or printed page, the decorated or decorative letters, the pictures, set amid the text, and finally the binding, holding the whole in its strong grip and for very love again itself becoming beautiful because in company with the idea.

8

This is the supreme Book Beautiful or Ideal Book, a dream, a symbol of the infinitely beautiful in which all things of beauty rest and into which all things of beauty do ultimately merge. The Ideal Book

¶ The Book Beautiful, then, should be conceived of as a whole, & the self-assertion of any Art beyond the limits imposed by the conditions of its creation should be looked upon as an Act of Treason. The proper duty of each Art within such limits is to co-operate with all the other arts, similarly employed, in the production of something which is distinctly Not-Itself. The wholeness, symmetry, harmony, beauty without stress or strain, of the Book Beautiful would then be one in principle with the wholeness, symmetry, harmony, and beauty without stress or strain, of that WHOLE OF LIFE WHICH IS CONSTITUTED OF OURSELVES & THE WORLD, THAT COMPLEX AND MARVELLOUS WHOLE WHICH, AMID THE STRIFE OF COMPETITIVE FORCES, SUPREMELY HOLDS ITS OWN, AND IN THE LANGUAGE OF LIFE WRITES, UPON THE ILLUMINED PAGES OF THE DAYS, THE VOLUMES OF THE CENTURIES, & THROUGH THE INFINITUDES OF TIME & SPACE MOVES RHYTHMICALLY ONWARD TO THE FULL DEVELOPMENT OF ITS ASTONISHING STORY THE TRUE ARCHETYPE OF ALL BOOKS BEAUTIFUL OR SUBLIME.

Figure 89 The typography of the Doves Press was directly modelled on that of the Kelmscott Press, particularly in its emphasis on close and even word-spacing. The sentiments of this essay on *The Book Beautiful* by Cobden-Sanderson, however, would doubtless have struck Morris as too pretentiously phrased.

printing with a mind set on decoration, and with a mind overscored with tapestry and woven effects, all which he reproduced where they were not wanted, on the pages of his books. And many of his effects, as I told him at the time, are "typographical impertinencies", and utterly destructive of the page as an expression of the author's thought.'[11]

Cobden-Sanderson brought his partnership with Walker to an end in 1909, and the Press itself closed in 1916; in order to prevent Walker from using the Doves fount in future books, Cobden-Sanderson melodramatically tossed the punches, matrices, and types into the Thames under cover of darkness. The antagonism that erupted between Walker and Cobden-Sanderson is, however, only a partial explanation of the ultimate collapse of the Doves Press: a deeper reason lies in the rigidity of Cobden-Sanderson's vision of what he termed 'the Book Beautiful' (a phrase sometimes erroneously attributed to Morris but presumably derived from Oscar Wilde's 'House Beautiful'). The unhappy truth is that the Doves books are boring in their cold perfection, each one an immaculate clone of the other. Albeit stunningly lovely, they disclose none of the vitality that one finds in even the least satisfactory Kelmscott books. Yet this extreme simplicity of the Doves Press volumes mediated between the ornate pages of the Kelmscott Press and the more austere tastes of twentieth-century typographers. By demonstrating that Morris's principles of book-design could be applied with equal success to a completely undecorated text, Cobden-Sanderson and Walker translated the Morrisian Ideal Book into terms acceptable to printers and publishers who found Morris's own style intolerable. Giovanni Mardersteig, whose Officina Bodoni has been regarded by some as the most outstanding private press of the twentieth century, is only one notable example of a young man who first became excited about the possibilities of fine printing through the example of the Doves Press.[12] What we must always bear in mind, however, is that the Doves Press could not have existed without the theories and practice of Morris lying behind it.

C. H. St John Hornby's Ashendene Press, on the other hand, had already been launched before Hornby met Morris, and his first books resemble, if anything, the Revd Daniel's little private productions at Oxford. Hornby began printing on a modest scale in 1894, but in March 1895 he made a pilgrimage to the Kelmscott Press, where he saw the *Chaucer* in production, talked to Morris in his study, and had tea with both of the Morrises in their dining-room.[12] Hornby, like

Cobden-Sanderson and Walker, was sufficiently strong-minded to avoid a direct imitation of the Kelmscott books, for, as he said, the *Chaucer*, though 'in many ways the greatest example of a printed book of all time', is 'inimitable'.[13] 'But', Hornby observed in 1935, 'we can follow the spirit of his work, and that spirit has without a doubt exercised an enormous influence during the past 40 years upon printing throughout the world. Even those who profess to make light of Morris as a printer are themselves largely influenced by what was best in the Morris tradition. . . .'[14]

In 1900 Hornby met Cockerell and through him Walker (Cockerell was at this time a partner in Walker's business), and they persuaded him that it would be possible to create his own type fount for the Ashendene Press. After consulting Robert Proctor, they decided to turn to the semi-Gothic of Sweynheym and Pannartz as their model: Hornby was of course aware that Morris had once begun designing a type based on the Sweynheym–Pannartz face, and this no doubt added to its appeal. The resulting Subiaco type was so successful that he continued to use it regularly until the Ashendene Press ceased operation in 1935. The Press had a long, illustrious history, much of it falling outside the boundaries of this book; but what Hornby's career as a printer especially demonstrates is that it was (and is) possible to be a disciple of Morris yet to chart one's own course in book-design. The Ashendene volumes convey the impression that Hornby still speaks Morris's language but has translated it into a new and unexpected dialect.

The Vale Press (1896–1904) of Charles Ricketts offers another instructive example of how the Morrisian influence could be diluted by other forces. Ricketts admired Morris's books immensely, but he recognized that it was only too easy for the Kelmscott style to be debased by less talented designers. In an essay entitled *A Defence of the Revival of Printing* (1899), Ricketts commented on the 'ornamental tendency' in Morris which led him to create books 'ornamentally conceived and of ornamental use. With such pomp of circumstance', he observed, 'they became more and more large, necessitating ornamental type such as the Troy type; certain books like the Chaucer should be perused at a lectern.' When the designing of such a book fell into the hands of a typographer who only half-understood Morris's principles, the results were disastrous: 'deprive ornament of its excellence, relate it to some casual fount, throw in with this a photogravure frontispiece, badly blocked boards, various papers, &c., and the new usual ornamental book becomes unbearable; instead of a revival in the shaping of books we have a new cumbersome trade article.'[15]

The type that Ricketts designed for the Vale Press is a bold, slab-serifed Roman which openly proclaims its indebtedness to the Golden type, and Ricketts's borders also owe something to Morris's, though the general effect of a Vale page is much lighter, more delicate and open. Ricketts, like Cobden-Sanderson, found the dark coloration of the Kelmscott books unattractive and sought inspiration instead in the less cluttered pages of the early Venetian printers. 'I would define the page of a fine Kelmscott book as full of wine, an Italian book as full of light,' he declared.[16] Yet if Hornby veers off into a different dialect of the Morrisian language of bookmaking, Ricketts seems to be speaking in a different tongue altogether: the earnestness and solidity of the Kelmscott style have, through some strange sea-change, become mannered and precious.

Ricketts's deepest links, despite his professed emulation of Morris, are in fact with the art-nouveau book that coexisted with the Morrisian style during the final decade of the century. John Russell Taylor, in a lengthy study of the art-nouveau fashion in typography, has offered a useful and concise distinction between the two traditions: 'Arts and Crafts [i.e. the Kelmscott-inspired style] is essentially, as the label suggests, direct, unaffected, simple (or at least simpliste); it is fairly solid and robust; and it is solidly traditional, often with a sort of revivalist fervour. Art nouveau, on the other hand, is essentially sophisticated and eclectic, accepting and assimilating influences from here, there and everywhere – Japanese, Greek, Celtic, Pre-Raphaelite. . . , Rococo, Dutch, Blake, Victorian industrial design, etc., etc. Moreover, art nouveau is above all highly mannered.'[17] The origins of art nouveau are obscure and complex; suffice it to say here that Aubrey Beardsley was responsible for introducing the style in book illustrations and ornaments, and, as we saw earlier, this came about as a direct result of his failed attempt to illustrate a Kelmscott book. When Morris peremptorily refused his frontispiece for *Sidonia the Sorceress*, Beardsley saw Dent's edition of *Le Morte Darthur* (1893–4) as a wonderful opportunity to achieve revenge by parodying the Morris–Burne-Jones conventions and modifying them in a sensual, decadent fashion. *Figure 90* The *Morte Darthur* was a great commercial success, and it effectively launched Beardsley's career as an illustrator. 'William Morris has sworn a terrible oath against me for daring to bring out a book in his

KING ARTHUR.

Book i. Chapter i.

HOW UTHER PENDRAGON SENT FOR THE DUKE OF CORNWALL AND IGRAINE HIS WIFE, AND OF THEIR DEPARTING SUDDENLY AGAIN.

IT befell in the days of Uther Pendragon, when he was king of all England, and so reigned, that there was a mighty duke in Cornwall that held war against him long time. And the duke was called the duke of Tintagil. And so by means King Uther sent for this duke, charging him to bring his wife with him, for she was called a fair lady, and a passing wise, and her name was called Igraine. So when the duke and his wife were come unto the king, by the means of great

Figure 90 Aubrey Beardsley's ornaments in the Dent edition of *Le Morte Darthur* (1893–4) represent a savage parody of the Kelmscott Press manner.

manner,' Beardsley exulted in 1893. 'The truth is that, while his work is a mere imitation of the old stuff, mine is fresh and original.'[18]

Beardsley's clever travesty of the Kelmscott books was both a personal rebellion against his respectable elders and a *fin-de-siècle* response to Morris's medievalism, which seemed to Beardsley's generation a tiresome symptom of mid-Victorian values. When Leonard Smithers proposed establishing a magazine with the very ninetyish name of *Peacock*, Beardsley wrote to him in 1897, 'On the art side I suggest that it should attack *untiringly and unflinchingly* the Burne-Jones and Morrisian medieval business, and set up a wholesome seventeenth and eighteenth-century standard of what picture making should be.'[19] Beardsley's crusade was not without effect. The *Studio* during the latter part of the decade sponsored periodic competitions for illustrated title-pages, and the winning pages were almost always more in the Beardsley vein than in a style of which Morris or Burne-Jones would have approved.

There was another emerging typographical fashion in the 1890s, most brilliantly represented in the skilfully designed books of Whistler, which emphasized asymmetric layout, minimal ornamentation, and mechanical methods of printing: the 'Aesthetic book' (as one scholar has called it)[20] seems to owe almost nothing to Morris's example, except perhaps a shared preoccupation with minute and careful planning of the printed page. Some historians are persuaded that there would have been a great revival of printing in England at the end of the nineteenth century even without the Kelmscott Press; this overstates the case, since Morris went well beyond the achievements of artists like Beardsley and Whistler by reaffirming the primary importance of fine materials (paper, ink, bindings) and by asking searching questions about the structure of the book that never entered their minds. There is no doubt, however, that other views of book-making were clamouring to be heard at the time, and that in many instances people accepted only selected portions of the typographical gospel according to Morris. Bernard Shaw, for example, was strongly impressed by *The House of the Wolfings* and *The Roots of the Mountains* (Morris's two pre-Kelmscott Press experiments at the Chiswick Press), became a Caslon-addict, and for decades badgered his printers whenever they failed to produce the closely spaced, evenly textured page that Morris had advocated.[21] Shaw never aspired to emulate the Kelmscott Press volumes, yet he had an intelligent grasp of Morris's typographical principles, which he attempted to follow while at the same time ignoring the private mannerisms of Morris's own books.

Shaw's case exemplifies the difficulties in tracing the reputation and influence of the Kelmscott Press during the final decade of the century, for Morris's ideas (which seem distinctive enough in retrospect) about book-design mingled subtly and confusingly with the notions of other practitioners of bookmaking. Nor was there any unanimity of opinion about the Kelmscott Press, which evoked astonishingly varying responses from readers throughout the decade. The earliest comments on the Press that appeared in newspapers and magazines were on the whole very favourable, and even the printers' trade journals printed occasional enthusiastic notices about Morris's activities in Hammersmith. The death of Morris in 1896 and the closing of the Kelmscott Press in 1898 stimulated further public praise of his work as a printer. Herbert P. Horne offered a judicious and lengthy appraisal of Morris's achievement in the *Saturday Review* (24 October 1896), finding the key to his success in the ideal of good craftsmanship; the *Library*, in commenting on an exhibition of the Kelmscott Press books mounted at the British Museum in 1898, remarked that 'the Kelmscott books may fitly take their place with those of Gutenberg and Schoeffer, of Zainer and Koberger, of Jenson and Wendelin of Speier, among the masterpieces of the printer's art'.[22] A. W. Pollard, reviewing the same exhibition in the *Guardian* (8 June 1898), found encouraging evidence of the spread of Morris's ideals, for 'three great firms – the Chiswick Press, Messrs. Constable, and the Clarendon Press – have already begun to use richer founts of type than have been seen in English books since black-letter went out of fashion, and with admirable effect'. When Henry R. Plomer wrote *A Short History of English Printing* for the English Bookman's Library (1900), the terminal date of his survey was, significantly, 1898, since he regarded the 'brilliant but short-lived career of the Kelmscott Press' as the culminating episode in a story that had begun with Caxton in 1476.

Figure 91

Yet this chorus of praise for the Kelmscott Press was by no means unchallenged. The most serious attack on Morris's principles of design grew out of an interview with him entitled ' "Master Printer Morris": A Visit to the Kelmscott Press' which was published in the *Daily Chronicle* on 22 February 1893; shortly thereafter the same newspaper printed a stinging reply called 'The Kelmscott Press: A Master Critic on a Master Printer'. The 'Master Critic' was apparently Morris's old adversary, Joseph Pennell, and the article was reprinted and discussed in at least two printers' journals.[23] It certainly pulled no punches: the

Figure 91 By the end of the last century, Morris had become linked in the popular imagination with Gutenberg, as the façade of the Edward Everard Building in Bristol demonstrates.

anonymous critic proclaimed that De Vinne in New York was printing finer books than Morris, assaulted the Troy and Chaucer types, and insisted that the Kelmscott volumes 'are filled with blemishes which would be ascribed to carelessness even in printers of lesser pretensions'.

The Kelmscott Press books did not again come under such strong attack until Morris's death, when Pennell once more made a brutal assault in the *Daily Chronicle*, the *British Printer* revived the old charge that Morris divided the lines of verse arbitrarily in his books, and the *British and Colonial Printer and Stationer* declared that the Kelmscott volumes were characterized by poor inking and register.[24] After the hyperbole lavished upon the Kelmscott Press in the preceding years, a reaction of this sort in 1896 was not unexpected. The *Academy* (12 December 1896) remarked that 'certain journalist writers' had used the occasion of Morris's death to 'air their opinions as to the superiority of the Continental and Transatlantic fashions of illustration over the style [of Morris] . . .; their opinions also as to the vast superiority of process blocks to hand engraving, machine-made papers and printing to hand-made papers and hand-printing, and so on'.

Other indications of the Kelmscott Press's status were equally ambiguous. The market value of the Kelmscott books – a good rough index of the esteem in which they were held by collectors – rose steadily until late 1903, when their prices inexplicably dipped. 'The truth is, collectors have tired of [the Kelmscott volumes] and will have none of them,' W. G. Menzies wrote in the *Connoisseur* (May 1905). 'That they may regain some of their lost prestige is possible, but that they will ever reach the high values of four or five years ago is improbable.' This prediction was wide of the mark, yet the pattern of prices is suggestive indeed. Within the printing trade there were also mixed responses to the Kelmscott experiment. Trade journals that had at first welcomed Morris's attempt at private printing gradually grew more hostile as his condemnations of commercial typography became more widely known through lectures and interviews. The Caslon typefoundry was reflecting a substantial body of opinion when it declared in 1897 that it could 'only condemn the fantastic types and methods adopted by Mr. Morris'.[25] John Southworth, in his book *Modern Printing* (1900), reluctantly admitted that the Kelmscott Press had attracted a great deal of attention, but 'few printers think there is much likelihood that what they regard as its peculiar characteristics will permanently influence present styles of book printing' (p. 280).

Yet printers continued to pay at least lip-service to Morris's high standards of craftsmanship even while they scorned his isolation from commercial realities. The Printing Number of *The Times* published on 10 September 1912, for example, though it was largely devoted to the burgeoning technology of the trade, prominently displayed a huge reproduction of a page of the Kelmscott *Chaucer*. An amusing example of this ambivalence can be seen on the cover of the *Monotype Recorder* of April 1913, which featured a facsimile of one of Morris's decorated initials; the *Recorder* had of course always been an enthusiastic advocate of increased mechanization, with early covers portraying a Monotype machine rising triumphantly into the sky above an old-fashioned compositor, often accompanied by such edifying slogans as 'The old order changeth, yielding place to new'.

Paradoxically, at the very time when Morris's reputation as a printer had declined, particularly in the printing trade, there were signs on every side of his influence making itself felt in more oblique ways. Caslon and Old Style type continued to gain ground in book-work, though Shaw, Walker, and the Chiswick Press as well as Morris's example contributed to the growing revival of old-face types. Founts based on the Golden type still proliferated in the first decade of this century. An examination of any representative group of English books published at that time reveals a slight but perceptible darkening of the page through the use of heavier faces and less leading. Volumes in Everyman's Library series, launched by Dent in 1905, displayed a Venetian typeface and ornaments that were intended to evoke memories of the Kelmscott Press. The Oxford University Press began issuing a handsome series of English classical texts in 1906, printed in the Fell types on pure linen rag paper; the Cambridge University Press a few years earlier acquired a new fount bearing a family resemblance to the Golden type which it used for printing limited editions of English classics.

But whereas the two university presses had special resources at their disposal, most printers before the First World War possessed unsatisfactory collections of typefaces. Harold Curwen, at the Curwen Press, was delighted when a shortage of lead for the munitions factories in 1915 enabled him to dispose profitably of vast quantities of types accumulated since the firm's founding: he cheerfully got rid of more than 200 founts, retaining only Caslon Old Face, Monotype Old Style No. 2, and two other faces.[26] Few other proprietors were willing to adopt such a drastic solution, if only because Caslon still seemed unsuitable for many varieties of commercial printing. As Morris had seen so clearly in 1888, English book-design could not be rejuvenated until the tired nineteenth-century modern-face types were supplanted by new founts modelled on classic old-face types. The earliest really successful effort to do this in England – and simultaneously to make the face generally available on the Monotype machine – came when Gerard Meynell and several associates commissioned an altered version of Caslon, named after their journal in which it was first used, the

Figure 92 *Imprint*, with the first issue appearing in January 1913. It is symptomatic that the *Imprint*, while saturated with the ideal of technological progress in printing, is also pervaded by the presence of Morris: the first article, 'Art and Workmanship' by W. R. Lethaby, quotes Morris in its second paragraph; the frontispiece of the Kelmscott Press *A Dream of John Ball* is reproduced; an editorial note declares, 'The Kelmscott Press, the Doves Press, and the Ashendene Press have done a great deal towards bringing about a Renaissance of Printing, and though the commercial printer is prone to speak slightingly of private presses, much of the improvement in his own work, especially in the design of the types he uses, is to be derived from this source.'

The *Imprint* survived for less than a year but is a harbinger of the resurgence of fine printing, by both private and commercial presses, that was to take place between the wars, a period that deserves to be regarded as the golden age of English printing. The transformation of the mass-produced book was especially remarkable: while the private presses, nearly all exhibiting Morrisian impulses, continued to pour forth handsome volumes of every variety, the renewal of printing now frequently took the form of books produced under normal trade conditions yet displaying a refinement of typography and illustration hitherto associated only with small, experimental presses. Morris's principles, after lying almost dormant for two decades, suddenly emerged during the 1920s into the different world of commercial publishing, and were applied, with modifications, to large editions of machine-printed books. What complicates this pattern of influence is that the typographical pioneers of the 1920s, committed to modern printing technology, in general were reluctant to acknowledge their ties to Morris or the private-press movement; they prided themselves on being forward-thinking men of a brave new century and professed scorn for the preciousness of Kelmscott-style ventures.

Uncertainty about Morris's position in the history of printing became, therefore, even more acute after 1918, and exercising his

The Imprint

ART AND WORKMANSHIP: By Professor W. R. LETHABY

WE have been in the habit of writing so lyrically of art and of the temperament of the artist that the average man who lives in the street, sometimes a very mean street, is likely to think of it as remote and luxurious, not " for the likes of him." There is the danger in habitual excess of language that the plain man is likely to be frightened by it. It may be feared that much current exposition of the place and purpose of art only widens the gap between it and common lives.

A proper function of criticism should be to foster our national arts and not to frighten timid people off with high-pitched definitions and far-fetched metaphors mixed with a flood of (as Morris said) " sham technical twaddle." It is a pity to make a mystery of what should most easily be understood. There is nothing occult about the thought that all things may be made well or made ill. A work of art is a well-made thing, that is all. It may be a well-made statue or a well-made chair, or a well-made book. Art is not a special sauce applied to ordinary cooking; it is the cooking itself if it is good. Most simply and generally art may be thought of as THE WELL-DOING OF WHAT NEEDS DOING. If the thing is not worth doing it can hardly be a work of art, however well it may be done. A thing worth doing which is ill done is hardly a thing at all.

Fortunately people are artists who know it not—bootmakers (the few left), gardeners and basketmakers, and all players of games. We do not allow shoddy in cricket or football, but reserve it for serious things like houses and books, furniture and funerals.

If it is necessary that everything must be translated into words, our art critics might occupy quite a useful place if they would be good enough to realise that behind the picture shows of the moment is the vast and important art of the country, the arts of the builder, furniture maker, printer and the rest, which are matters of national well-being.

It is doubtful if we have it in us to form a leading school of painting at

1 B

Figure 92 The initial number (January 1913) of the *Imprint*, a journal that signalled the beginning of the revival of commercial typography in twentieth-century Britain. It is significant that the first article is by W. R. Lethaby, who was deeply influenced by Morris's ideals, and contains a direct reference to Morris in the second paragraph.

ghost was a common activity. The publication of Sparling's history of the Kelmscott Press in 1924 served as the occasion for much energetic debunking, as in this review in *The Times Literary Supplement* (22 January 1925): 'most students of typography will find this a profoundly depressing book. The book itself – the way in which it is printed – the Morris designs and the Burne-Jones illustrations – are all of a piece, and all have the pathos of dead enthusiasms. . . . It is certain that the reaction against Morris's canons of printing is not the outcome either of moral baseness or of our system of society; it is due to the discovery that in many things Morris was wrong – wrong as an antiquary, wrong as a practitioner.' During the interwar period Holbrook Jackson, one of this century's most influential writers on printing history, kept up a series of attacks upon the Kelmscott Press in articles, lectures, and books: reading a paper at the Morris Centenary Dinner of the Double Crown Club on 2 May 1934, he seized the occasion to describe Morris's books as pretentious, ill conceived, and unreadable. They were, he said, 'museum pieces, typographical monuments – beautiful and ineffectual angels beating in the void their luminous wings in vain'.[27] At the same dinner James Shand paid tribute to 'Morris's magnificent technical achievements, yet found that he meant nothing to the printers of today'.[28] Paul Johnston's *Biblio-typographical: A Survey of Contemporary Fine Printing Style* (New York, 1930), began: 'The fallacy that William Morris is the patriarch of modern fine printing is as common as it is inaccurate.'

Naturally Morris had his defenders after the war, chief of whom was Bernard Newdigate, the accomplished printer at the Shakespeare Head Press, who wrote a regular column entitled 'Book Production Notes' for the *London Mercury* between 1920 and 1937. Newdigate repeatedly reminded his readers of Morris's achievement at the Kelmscott Press, as in this representative passage from a column of 1934: 'A few years since it came to be the fashion to disparage Morris's work as a printer, or, at best, to damn it with faint praise; and this notwithstanding that in the eight-and-thirty years which have passed since his death, no books have surpassed his in excellence and sheer beauty, and none have been printed anywhere save at the Doves and the Ashendene presses which are worthy to rank with them.'[29] Newdigate was one of several typographical enthusiasts – the others were Francis Meynell, Stanley Morison, Oliver Simon, and Holbrook Jackson – who met in London during 1922 in order to establish a publishing society that would issue one book a year, set and printed by mechanical means yet produced to the highest standards. The would-be founders immediately discovered themselves divided ideologically: 'There were disputes on the texts suggested and the manner of printing them and Newdigate was felt to have too strong a tendency towards the ideas and practice of William Morris,' Simon recalled.[30] The society was disbanded before it was even organized. Though no more than a minor episode, the stillborn Fleuron Society represented emblematically the struggle between Morrisian handicraft methods and the modern mechanization of printing.

This was in fact the very issue that rendered Morris's accomplishments so problematic during the second and third decades of our century. Among the intelligentsia of the day there were urgent calls to harness the machine to aesthetic and humane ends, emphasizing the importance of design in industrial production, and Morris's Luddite tendencies now seemed distinctly embarrassing. In England the most eloquent expression of this viewpoint came from Aldous Huxley, who wrote in 1928: 'It has become obvious that the machine is here to stay. Whole armies of William Morrises and Tolstoys could not now expel it. . . . The sensible thing to do is not to revolt against the inevitable, but to use and modify it, to make it serve your purposes. Machines exist; let us then exploit them to create beauty – a modern beauty, while we are about it. For we live in the twentieth century; let us frankly admit it and not pretend that we live in the fifteenth. . . . The printer who makes a fetish of hand-work and medieval craftsmanship, who refuses to tolerate the machine or to make any effort to improve the quality of its output, thereby condemns the ordinary reader to a perpetuity of ugly printing.'[31] Huxley thus transforms Morris's profoundly moral rejection of industrialism into a conspiracy against the reader of modest means; the same point was often made during Morris's lifetime (Talbot Baines Reed, for example, said something very like this in a letter to Emery Walker written in 1888[32]), yet Huxley's polemic is striking both because of its cleverness and because of its assumption that the benefits of mechanization and mass-production are so self-evident that they no longer require defence.

Even those whose fundamental attitudes were close to Morris's acknowledged the supremacy of the machine in the contemporary world. Eric Gill's brilliant but eccentric *Essay on Typography* (1931) illustrates how Morrisian (and Ruskinian) ideas could, with the passing of time, be metamorphosed into mysterious new shapes. Gill, a master of *reductio ad absurdum*, argues that, since industrialism and handicrafts represent two separate worlds of human experience, machine-made goods should be plain and simple, not imitations of craft artefacts. Once we have decided to make the machine our master, we must learn to accept the only kind of beauty it can create. On the Continent the Bauhaus movement had been advancing the same case, but what sets his reasoning apart from theirs is that it is coloured by intense hatred of our industrial civilization. When one probes beneath the surface of Gill's writings, his unlikely advocacy of functionalism proves to be not very different from the position of Morris, whose medievalism Gill claimed to despise.[33]

Morris's understanding of machinery was certainly more complex than Huxley's caricature of him as a regressive escapist might suggest. The confusion about his perspective was compounded by the simple-minded propaganda of some of his disciples, particularly Sparling, who offered the world a one-dimensional version of Morris's ideas on handicrafts: Sparling could not even bring himself to admit that an Albion press is a machine.[34] When the restoration of Morris's reputation came about in the following decades, therefore, it signalled a growing realization that Morris had anticipated many twentieth-century notions of functionality and faithfulness to the nature of materials. The fullest exposition of this revisionist view of Morris is to be found in Pevsner's *Pioneers of the Modern Movement* (1936), which was retitled in the second edition *Pioneers of Modern Design: From William Morris to Walter Gropius*; but one can see the same pattern of coming to terms with Morris in the public and private pronouncements of Stanley Morison, who was largely responsible for the important programme of historic type revivals at the Monotype Corporation. Morison, like many of his contemporaries, viewed Morris's printing with distrust and delivered a steady stream of criticism of it during the 1920s and 1930s. In a 1944 lecture delivered in Edinburgh (published in Morison's *The Typographic Arts: Two Lectures*, 1950), however, he adopts a more Pevsnerian view of Morris, arguing that he was not really opposed to the machine-made book. Morison is able to make a persuasive case because Morris does indeed refer occasionally in his typographical writings to the application of good principles of design to commercially produced books. Yet to make Morris a crypto-modernist who had the misfortune to be born in the nineteenth century, as both Pevsner and Morison imply, is a form of special pleading that brings us no closer to the core of Morris's ideas than do Sparling's purple passages attacking machinery.

Another issue on which Morris was seen to be in the wrong was the ornamentation of books; one of the most widespread manifestations of anti-Victorian sentiment in the 1920s was the conviction that ornament, especially of the historicist kind, was morally and aesthetically inappropriate for a machine-made product. The application of this doctrine to book-design led to what one French critic later described, in a different context, as a 'style calviniste',[35] characterized by a puritanical fear of any decoration of the text. Holbrook Jackson offered a trenchant summary of the doctrine in 1924: 'Decadent printing is always ornamental printing. When typographical imagination fails it becomes decorative.'[36] It was the Bauhaus school that adopted the idea most zealously, eschewing all ornamental patterns except for the functionalist structure of blocks of plain sanserif type. In England the hostility to ornament rarely reached such a pitch; even Morison, who espoused a different variety of typographical austerity, admitted that the taste for ornament is an essential human need.[37] So long as ornament was seen as a symptom of moral weakness, however, it was impossible for Morris's contributions to fine printing to be acknowledged fully.

Whereas Morris's contemporaries had been startled by his rejection of typographical orthodoxies, in the early decades of this century the Kelmscott books were seen as the final, overripe expression of Victorian ornamental eclecticism. Under these circumstances, how can one claim any connection at all between Morris's theories of book-design and the renewal of English printing that took place after 1918? There is indeed a direct line of descent from the Kelmscott Press to the 1920s and 1930s, but, because Gill's and Morison's generation found it psychologically impossible to acknowledge the ancestry of their typographical views, the line is not always easy to trace; we can glimpse the connection, however, by looking briefly at the careers of two men who, with Emery Walker, were largely responsible for bringing the Morrisian vision of the Ideal Book into the market-place.

J. H. Mason, an admirer of Morris and compositor at the Doves Press, began teaching evening classes in printing at London's Central School of Arts and Crafts in 1905, and four years later, on the recommendation of Walker and Cobden-Sanderson, he was appointed the superintendent of the Printing School.[38] In that position he was responsible for training many of the next generation of British printers. The typographical style that he imparted to his students was a simple adaptation of the Kelmscott/Doves manner: closely spaced

Caslon Old Face (almost the only type of which he approved) printed on handmade, dampened paper on an Albion press. Mason was aware that such conditions could not be duplicated in most twentieth-century printing firms, but he argued that, in order to operate complex modern machinery, a young printer first had to master the fundamental elements of his craft – a principle of which Morris would surely not have disapproved.

Leonard Jay, who enrolled as Mason's first pupil and in 1912 became his assistant at the Central School, eventually grew restless under Mason's regime of Morrisian orthodoxy and in 1925 left to become head of the School of Printing in Birmingham. There he continued his master's tradition of exacting craftsmanship and careful design, but his students in Birmingham were also encouraged to study advertising typography and to make use of the Linotype and Monotype machines. Jay regarded these innovations not as a lapse from the purity of the Kelmscott/Doves faith but as an attempt to prepare students for a world in which handicraft methods no longer prevailed. In a lecture delivered in 1929, Jay declared, 'The product of machinery can only be improved when its producers are imbued with the same spirit which produced fine work in the past, and every effort should be directed towards breathing life and beauty into the products of machinery. This is a vital necessity in this mechanical age; for the more machinery that is introduced into the printing trade the greater the need for more alert, intelligent, and enlightened craftsmen. Perfect machinery is available everywhere; our saving grace lies in better quality of production.'[39] In the differences between Mason and Jay one sees the process by which Morris's ideas were adapted to mechanized printing and stripped of their Ruskinian implications. Jay and his students were just as fastidious in their search for high standards of craftsmanship as Morris was, and they nearly always followed his advice on such crucial matters as word-spacing, leading, and margins; but Morris's fear of the dehumanizing effects of machinery upon workers is no longer in evidence.

Francis Meynell's highly successful Nonesuch Press is another example of how the Morrisian principle of hand-production could be gradually assimilated to more modern methods, though Meynell, unlike Mason and Jay, thought that Morris was 'something of a calamity' to the development of printing.[40] Meynell recognized the relationship between the Nonesuch Press and Jay's work in Birmingham, because in both instances there was an abandonment of the handicraft ideal, accompanied by an abiding belief in the importance of painstaking design. In 1930 Meynell wrote, 'My previous experience had shown me quite clearly that, in order to avoid monotony and to produce desirable editions at a reasonable cost, one must intelligently exploit the best mechanical equipment and the highest technical skill available. To-day there is more fine typographical material to be had than even the largest printing house in the country could possess; and the various "mercenary" presses have developed technical skill and variety along various lines. There was therefore no good reason, we thought, for a new private press in the old style, arrogantly self-contained, and with but one type and obsolete "hand" machinery. . . . In this faith we began to mobilise other people's resources; to be designers, specifiers, rather than manufacturers; architects of books rather than builders. In this faith we proceed to compel the fully mechanical processes to serve for the first time the end of Fine Book making.'[41]

Though this stridently anti-Morris blast sounds like a complete rejection of all that the Kelmscott Press stood for, in reality Morris's principles of book-design had simply been detached from their original intellectual context and were now being claimed as their own by a new generation which believed that machinery could be tamed and civilized.

Morris's impact on printing and book-design outside England was, for at least a decade or two, unparalleled, with the force of his influence being felt most dramatically in northern Europe and English-speaking countries. In the United States imitating the Kelmscott books became a temporary craze towards the end of the century, as Professor Susan Thompson has shown in her exhaustive study *American Book Design and William Morris* (New York, 1977). Roberts Brothers of Boston published a pirated facsimile of *The Story of the Glittering Plain* a few months after it was issued by the Kelmscott Press in 1891, and within a few years the New York periodical the *Bookman* (November 1896) was complaining that 'Kelmscott–Jenson type, more or less accurately reproduced, has spread from Portland to San Francisco, and seems especially dear to the heart of advertisers of bicycles and malt extracts'. The more responsible American printers and book-designers were alarmed by this commercial degradation of Morris's ideals. 'Everyone who cared for William Morris must regret the extent to which his name and work are being used in the matter of printing,' wrote

Theodore L. De Vinne. 'Tasteless imitations of his types and ornaments threaten us on every side, and no one seems able to start a new press or get some type without taking Mr. Morris's name in vain.'[42] Bruce Rogers, looking back on the phenomenon in 1912, recalled that 'for a period of ten years the work of William Morris had an enormous influence on American printing – and not, on the whole, a beneficial one, except in its insistence upon sound craftsmanship'.[43]

The Kelmscott style in the United States was initially promulgated by a handful of periodicals and publishers of anglophile or Arts and Crafts tendencies. The most ambitious of the periodicals was the *Knight-Errant* of Boston, which survived for only four issues during 1892–3; set in Old Style type on handmade paper and ornamented with initials and tailpieces designed by Bertram Grosvenor Goodhue, the *Knight-Errant* was self-consciously Morrisian and aesthetic. In the July 1892 issue an essay by Francis Watts Lee, 'Some Thoughts Upon Beauty in Typography, Suggested by the Work of Mr. William Morris at the Kelmscott Press', provided facsimiles of three Kelmscott pages and the printer's mark designed by Morris, and described his books as 'the crowning glory of the nineteenth century'. Another Boston periodical was *Modern Art*, a quarterly published between 1893 and 1897. Its design, though more satisfactory than that of the *Knight-Errant*, was equally Morrisian (Kelmscott-style borders and initials, a replica of the Golden type, shoulder titles and notes in red), and it had the distinction of publishing 'A Note by William Morris on His Aims in Founding the Kelmscott Press', an essay reprinted in the last Kelmscott book. A considerably less imposing journal was the *Craftsman*, produced by a crafts guild called the United Crafts of Eastwood, N.Y., the first issue of which (October 1901) was devoted entirely to Morris, with vaguely Morrisian borders and indifferent typography.

Figure 93

The publishing houses that fell under Morris's sway in the 1890s displayed his influence at a relatively sophisticated level, since they were responding to a variety of styles in book-design that floated across the Atlantic. Copeland and Day of Boston, for example, occasionally imitated the Kelmscott manner, most conspicuously in their edition of Rossetti's *House of Life* (1894), but many of their books possessed a strongly aesthetic and decadent quality reminiscent of John Lane's characteristic style. W. Irving Way of Way and Williams (Chicago) made a pilgrimage to Hammersmith in the spring of 1895 and persuaded Morris to print *Hand & Soul*, the only Kelmscott book co-published in the United States (though Copeland and Day nearly

Figure 94

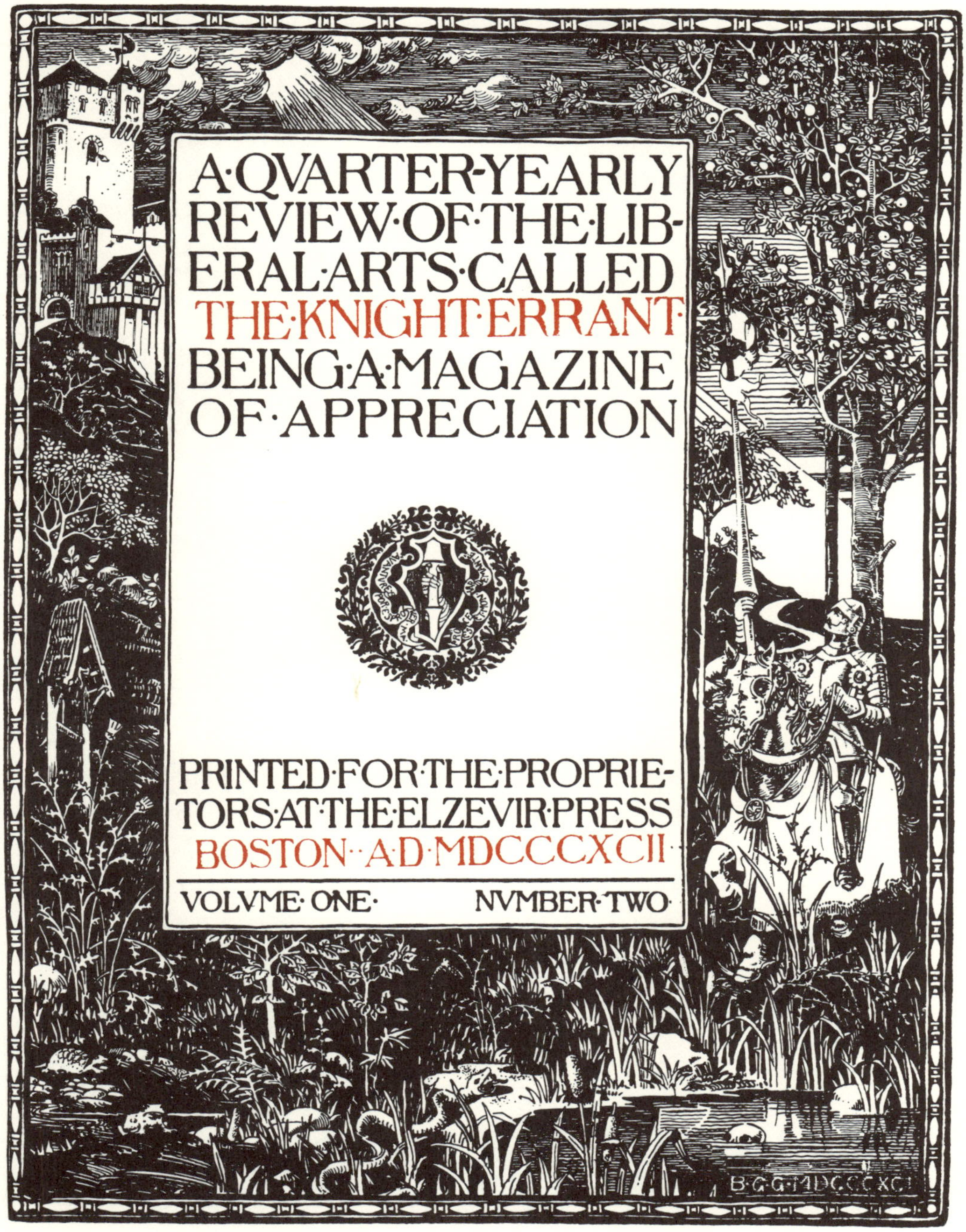

Figure 93 The Boston magazine *Knight-Errant* gave clear evidence of being under Morris's spell in 1892. Pages of some of the Kelmscott books were reproduced in it for the edification of American readers.

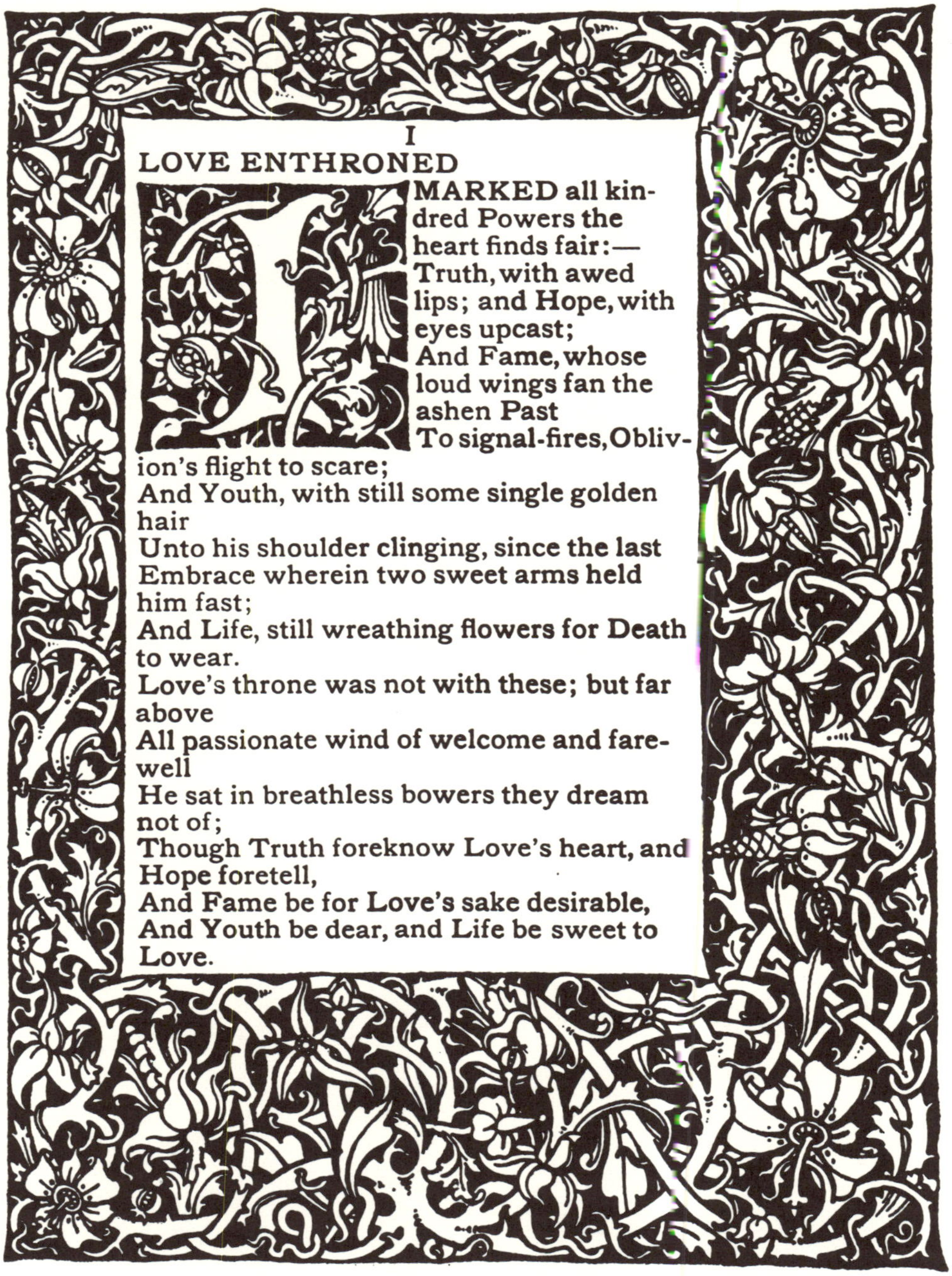

I

LOVE ENTHRONED

I MARKED all kindred Powers the heart finds fair:—
Truth, with awed lips; and Hope, with eyes upcast;
And Fame, whose loud wings fan the ashen Past
To signal-fires, Oblivion's flight to scare;
And Youth, with still some single golden hair
Unto his shoulder clinging, since the last
Embrace wherein two sweet arms held him fast;
And Life, still wreathing flowers for Death to wear.
Love's throne was not with these; but far above
All passionate wind of welcome and farewell
He sat in breathless bowers they dream not of;
Though Truth foreknow Love's heart, and Hope foretell,
And Fame be for Love's sake desirable,
And Youth be dear, and Life be sweet to Love.

Figure 94 The American publishing house of Copeland and Day in Boston closely imitated the Kelmscott Press style in this edition of Dante Gabriel Rossetti's *The House of Life* (1894). Both the initial and border, however, are more cluttered than Morris's ever were.

convinced Morris to print an edition of Emerson's poems for them). On the same trip to London, however, Way also secured nine books from John Lane. A third American publisher, Thomas B, Mosher of Maine, produced beautiful pirated editions of English books rather more in the Whistlerian than Morrisian vein, but in 1900 he issued a *Hand & Soul* that was a thoughtful imitation of the Kelmscott version. Unlike many of his American contemporaries, Mosher realized that the secret of Morris's success as a printer lay not in his types and ornaments (which were so easily facsimiled by photoengraving methods) but in patient craftsmanship. 'Mr. Irving Way who brought out the Morris edition in this country recently wrote me desiring to know how I succeeded in getting so close a duplicate [of *Hand & Soul*],' Mosher said. 'It was done by using our best efforts on the press work and by procuring special inks, which I think on comparison fully come up to the Morris ink.'[44]

Copeland and Day, Way and Williams, and Mosher were all reputable publishers (Mosher, known as 'the moral pirate', actually sent royalties to English authors) who responded briefly and with restraint to the enthusiasm for Kelmscott-style typography in the United States; to see the Morris mania in a less inhibited form, we must turn to Elbert Hubbard, who, after meeting Morris, created the Roycrofters, a self-described crafts community, in East Aurora, N.Y. ('Take the train for East Aurory | Where we work for Art and Glory' was one of Hubbard's many slogans.) There the 'P. T. Barnum of art', as one newspaper justly called him, built the Roycroft Inn for visitors (with one of the rooms named after Morris), a printing plant, a book bindery, and a chapel, all in a Ye Olde Worlde fashion. The Morrisian doctrine of handicrafts was oddly blended with inspirational capitalist sentiments ('I am not in the business of defaming America . . . I believe in big business and more of it,' Hubbard announced[45]), and at annual conventions in East Aurora the numerous disciples of 'Fra Elbertus' listened to band concerts on the lawn and pondered such themes as 'HEALTH, HAPPINESS, TRUTH'. May Morris, visiting Buffalo in 1904, was handed an invitation from Hubbard to visit the Roycrofters; angrily crumbling the letter in her hand, she replied, 'I most certainly will not go to East Aurora, nor do I have any desire to see that obnoxious imitator of my dear father.'[46]

Figure 95

Hubbard, who had called at Kelmscott House a second time in June 1896 when Morris was too ill to see him, always regarded himself as Morris's chief American disciple. The sentimental, self-indulgent tone

Figure 95 Elbert Hubbard (known as 'Fra Elbertus' to his numerous followers), Morris's flamboyant American disciple, printed a series of Kelmscott-style books, often with such odd features as limp chamois covers and halftone illustrations. May Morris described him as 'that obnoxious imitator of my dear father'.

of the Roycroft propaganda is well represented by this typical effusion from Hubbard's son after his father's death: 'When the Roycroft workers have departed and the still serenity of deep night has settled down like a blanket upon the shops, the kindred spirits of William Morris and Elbert Hubbard often glide noiselessly through the print shop, the chapel and the book-bindery to learn how we, their earth-bound successors, are carrying on with the heritage of fine craftsmanship which they have left to us.'[47]

The first Roycroft books published in the 1890s were printed on a hand-press, but 'gradually we got over that idea', Hubbard confessed,[48] and his shop in time was filled with rotary presses, many of them used to produce Hubbard's popular magazine the *Philistine*. The books offered a bastardized rendition of the Kelmscott style, with coarse paper, Morris-like ornaments, incongruous halftone illustrations, abundant red ink, conventional typefaces, and limp chamois covers. Yet, unappealing as the Roycroft volumes may seem to us today, they had a strong influence on some young American book-lovers such as Dard Hunter and Will Ransom. Through the improbable medium of Hubbard's tasteless imitations, the Morrisian idea of fine printing was translated into a vernacular easily grasped by a mass audience. 'That is how the tradition first came to America,' wrote Ransom, 'was expressed in terms of the national characteristic, publicity, by a master of that delicate art, and set many of us on the road to our own goal.'[49]

The usual superficial indications of Morris's presence in American book-design at the turn of the century, apart from a predilection for the ATF Jenson type, were title-pages surrounded by dark Kelmscott-like borders, often printed separately on coated paper and merely inserted in a book that had been produced earlier. Such affectations do not suggest a deep understanding of Morris's views on bookmaking, and, if we wish to assess his more permanent contribution to American typography, we must consider his influence on the most accomplished printers and designers of the United States. To cite only a few random examples, Frederic Goudy, the type-designer, bought Morris's *Chaucer* press, issued Morris and Walker's essay 'Printing' as the first publication of the Village Press, and made his Village type remarkably similar to the Golden fount; John Henry Nash, the celebrated California printer, commissioned a portrait of Morris for his library; Carl Purington Rollins, for many years the director of the Yale University Press, began collecting Kelmscott books when still a young man and,

Figure 96

NO one goes there now:
For what is left to fetch
away
From the desolate battle-
ments all arow,
And the lead roof heavy
and grey?
"Therefore," said fair
Yoland of the flowers,
"This is the tune of Seven Towers."

NO one walks there now;
Except in the white moonlight
The white ghosts walk in a row;
If one could see it, an awful sight,—
"Listen!" said fair Yoland of the flowers,
"This is the tune of Seven Towers."

BUT none can see them now,
Though they sit by the side of the moat,
Feet half in the water, there in a row,
Long hair in the wind afloat.
"Therefore," said fair Yoland of the flowers,
"This is the tune of Seven Towers."

Figure 96 A sadly characteristic example of a debased Morrisian style in American bookmaking at the turn of the century: an edition of *Pre-Raphaelite Ballads* by Morris (New York, 1900), with illustrations and ornaments by H. M. O'Kane.

after arriving at Yale, rescued the *Chaucer* which he found being used as a doorstop in the Art School library.[50] Reading the memoirs and biographies of numerous important figures in American typography who came of age in the early decades of the twentieth century, one can discover a curiously similar pattern of influence: it was the Kelmscott books that first opened their eyes to the possibilities of fine printing, and, long after they stopped imitating Morris's typographical peculiarities, they continued to revere the founder of the Kelmscott Press as a symbol of craftsmanship and moral integrity.

One can trace this evolution of attitudes in the lives of Daniel Berkeley Updike and Bruce Rogers, the most distinguished printer and book-designer respectively that the United States has yet produced. The first ambitious book undertaken by Updike during the 1890s was the *Altar Book* for the Episcopal Church, and from the beginning it was conceived as a homage to the Kelmscott Press. In August 1893 Updike wrote to Morris, 'I am engaged in the printing of a devotional book which I sincerely desire to make the most perfect piece of printing that is possible. And to this end I take the liberty of addressing you to know if through you, I can get a Roman type for the printing of it that has enough blackness and strength to allow the use of decoration of a solid yet brilliant kind.' Would Morris sell him founts of the Golden type or allow it to be recut in Great Primer size?[51] It is not known whether Morris replied to this request, and if he did his answer was no doubt negative. Updike therefore turned to the unauthorized adaptation of the Golden type produced at the Dickinson Type Foundary in Boston but decided not to use it when it was supplied to another printing house. The only remaining solution was to commission his own design, a task put into the hands of Bertram Goodhue. The *Altar Book*, published in 1896, was perhaps the most stupendous tribute to the influence of the Kelmscott Press ever published in the United States (or in any other country): an enormous folio, it was embellished with ornaments by Goodhue and illustrations by Robert Anning Bell (an Englishman) and was printed by De Vinne in New York. *Figure 97*

This remarkable pastiche signals the high-water mark of Morrisian enthusiasm in both American book-design and Updike's career, for the majority of the early books printed at his Merrymount Press in Boston are set in Caslon and resemble the exercises in puritanical severity produced by Mason's students at the Central School of Arts and Crafts. Nor, dazzled though he was by Morris's ornaments, did

Figure 97 Daniel Berkeley Updike's folio *Altar Book* (1896) for the Episcopal Church in the United States is perhaps the most sumptuous imitation of the Kelmscott style ever produced on either side of the Atlantic.

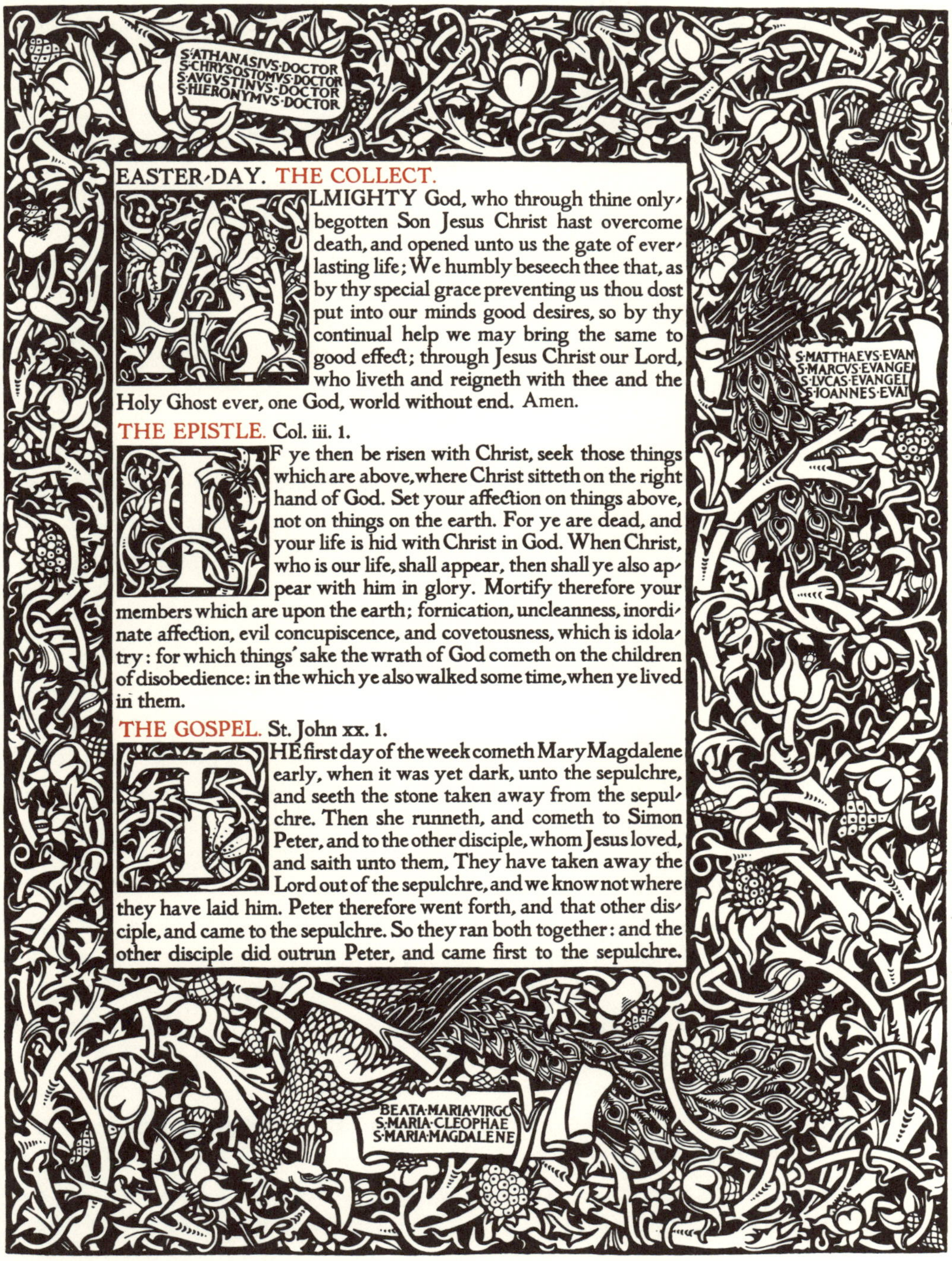

EASTER-DAY. THE COLLECT.

ALMIGHTY God, who through thine only-begotten Son Jesus Christ hast overcome death, and opened unto us the gate of everlasting life; We humbly beseech thee that, as by thy special grace preventing us thou dost put into our minds good desires, so by thy continual help we may bring the same to good effect; through Jesus Christ our Lord, who liveth and reigneth with thee and the Holy Ghost ever, one God, world without end. Amen.

THE EPISTLE. Col. iii. 1.

IF ye then be risen with Christ, seek those things which are above, where Christ sitteth on the right hand of God. Set your affection on things above, not on things on the earth. For ye are dead, and your life is hid with Christ in God. When Christ, who is our life, shall appear, then shall ye also appear with him in glory. Mortify therefore your members which are upon the earth; fornication, uncleanness, inordinate affection, evil concupiscence, and covetousness, which is idolatry: for which things' sake the wrath of God cometh on the children of disobedience: in the which ye also walked some time, when ye lived in them.

THE GOSPEL. St. John xx. 1.

THE first day of the week cometh Mary Magdalene early, when it was yet dark, unto the sepulchre, and seeth the stone taken away from the sepulchre. Then she runneth, and cometh to Simon Peter, and to the other disciple, whom Jesus loved, and saith unto them, They have taken away the Lord out of the sepulchre, and we know not where they have laid him. Peter therefore went forth, and that other disciple, and came to the sepulchre. So they ran both together: and the other disciple did outrun Peter, and came first to the sepulchre.

Bertram Goodhue furnished the ornaments, and the illustrations were by Robert Anning Bell.

Updike share his dislike of machinery. He did not hesitate to install cylinder presses, though he operated them at slower speeds than most printers did, and he was an advocate of Monotype composition, especially after more attractive typefaces became available from Monotype in the 1920s. Updike, in short, moved from a youthful phase of uncritical adulation of Morris to a more judicious, mature appreciation of his achievement. Updike in later years criticized Morris's types and his failure to follow always his own typographical principles, but the American never abandoned his conviction that Morris 'was a great printer because he was a great man who printed greatly, as he did much else'.[52]

Never as bewitched by Morris as Updike was, Bruce Rogers none the less discovered his vocation as a book-designer when J. M. Bowles, the founder of *Modern Art*, showed him several Kelmscott volumes. Only in one of Rogers's youthful productions, R. B. Gruelle's *Notes: Critical and Biographical. Collection of W. T. Walters*, published by Bowles in 1895, did he attempt a frank imitation of the Kelmscott style, and, like Updike, he subsequently expressed serious misgivings about Morris's books. Though he remained primarily a 'limited edition' designer, Rogers faced directly (and solved) many typographical problems – such as the designing of footnotes and indexes – that Morris, with his disdain for scholarly apparatus, never had to confront, and he went beyond Morris also in his ready acceptance of mechanical typesetting equipment and power presses. Rogers's work occasionally rather reminds one of Francis Meynell's books at the Nonesuch Press, for in both instances it is the sheer versatility and inventiveness of the designer that astonishes. And, as in Meynell's case, the impact of Morris is sometimes difficult to discern on the brilliant surface of Rogers's books, yet, as Beatrice Warde suggested, we can discover just beneath that surface the surprising truth that Rogers 'managed to steal the Divine Fire which glowed in the Kelmscott Press books, and somehow to be the first to bring it down to earth'[53] – a phrase which adroitly summarizes the transformation of Morrisian principles in both the United States and England during the first half of the twentieth century.

In Germany there was rapid recognition of the importance of the Kelmscott Press, which inspired a flourishing private-press movement in the first decade of the new century. As early as 1898, when the Press closed, an account of Morris and his accomplishments as a printer was published in *Zeitschrift für Bücherfreunde* which included numerous facsimile reproductions of his types and ornaments.[54] The centre of German interest in the Kelmscott Press was Weimar, where in May 1905 there was an exhibition of the Kelmscott and Doves books and where Count Kessler, an admirer of the English private presses, set up one of his own, the Cranach-Presse. J. H. Mason was brought over to instruct the German compositors; the services of Edward Johnston were enlisted; Emery Walker gave advice and assistance. A few years earlier Walker, Johnston, and Eric Gill also contributed their skills to a series of German classics issued by the Insel Verlag of Leipzig, a commercial publisher. The Morrisian influence, as mediated through his associates and disciples, was strong in Germany until 1914, and after the war, in a backward flow of ideas, some English printers received their first exposure to Morris's typographical principles through German books.

Other European countries also reacted enthusiastically to the Kelmscott Press books. Switzerland, at this time under German cultural sway, tended to reflect the typographical fashions of its larger neighbour. In 1894 visitors to the 'Libre Esthétique' exhibition in Brussels were able to see the Kelmscott books assembled for the first time, and a correspondent for the *Studio* (April 1894) reported: 'It is with the most lively admiration that connoisseurs are arrested before the books of William Morris – *The Defence of Guenevere*, *The History of Troy*, and *A Dream of John Ball*, of which the frontispiece has been designed with such distinction by Sir Edward Burne-Jones. One recognizes in them the perfection of quality; and the first appearance of these works of the Kelmscott Press before the Brussels public, one can say, has been a real triumph.' For the opening of a similar exhibition in Antwerp the same year, Emmanuel de Bom delivered a lecture on Morris and his influence on typography which was later published. On the Continent, however, the art-nouveau influence was strong, and at that distance from England both readers and book-designers found it difficult to distinguish between the rival styles; thus Henry Van de Velde, the most influential figure in Belgian book-production in the early years of the twentieth century, may seem a very Morrisian figure, printing in Caslon on a hand-press and designing all his own ornaments, but the latter prove, on closer inspection, to be more in the manner of Beardsley than of Morris. In the Netherlands there was also a Morris-like personality, Dr P. J. H. Cuypers, an ardent medievalist architect, not to mention a book-designer of the first rank, S. H. de Roos, whose earliest volume was a translation of some Morris essays (1903).

The Legacy of Morris and his Press

Morrisian ideas were spread in Scandinavia primarily by a few printers such as Waldemar Zachrisson in Sweden and Simon Bernsteen in Denmark. Frederik Hendrikson, a Dane, had even anticipated Morris by writing an article on the deficiencies of modern book-design for the newspaper *Politiken* in 1884, and in 1888 he founded Forening for Boghaandvaerk (the Danish Society for Book Craft). But if northern Europe was receptive to Morris's theories, this was not the case in France and Italy, both of which were dominated by strong native traditions of typography and illustration. The French especially had their own views on the Ideal Book, and their ideal bore scarcely any resemblance to Morris's. French bibliophiles, moreover, were so convinced of the superiority of their own *éditions de luxe* that they were not at all receptive to an Anglo-Saxon standard. Octave Uzanne, in *Nos amis les livres* (Paris, 1886), declared unhesitatingly: 'The art of producing beautiful books is the only one that the foreigners have not been able to imitate. Essentially it is a purely French art and one could parody Villon and say: "Il n'est bon livre que de Paris!" Printing, illustration, stitching, binding make us the absolute superior of all others in Europe or America, and this applies to everything, not only to the talent of our artists, but to the eye and perfect taste of our main publishers.'[55] In Italy the only minor evidences of Morris's influence are some volumes published by the Fratelli Treves of Milan early in the century in an imitation of the Golden type, and a temporary revival of darker typefaces and wood-engraved title-pages.

A. W. Pollard, long an astute observer of the contemporary printing scene as well as a bibliographer of early English books, in 1934 offered this summary of the meaning of Morris's typographical adventure: 'The beauty of all these books [by the Doves, Vale, and Ashendene presses and Robert Proctor] reinforced the influence of the Kelmscott Press ones, by proving that what Morris had done on his own lines could be done by lesser men with the variations suggested by their individual tastes. They reinforced also the proof which Morris had given that so long as it is regarded as a hobby (or in a commercial house as an advertisement) the production of really fine specimens of printing is not an impossibly expensive one. Morris made no profit from the Kelmscott books as a publisher. . . . But he proved the existence of a public willing to pay for the cost of print and paper, even when print and paper were the best which money could buy.'[56]

Morris's discovery of a market for limited-edition books was indeed significant, and his customers were for the most part the same middle-class bibliophiles who were caught up in the first-edition mania fostered at the end of the nineteenth century by T. J. Wise and Harry Buxton Forman; it is no accident that the first (and only) full-length bibliographical account of Morris's books was written by Forman. In order for private presses to move beyond the Victorian parlour, it had to be demonstrated that there was a substantial number of book-lovers who were prepared to purchase handsomely printed volumes. The success of the Kelmscott Press proved this was so. At the same time Morris supplied an ideology, rooted in Ruskinian thought and the Arts and Crafts movement, which justified the hand-labour required by such undertakings.

The nature of Morris's contribution to modern book-design may still inspire debate, but it is indisputable that he is the progenitor of nearly all twentieth-century private presses that have emphasized fine printing. Naturally these presses have evolved in various ways, and few of them today are similar in every respect to the Kelmscott or Doves model. After the economic collapse of 1929 the bibliophilic market shrank dramatically, making it unthinkable to produce works on the scale of the Kelmscott *Chaucer*, the Doves Bible, or the Ashendene Dante. Many of the private presses in Britain and the United States in the period following the Second World War embodied, in fact, a temporary reversion to the Daniel ideal of the hobbyist printer distributing slender little volumes to his family and friends. But that pattern has changed yet again in recent decades, and today the private presses seem to be prospering once more, flourishing with amazing vitality in such centres as Leicester, Iowa City, and San Francisco.

One major difference between the earlier Kelmscott-type presses and contemporary private presses, however, is that the latter are less likely to have any direct influence on the world of commercial printing. The reason for this appears to be that their respective printing technologies have diverged so abruptly in the last two or three decades that they no longer have much in common except certain problems of designing. The hand-press methods still used by the majority of private printers and the automated realm of offset presses and computerized typesetting simply represent two separate worlds, and, despite occasional attempts to bridge the gap – such as the University of California's admirable lithographic reprints of private-press books like the Arion Press *Moby Dick* and the Pennyroyal *Alice* volumes –

it seems unlikely that any private press in the future will ever have the impact on printing and publishing at large that the Kelmscott and Doves Presses and their ilk once did. This is an unhappy development, since there is unmistakable evidence of deteriorating typographical standards at the moment as a result of the new technologies.

Except for Blake, who rejected letterpress printing altogether, Morris was the first English artist of major stature who succeeded in controlling every aspect of the printing of a book from types to bindings. He did so not only because he had a passion for thoroughness but also because it seemed to him the only way to achieve that unity of effect which he and Walker regarded as the primary characteristic of fine printing. Two centuries earlier Joseph Moxon, in his *Mechanick Exercises on the Whole Art of Printing*, had offered a definition – often since quoted – of the typographer (or, as we might say today, the book-designer) as 'such a one, who by his own Judgement, from solid reasoning with himself, can either perform, or direct others to perform from the beginning to the end, all the Handy-works and Physical Operations relating to *Typographie*'.[57] With increasing automation and specialization in modern printing, there has been an ever greater need for one man or woman to orchestrate all the multitudinous activities now required to produce a book: Morris, it might be argued, invented (or reinvented, if we take Moxon into account) the role of book-designer.

Yet such a claim on his behalf has troubling implications, because it suggests an acceptance of the division of labour which Morris, as a disciple of Ruskin, found abhorrent. Peter Floud, in a talk on Morris delivered on BBC radio in 1954 and later printed in the *Listener*, made the point that 'though Morris's designs [for fabrics and wallpapers] were certainly executed by hand, they are not examples of individual craftwork, but rather what one can call serial hand-production on a factory basis, which makes full use of the division of labour and differs very little in its moral and social implications from straightforward machine-production'.[58] Floud may have put his case in strong language, but there can be no doubt that he has discovered the essential discrepancy between Morris's theory and practice. A similar analysis of the Kelmscott Press would show that Morris's printers were compelled to follow the typographical patterns and designs he devised in his study at Kelmscott House rather than to express the roughness and majesty of their own souls (as Ruskin phrased it in *The Nature of Gothic*). On the rare occasions when they were allowed complete freedom of choice, the Kelmscott printers produced something conspicuously hideous like the Wayzegoose menus.

But it is almost too easy to point out the inconsistencies of Morris's system of ideas. If we find Morris the thinker irritating, we can do what Pevsner, Morison, and others have done: ignore the intellectual foundations of his medievalism and concentrate instead on Morris the precursor of twentieth-century functionalism. But to do so is surely to ignore the richness and complexity of the man and to isolate only a small part of his mind that happens to suit our modern tastes. In the end the Kelmscott Press is important to us for reasons that go beyond the technical and aesthetic innovations described in earlier chapters of this book. The Press still fascinates us today because of the profound questions that Morris posed about the triumph of the machine.

Morris saw that more was at stake than merely the revival of attractive printing, and it is the reverberation of his ideas, quite apart from whether one is personally drawn to the Kelmscott Press books, that makes the books so extraordinarily compelling in their appeal. Although he was unable to foresee the specific technology of the next century – films, radio, television, video cassettes, computers – Morris once expressed to Burne-Jones, in a characteristically jocular way, a dark vision of the future of the printed book in modern civilization: 'he railed fiercely against the invention of printing and gave it a hundred years at most to come to an end. Already said he the magazines are driving books out of the field – presently newspapers will have killed magazines – then the telephone will come – bottles of talk – & the newspapers will be ended, & thats a comfort – as a centipede eats a cockroach & a cockroach eats a bug – a hundred years will do it said he. . . .'[59]

The Kelmscott books are imperfect creations, produced by occasionally uneasy compromises with modern technology and often flawed by Morris's hatred of all post-medieval thought and art. But from our point of view today, a century later, they represent a logical and, on the whole, satisfying culmination of the art of fine printing that began with Gutenberg. Of course Morris was wrong in a literal sense: books that are in themselves works of art are still being printed and will continue to be printed in the future. But Morris's astonishingly accurate prediction of the death of the book as a central cultural artefact in the twentieth century, with moral and social consequences we cannot yet gauge, makes the Kelmscott Press seem, more and more, like the end rather than the beginning of something good.

APPENDIX A ❦ CHECKLIST OF THE KELMSCOTT PRESS BOOKS

THE following list includes all the books printed at the Kelmscott Press but not the ephemera such as advertising circulars. My *A Bibliography of the Kelmscott Press* (Oxford, 1984) supplies much more detailed descriptions of each of the books.

The books were all produced in one of several buildings in the Upper Mall, Hammersmith, except for *Gothic Architecture* (no. 18), which was printed during the Arts and Crafts Exhibition at the New Gallery, London. Unless otherwise indicated, the books are printed in black ink only. The two sizes of Flower paper (16 × 11 inches and 16 × 22 inches) are designated Flower (1) and Flower (2) respectively. I have made no attempt to describe the initials, borders, and engraved title-pages in this brief list; suffice it to say that every Kelmscott book was decorated with wood-engraved ornaments designed by Morris, some much more fully than others.

❦

1. *The Story of the Glittering Plain. Which has been also Called the Land of Living Men or the Acre of the Undying. Written by William Morris.* Small 4°: Flower (1) paper. 192 pages. Golden type. Stiff vellum binding with washleather ties. 200 paper, 6 vellum copies. Colophon dated 4 April 1891. Published by Reeves and Turner, 8 May 1891, at 2 guineas (paper) and 12 and 15 guineas (vellum).

2. *Poems by the Way. Written by William Morris.* Small 4°: Flower (1) paper. 204 pages. Golden type. Black and red ink. Stiff vellum binding with silk ties. 300 paper, 13 vellum copies. Colophon dated 24 September 1891. Published by Reeves and Turner, 20 October 1891, at 2 guineas (paper) and about 12 guineas (vellum).

3. *The Love-Lyrics & Songs of Proteus by Wilfrid Scawen Blunt with the Love-Sonnets of Proteus by the Same Author now Reprinted in their Full Text with Many Sonnets Omitted from the Earlier Editions.* Small 4°: Flower (1) paper. 264 pages. Golden type. Red and black ink. Stiff vellum binding with silk ties. 300 paper copies. Colophon dated 26 January 1892. Published by Reeves and Turner, 27 February 1892, at 2 guineas.

4. *The Nature of Gothic* [:] *A Chapter of the Stones of Venice. By John Ruskin.* Small 4°: Flower (1) paper. 136 pages. Golden type. Line-block diagrams. Stiff vellum binding with silk ties. 500 paper copies. Colophon undated. Published by George Allen, 22 March 1892, at 30*s.*

5. *The Defence of Guenevere, and Other Poems. By William Morris.* Small 4°: Flower (1) paper. 176 pages. Golden type. Black and red ink. Limp vellum binding with silk ties. 300 paper, 10 vellum copies. Colophon dated 2 April 1892. Published by Reeves and Turner, 19 May 1892, at 2 guineas (paper) and about 12 guineas (vellum).

6. *A Dream of John Ball and A King's Lesson. By William Morris.* Small 4°: Flower (1) paper. 132 pages. Golden type. Black and red ink. Wood-engraved frontispiece by Edward Burne-Jones. Limp vellum binding with silk ties. 300 paper, 11 vellum copies. Colophon dated 13 May 1892. Published by Reeves and Turner, 24 September 1892, at 30*s.* (paper) and 10 guineas (vellum).

7. Jacobus de Voragine, *The Golden Legend.* Translated by William Caxton and edited by F. S. Ellis. 3 volumes. Large 4°: Flower (2) paper. 1310 pages. Golden type. 2 wood-engraved illustrations by Burne-Jones. Quarter holland binding. 500 paper copies. Colophon dated 12 September 1892. Published by Bernard Quaritch, 3 November 1892, at 5 guineas.

8. Raoul Lefevre, *The Recuyell of the Historyes of Troye.* Translated by William Caxton and edited by H. Halliday Sparling. 2 volumes. Large 4°: Flower (2) paper. 740 pages. Troy and Chaucer types. Black and red ink. Limp vellum binding with silk ties. 300 paper, 5 vellum copies. Colophon dated 14 October 1892. Published by Bernard Quaritch, 24 November 1892, at 9 guineas (paper) and £80 (vellum).

9. *Biblia innocentium: Being the Story of God's Chosen People before the Coming of Our Lord Jesus Christ upon Earth, Written Anew for Children by J. W. Mackail, Sometime Fellow of Balliol College, Oxford.* 8°: Flower (2) paper. 268 pages. Golden type. Stiff vellum binding with silk ties. 200 paper copies. Colophon dated 22 October 1982. Published by Reeves and Turner, 9 December 1892, at 1 guinea.

10. *The History of Reynard the Foxe by William Caxton.* Edited by Sparling. Large 4°: Flower (2) paper. 172 pages. Troy and Chaucer types. Black and red ink. Limp vellum binding with silk ties. 300 paper, 10 vellum copies. Colophon dated 15 December 1892. Published by Bernard Quaritch, 25 January 1893, at 3 guineas (paper) and 15 guineas (vellum).

11. *The Poems of William Shakespeare, Printed after the Original*

Copies of Venus and Adonis, 1593. The Rape of Lucrece, 1594. Sonnets, 1609. The Lover's Complaint. Edited by Ellis. 8°: Flower (2) paper. 232 pages. Golden type. Black and red ink. Limp vellum with silk ties. 500 paper, 10 vellum copies. Colophon dated 17 January 1893. Published by Reeves and Turner, 13 February 1893, at 25*s.* (paper) and 10 guineas (vellum).

12. *News from Nowhere: Or, An Epoch of Rest, Being Some Chapters from a Utopian Romance, by William Morris.* 8°: Flower (2) paper. 320 pages. Golden type. Black and red ink. Wood-engraved frontispiece by C. M. Gere. Limp vellum binding with silk ties. 300 paper, 10 vellum copies. Colophon dated 22 November 1892. Published by Reeves and Turner, 24 March 1893, at 2 guineas (paper) and 10 guineas (vellum).

13. Ramon Lull, *The Order of Chivalry.* Translated by Caxton. With *L'Ordene de chevalerie.* Translated by Morris. Small 4° and 8°: Flower (1) and Flower (2) paper. 156 pages. Chaucer type. Black and red ink. Wood-engraved frontispiece by Burne-Jones. Limp vellum binding with silk ties. 225 paper, 10 vellum copies. Colophon dated 24 February 1893. Published by Reeves and Turner, 12 April 1893, at 30*s.* (paper) and 10 guineas (vellum).

14. *The Life of Thomas Wolsey, Cardinal Archbishop of York Written by George Cavendish.* Edited by Ellis. 8°: Flower (2) paper. 296 pages. Golden type. Limp vellum binding with silk ties. 250 paper, 6 vellum copies. Colophon dated 25 December 1892. Published by Reeves and Turner, 3 May 1893, at 2 guineas (paper) and 10 guineas (vellum).

15. Guilelmus, Archbishop of Tyre, *The History of Godefrey of Boloyne and of the Conquest of Iherusalem.* Edited by Sparling. Large 4°: Flower (2) paper. 476 pages. Troy and Chaucer types. Black and red ink. Limp vellum binding with silk ties. 300 paper, 6 vellum copies. Colophon dated 27 April 1893. Published by the KP, 24 May 1893, at 6 guineas (paper) and 20 guineas (vellum).

16. *Utopia Written by Sir Thomas More.* Translated by Ralph Robinson and edited by Ellis. 8°: Flower (2) paper. 300 pages. Chaucer and Troy types. Black and red ink. Limp vellum binding with silk ties. 300 paper, 8 vellum copies. Colophon dated 4 August 1893. Published by Reeves and Turner, 8 September 1893, at 30*s.* (paper) and 10 guineas (vellum).

17. *Maud, a Monodrama, by Alfred Lord Tennyson.* 8°: Flower (2) paper. 80 pages. Golden type. Black and red ink. Limp vellum binding with silk ties. 500 paper, 5 vellum copies. Colophon dated 11 August

1893. Published by Macmillan and Co., 30 September 1893, at 2 guineas (paper). (Vellum copies not for sale.)

18. *Gothic Architecture: A Lecture for the Arts and Crafts Exhibition Society by William Morris.* 16°: Flower (2) paper. 74 pages. Golden type. Red and black ink. Quarter holland binding. Three impressions. 1,500 paper, 45 vellum copies. Colophon dated 1893. Published by the KP, 21 October 1893, at 2*s.* 6*d.* (paper) and 10*s.* and 15*s.* (vellum).

19. *Sidonia the Sorceress by William Meinhold Translated by Francesca Speranza Lady Wilde.* Large 4°: Flower (2) paper. 472 pages. Golden type. Black and red ink. Limp vellum binding with silk ties. 300 paper, 10 vellum copies. Colophon dated 15 September 1893. Published by the KP, 1 November 1893, at 4 guineas (paper) and 20 guineas (vellum).

20. *Ballads and Narrative Poems by Dante Gabriel Rossetti.* Edited by William Michael Rossetti. 8°: Flower (2) paper. 240 pages. Golden type. Black and red ink. Limp vellum binding with silk ties. 310 paper, 6 vellum copies. Colophon dated 14 October 1893. Published by Ellis and Elvey, November 1893, at 2 guineas (paper) and 10 guineas (vellum).

20a. *Sonnets and Lyrical Poems by Dante Gabriel Rossetti.* Edited by William Michael Rossetti. (The second volume of no. 20.) 8°: Flower (2) paper. 216 pages. Golden type. Black and red ink. Limp vellum binding with silk ties. 310 paper, 6 vellum copies. Colophon dated 20 February 1894. Published by Ellis and Elvey, 21 April 1894, at 2 guineas (paper) and 10 guineas (vellum).

21. *The Tale of King Florus and the Fair Jehane.* Translated by Morris. 16°: Perch paper. 106 pages. Chaucer type. Black and red ink. Quarter holland binding. 350 paper, 15 vellum copies. Colophon dated 16 December 1893. Published by the KP, 28 December 1893, at 7*s.* 6*d.* (paper) and 30*s.* (vellum).

22. *The Story of the Glittering Plain which has been also Called the Land of Living Men or the Acre of the Undying. Written by William Morris.* 4°: Perch paper. 192 pages. Troy and Chaucer types. Red and black ink. 23 wood-engraved illustrations by Walter Crane. Limp vellum binding with silk ties. 250 paper, 7 vellum copies. Colophon dated 13 January 1894. Published by the KP, 17 February 1894, at 5 guineas (paper) and £20 (vellum).

23. *Of the Friendship of Amis and Amile.* Translated by Morris. 16°: Perch paper. 80 pages. Chaucer type. Black and red ink. Quarter holland binding. 500 paper, 15 vellum copies. Colophon dated

13 March 1894. Published by the KP, 4 April 1894, at 7*s.* 6*d.* (paper) and 30*s.* (vellum).

24. *The Poems of John Keats.* Edited by Ellis. 8°: Flower (2) paper. 400 pages. Golden type. Black and red ink. Limp vellum binding with silk ties. 300 paper, 7 vellum copies. Colophon dated 7 March 1894. Published by the KP, 8 May 1894, at 30*s.* (paper) and 9 guineas (vellum).

25. *Atalanta in Calydon: A Tragedy. By Algernon Swinburne.* Large 4°: Perch Paper. 92 pages. Golden and Troy types and Selwyn Image's Greek type. Black and red ink. Limp vellum binding with silk ties. 250 paper, 8 vellum copies. Colophon dated 4 May 1894. Published by the KP, 24 July 1894, at 2 guineas (paper) and 12 guineas (vellum).

26. *The Tale of King Coustans and of Over Sea.* Translated by Morris. 16°: Perch paper. 144 pages. Chaucer type. Black and red ink. Quarter holland binding. 525 paper, 20 vellum copies. Colophon dated 30 August 1894. Published by the KP, 26 September 1894, at 7*s.* 6*d.* (paper) and 2 guineas (vellum).

27. *The Wood beyond the World. By William Morris.* 8°: Flower (2) paper. 272 pages. Chaucer type. Black and red ink. Wood-engraved frontispiece by Burne-Jones. Limp vellum binding with silk ties. 350 paper, 8 vellum copies. Colophon dated 30 May 1894. Published by the KP, 16 October 1894, at 2 guineas (paper) and 10 guineas (vellum).

28. Sulkhan-Saba Orbeliani, *The Book of Wisdom and Lies.* Translated by Oliver Wardrop. 8°: Flower (2) paper. 276 pages. Golden type. Black and red ink. Limp vellum binding with silk ties. 250 paper copies. Colophon dated 29 September 1894. Published by Bernard Quaritch, 29 October 1894, at 2 guineas.

29. *The Poetical Works of Percy Bysshe Shelley.* Edited by Ellis. 3 volumes. 8°: Flower (2) paper. 1,264 pages. Golden type. Black and red ink. Limp vellum binding. 250 paper, 6 vellum copies. Colophon dated 21 August 1895. Published by the KP, 29 November 1894–28 November 1895, at 25*s.* (paper) and 8 guineas (vellum) per volume.

30. *Psalmi penitentiales.* Edited by Ellis. 8°: Flower (2) paper. 72 pages. Chaucer type. Black and red ink. Quarter holland binding. 300 paper, 12 vellum copies. Colophon dated 15 November 1894. Published by the KP, 10 December 1894, at 7*s.* 6*d.* (paper) and 3 guineas (vellum).

31. Girolamo Savonarola, *Epistola de contemptu mundi di Frate Hieronymo da Ferrara dellordine de frati predicatori la quale manda ad Elena Buonaccorsi sua madre, per consolarla della morte del fratello, suo zio.* Edited

by Charles Fairfax Murray. 8°: Flower (2) paper. 16 pages. Chaucer type. Black and red ink. Wood-engraved illustration by Murray. Quarter holland binding. 150 paper, 6 vellum copies. Colophon dated 30 November 1894. Privately printed for Murray (ready 12 December 1894); not for sale.

32. *The Tale of Beowulf.* Translated by Morris and A. J. Wyatt. 4°: Perch paper. 128 pages. Troy and Chaucer types. Black and red ink. Limp vellum binding with ties. 300 paper, 8 vellum copies. Colophon dated 10 January 1895. Published by the KP, 2 February 1895, at 2 guineas (paper) and £10 (vellum).

33. *Syr Perecyvelle of Gales.* Edited by Ellis. 8° Flower (2) paper. 112 pages. Chaucer type. Black and red ink. Wood-engraved frontispiece by Burne-Jones. Quarter holland binding. 350 paper, 8 vellum copies. Colophon dated 16 February 1895. Published by the KP, 2 May 1895, at 15*s.* (paper) and 4 guineas (vellum).

34. *The Life and Death of Jason, a Poem. By William Morris.* Large 4°: Perch paper. 368 pages. Troy type. Black and red ink. Wood-engraved frontispiece and illustration by Burne-Jones. Limp vellum binding with ties. 200 paper, 6 vellum copies. Colophon dated 25 May 1895. Published by the KP, 5 July 1895, at 5 guineas (paper) and 20 guineas (vellum).

35. *Child Christopher and Goldilind the Fair. By William Morris.* 2 volumes. 16°: Flower (2) paper. 512 pages. Chaucer type. Black and red ink. Quarter holland binding. 600 paper, 12 vellum copies. Colophon dated 25 July 1895. Published by the KP, 25 September 1895, at 15*s.* (paper) and 4 guineas (vellum).

36. *Hand & Soul by Dante Gabriel Rossetti.* 16°: Flower (2) paper. 66 pages. Golden type. Black and red ink. Stiff vellum binding. English edition: 225 paper, 10 vellum copies; American edition: 300 paper, 11 vellum copies. Both colophons dated 24 October 1895. Published by the KP, 12 December 1895, at 10*s.* (paper) and 30*s.* (vellum); and by Way and Williams (of Chicago) at $3.50 (vellum). (Price of American paper copies not known.)

37. *Poems Chosen out of the Works of Robert Herrick.* Edited by Ellis. 8°: Flower (2) paper. 312 pages. Golden type. Black and red ink. Limp vellum binding with silk ties. 250 paper, 8 vellum copies. Colophon dated 21 November 1895. Published by the KP, 6 February 1896, at 30*s.* (paper) and 8 guineas (vellum).

38. *Poems Chosen out of the Works of Samuel Taylor Coleridge.* Edited by Ellis. 8°: Flower (2) paper. 108 pages. Golden type. Black and red

ink. Limp vellum binding with silk ties. 300 paper, 8 vellum copies. Colophon dated 5 February 1896. Published by the KP, 12 April 1896, at 1 guinea (paper) and 5 guineas (vellum).

39. *The Well at the World's End by William Morris.* Large 4°: Flower (2) paper. 504 pages. Chaucer type. Black and red ink. 4 wood-engraved illustrations by Burne-Jones. Limp vellum binding with silk ties. 350 paper, 8 vellum copies. Colophon dated 2 March 1896. Published by the KP, 4 June 1896, at 5 guineas (paper) and 20 guineas (vellum).

40. *The Works of Geoffrey Chaucer.* Edited by Ellis. 2°: Perch paper. 564 pages. Chaucer and Troy types. Black and red ink. 87 wood-engraved illustrations by Burne-Jones. Quarter holland binding. (48 copies bound in pigskin at the Doves Bindery from a design by Morris.) 425 paper, 13 vellum copies. Colophon dated 8 May 1896. Published by the KP, 26 June 1896, at £20 (paper) and 120 guineas (vellum).

41. *The Earthly Paradise. By William Morris.* 8 volumes. Medium 4°: Apple paper. 1,560 pages. Golden type. Black and red ink. Limp vellum binding with silk ties. 225 paper, 6 vellum copies. Colophons dated 7 May 1896–10 June 1897. Published by the KP, 24 July 1896–27 September 1897, at 30*s.* (paper) and 7 guineas (vellum) per volume.

42. *Laudes Beatae Mariae Virginis.* Large 4°: Perch paper. 48 pages. Troy type. Black, red, and blue ink. Quarter holland binding. 250 paper, 10 vellum copies. Colophon dated 7 July 1896. Published by the KP, 7 August 1896, at 10*s.* (paper) and 2 guineas (vellum).

43. Sir Thomas Clanvowe, *The Floure and the Leafe, & the Boke of Cupide, God of Love, or the Cuckow and the Nightingale.* Edited by Ellis. Medium 4°: Apple paper. 56 pages. Troy type. Black and red ink. Quarter holland binding. 300 paper, 10 vellum copies. Colophon dated 21 August 1896. Published by the KP, 2 November 1896, at 10*s.* (paper) and 2 guineas (vellum).

44. Edmund Spenser, *The Shepheardes Calender: Conteyning Twelve Æglogues, Proportionable to the Twelve Monethes.* Edited by Ellis. Medium 4°: Apple paper. 112 pages. Golden type. black and red ink. 12 line-block illustrations by Arthur J. Gaskin. Quarter holland binding. 225 paper, 6 vellum copies. Colophon dated 14 October 1896. Published by the KP, 26 November 1896, at 1 guinea (paper) and 3 guineas (vellum).

45. *The Water of the Wondrous Isles by William Morris.* Large 4°: Flower (2) paper. 352 pages. Chaucer and Troy types. Black and red

ink. Limp vellum binding with silk ties. 250 paper, 6 vellum copies. Colophon dated 1 April 1897. Published by the KP, 29 July 1897, at 3 guineas (paper) and 12 guineas (vellum).

46. Specimen pages of *Froissart's Chronicles*. 2°: vellum. 4 pages. Chaucer and Troy types. Black and red ink. Not bound. 160 copies. Colophon dated September 1897. Published by the KP, 7 October 1897, at 7 guineas.

47. *Sire Degrevaunt*. Edited by Ellis. 8°: Flower (2) paper. 96 pages. Chaucer type. Black and red ink. Wood-engraved frontispiece by Burne-Jones. Quarter holland binding. 350 paper, 8 vellum copies. Colophon dated 14 March 1896. Published by the KP, 12 November 1897, at 15*s*. (paper) and 4 guineas (vellum).

48. *Syr Isambrace*. Edited by Ellis. 8°: Flower (2) paper. 52 pages. Chaucer type. Black and red ink. Wood-engraved frontispiece by Burne-Jones. Quarter holland binding. 350 paper, 8 vellum copies. Colophon dated 14 July 1897. Published by the KP, 11 November 1897, at 12*s*. (paper) and 4 guineas (vellum).

49. Morris, *Some German Woodcuts of the Fifteenth Century*. Edited by Sydney Cockerell. Large 4°: Perch paper. 72 pages. Golden type. Black and red ink. 35 line-block illustrations. Quarter holland binding. 225 paper, 8 vellum copies. Colophon dated 15 December 1897. Published by the KP, 6 January 1898, at 30*s*. (paper) and 5 guineas (vellum).

50. *The Story of Sigurd the Volsung and the Fall of the Niblungs by William Morris*. Small 2°: Apple paper. 220 pages. Chaucer and Troy types. Black and red ink. Wood-engraved frontispiece and illustration by Burne-Jones. Limp vellum binding with silk ties. 160 paper, 6 vellum copies. Colophon dated 19 January 1898. Published by the KP, 25 February 1898, at 6 guineas (paper) and 20 guineas (vellum).

51. *The Sundering Flood Written by William Morris*. 8°: Flower (2) paper. 516 pages. Chaucer type. Black and red ink. Line-block map by H. Cribb. Quarter holland binding. 300 paper, 10 vellum copies. Colophon dated 15 November 1897. Published by the KP, 25 February 1898, at 2 guineas (paper) and 10 guineas (vellum).

52. *Love is Enough, or the Freeing of Pharamond: A Morality. Written by William Morris*. Large 4°: Perch paper. 100 pages. Troy and Chaucer types. Black, red, and blue ink. Wood-engraved frontispiece and illustration by Burne-Jones. Limp vellum binding with silk ties. 300 paper, 8 vellum copies. Colophon dated 11 December 1897.

Published by the KP, 24 March 1898, at 2 guineas (paper) and 10 guineas (vellum).

53. *A Note by William Morris on his Aims in Founding the Kelmscott Press together with a Short Description of the Press by S. C. Cockerell, and an Annotated List of the Books Printed thereat.* 8°: Flower (2) paper. 80 pages. Golden, Troy, and Chaucer types. Black and red ink. Wood-engraved frontispiece by Burne-Jones. Quarter holland binding. 525 paper, 12 vellum copies. Colophon dated 4 March 1898. Published by the KP, 24 March 1898, at 10*s.* (paper) and 2 guineas (vellum).

APPENDIX B ❦ LECTURE BY EMERY WALKER

THE lecture by Emery Walker on 'Letterpress Printing and Illustration', delivered on 15 November 1888 to the Arts and Crafts Society, was a turning-point in the history of modern printing, for it inspired William Morris to design his own typeface and to establish the Kelmscott Press. Since Walker spoke (rather nervously, according to Morris) from notes, the lecture was never published in the form in which it was delivered; yet, because it had such far-reaching consequences, I have made an attempt here to reproduce the substance of the lecture. There are three important sources of information: Walker's notes (in the Humanities Research Center, University of Texas); Oscar Wilde's (unsigned) article, 'Printing and Printers', *Pall Mall Gazette*, 16 November 1888, p. 5; and 'The Art of Printing . . . A Lecture by Emery Walker', *British and Colonial Printer and Stationer*, 6 February 1890, p. 94, which reports a slightly different version of the talk, delivered at the City of London College in 1890. Of these, the notes and Wilde's article are more significant, and I have therefore transcribed them in their entirety, but I have also made use of the article in the *British and Colonial Printer and Stationer* (which is more detailed but filled with inaccuracies) to clarify certain doubtful allusions. Two other reports of essentially the same lecture by Walker were published in the *British and Colonial Printer and Stationer*, 2 January 1896, pp. 10–11, and 31 March 1898, pp. 205–6, both accompanied by illustrations based upon Walker's slides. In the later variants of the lecture Walker concluded by showing pages from the Kelmscott Press books. An important recent study of the lecture is John Dreyfus, 'Emery Walker's 1888 Lecture on "Letterpress Printing": A Reconstruction and a Reconsideration', *Craft History*, 1(1988), 118–30.

Since Walker used these lecture notes on more than one occasion, it is not always possible to distinguish precisely between the original text and later additions to it. The inconsistencies in numbering the slides reflect these revisions: only Walker's final renumberings have been transcribed here. Any necessary annotations of the two texts have been placed within square brackets; Walker's own square brackets in his notes are reproduced as arrow brackets.

Appendix B

Walker's lecture notes

Although, I suppose one may not unreasonably think that every person here, has at least some idea, of the methods by w[hi]ch books are made, I hope I may be pardoned, if for the sake of impressing the essentials of the various processes on your minds, I, with the assistance of my friend here, very briefly demonstrate the rudiments of the art. Type and paper, may be said to be to a printed book, what stones or bricks, and mortar, are to architecture. They are the essentials, without which there can be no book in the one case, and no architecture, in the other.

I do not want to dwell on the purely historical aspect longer than I can help, but it is necessary to say in what way modern, that is, type printing differs from the method which has been practised in China for centuries. Shortly it is the use of moveable letters instead of engraved blocks, and as Mr. Talbot Reed [*inserted in pencil*: whose book is certainly the best historical work in the language] has said, the man who invented typography, was doubtless he, who invented moveable type.

⟨Here show punch, matrix, adjustable mould and type, and Japanese wood block.⟩

⟨metal composed of lead[,] tin and antimony⟩ Early printers cast their own type

Describe and names [of] various parts of the type, face, serif, feet[,] shoulder[,] nick. Now generally cast by machines when large quantities are wanted

Compositor[']s Frame[.] Upper and Lower Case arrangement of letters[:] 'e' in centre of the lower case.

Composing stick. Galley. chase.

You will see that the labour of casting and 'setting up' is precisely the same whether the type is as beautiful as that of the early printers or as ugly as the latest American importation.

The earliest punch cutters naturally copied the current handwriting of the country in which they lived, and in the specimens which will now be shown, enlarged on the screen, we shall see how close the resemblance is to the written books

[*Inserted in pencil*: 1 Block book ars Moriendi.]

1 First edition of Bible known to Bibliographers as the 42 line[,] usually called the Mazarin or Gutenberg's[,] printed at Mentz

probably by John [Johann] Schoeffer between the years 1450 and 1455[.] Excellence of press work, paper and ink

2 German Missal written at Cologne in 145[;] size of page $10\frac{1}{2} \times 7$. [Owned by Morris.] Contractions make black[letter] of this period difficult to read

4 Missal printed at Bamberg in 1481 by [Johann] Sensenschmidt[.] part of a page of M.S. Missale written at Wurtzburg in 1470 or 80

5 Page of an Italian M.S. dated 1451 – an epitome of the History of Trogus Pompeius by Justin[,] [Justinianus] a Latin historian on [*sic*] the second or third century. page measures $7\frac{1}{8} \times 4\frac{3}{4}$[,] written [in] a hand similar to that from which the earliest type we call Roman was copied.

6 Livy. History of Rome [*Historiae Romanae decades*]. Printed at Rome by Sweynheim and Pannartz in 1469. The printers were Germans and had produced the first printed book in Italy in 1465. note margin ⟨read note on p. 5 verso[⟩]

7 Page of Italian M.S. of about 1480 or 1490. Columella Treatise on Agriculture $9 \times 5\frac{1}{2}$ inches [owned by Morris; now in Morgan Library]

8 Cicero's Familiar letters [*Epistolae ad familiares*]. printed in Venice by John of Spires [Johannes de Spira] in 1469 and the 1st book printed in Venice.

9 Cicero's letters to Brutus [*Epistolae ad Brutum*]. Printed by Nicolas Jenson at Venice 1470.

10 M.S. Triumphs and Sonnets of Petrarch [*Trionfi e canzoniere*] probably xvi century. Size of p. $5\frac{1}{2} \times 2\frac{5}{8}$

13 Small cheap book printed by Aldus Manutius in Venice. Type generally and incorrectly said to have been copies from the handwriting of Petrarch. The engraver is said to have been identified with Francesco Francia of Bologna. Many of the early punch cutters are said also to have been goldsmiths. No contractions

11 Vincentino [Ludovico Vicentino degli Arrighi]. a page from his copy-book [either *La operina* or *Il modo de temperare le penne*, both owned by Morris]. Roman letter. Venice 1523. Interesting looked at in connexion with Mr. Ruskin[']s remarks on handwriting in Fors Clavigera 1875 a paragraph of w[hi]ch has been set in three types now in use.

12 Italian hand from same book. C.f. with what are called 'jobbing' types[.] Show American specimen book.

14 The Ancient Abode of the Insubrian Gauls [Bonaventura

Castiglione, *Gallorum insumbrum antiquae sedes*]. Milan 1541[.] shows similar character, also decorative finish to a page

15 The Comedies of Plautus [*Comoediae viginti*]. printed by Junta [P. de Giunta] at Florence in 1514. 8vo. Copy from w[hi]ch photograph was taken was illuminated by hand for presentation to Lorenzo di Medici.

16 Polydorus Vergil's History of England [*Anglicae historiae libri virgini septum*] printed [by Palma] Isingrin of Basle in 1556 but type had been modelled by [Johann] Froben about 40 years previously. [Owned by Morris.] This is of the same character as Froben's and may be said to represent Roman type in perfection. The punch cutters up to the date of this type had fine handwriting as a model. After this, the period of decadence begins and the engravers having no longer a living reference copied one from the other[.] The Elzevirs['] types w[hi]ch were cut by Christopher [Christoffel] van Dijk are about the best of xvii century founts.

17 [Daniel] Heinsius. Leyden 1630. life of punches Inschede [Enschedé, Dutch typefounding firm] –

Caslon 1720[.] called 'old face['] on printed sheet[.] punches still in existence. cut in imitation of Elzevir.

18 Baskerville of Birmingham[,] Virgil 1757. His liking for glazed paper. Types said to have been designed by Hogarth

19 Bodoni[,] great printer and type founder at Parma to whom we are indebted for the ungraceful types now used in Blue books and newspapers ⟨c.f. with type called modern on specimen sheet[⟩]. The moral is if we want beautiful type we must [*crossed out*: teach children to write beautifully] write beautifully & to this end we may profitably study Mr. Ruskin[']s remarks[.] Miller & Richard's fount old style [.] *composition*

20 Page of Hobby Horse [periodical, the *Century Guild Hobby Horse*] leaded. objectionable appearance of white stripes

21 same solid[.] 'gets in' about the same amount and allows of the use of a larger face of type. More unity of effect in consequence.

22 The Hypnerotomachia of Poliphilo [Francesco Colonna, *Hypnerotomachia Poliphili*] printed Venice by Aldus [Manutius] in 1499. Advantage of setting headings and titles solid. Conducive to the simplicity and effectiveness to be desired in titles. Many types on a title page very objectionable. publishers['] colophons of a large size very decorative.

23 [*Added in pencil*: House of the Wolfings]

Black letter often used in titles. It shouldn't be mixed with roman. margins. Inner margin as small as possible so that it will allow the printed page to lie flat when the book is opened. Top margin may be the same width as inner margin. Fore-edge and tail large – especially tail[.] Reasons: – that the wear comes at bottom left hand corner – and the binder might cut fore-edge and tail but can't cut inner margin. When a book is opened in w[hi]ch margins have been well arranged the two pages are seen in the middle.

Headlines *may* be left out; where a short running abstract is used by authors this might be set [*deleted*: as a marginal note] in a small square in margin.

Paper. Type looks best when printed upon a moderately smooth Hand made paper which has been damped. If a machine made paper is used it should not be too smooth. Highly glazed paper very objectionable in appearance. Fashion of retaining rough edges objectionable on machine made paper. [*Added in margin*: Sizes of paper & books. Foolscap & post. water marks. laid & wove paper.]

Press Work[.] Type should not merely be brought in contact with paper[,] should be [*deleted*: driven] well impressed *into* the paper and should be quite black not grey.

⟨show page of type – forme, chase, furniture. print page of type and block with it. Show mechanical relation.[⟩]

Show American Wood cut full of admirable technical qualities – chiaroscuro. pull it on hand-made paper. Demonstrate want of mechanical relation. Show overlays and refer to the proper paper for printing such a block which is glossy & objectionable. [*Added in margin*: Early printing in England not as good as in other countries.]

24 Bewick cut[.] British Birds – perfect mechanical relation and in harmony

25 Rogers['] Italy with Turner's illustrations. Want of relation[,] mechanical or artistic. Mention Present Fashion shown in French Books with various coloured prints from copper plates

26 La Mer des Histoires. Paris[,] Pierre le rouge 1488[.] Example of harmonious relation of block to type [owned by Morris]

27 d[itt]o d[itt]o[.] show space left for arms[.] Fitness of Heraldry for book decoration when drawn without tinctures [*Inserted in pencil*: The arms of a country or of a person may be]

28 Boccaccio's 'Renowned and Noble ladies,' [*De claris mulieribus*] Ulm 1473[.] Finely designed woodcut with Gothic feeling. Type of a transitional character.

29 Supplementum Supplementi [by Jacopo Filippo Foresti.] Venice 1524. topographical view made decorative by the thorough knowledge possessed by the engraver of the mechanical limitations. These limitations will help rather than hinder when understood. No difficulty in the way of the modern book illustrator if he w[oul]d study the mechanical limitations of w[hic]h as a rule he never thinks. Many object[s] w[oul]d look very decorative[.] Botanical figures

30 Life of the Virgin[.] Durer [Nurnberg, 1511; owned by Morris]

31 [Hans] Burgmair's [Burgkmair] Emp[eror] Maximilian[']s procession [collection of woodcuts, 1516–18, to which Dürer also contributed]

Copper-plates and wood-engravings of a pictorial nature i.e. drawn without reference to there [*sic*] surroundings are best printed upon a separate page on paper of the same colour as the rest of the book [*deleted*: (illegible word) process blocks only substitutes for] and separate from the opposite page with a blank leaf [*Added in pencil*: Magazines & parallel with photographs.]

Oscar Wilde's article

Nothing could have been better than Mr. Emery Walker's lecture on Letterpress, Printing, and Illustration, delivered last night at the Arts and Crafts. A series of most interesting specimens of old printed books and manuscripts were displayed on the screen by means of the magic-lantern, and Mr. Walker's explanations were as clear and simple as his suggestions were admirable. He began by explaining the different kinds of type, and how they are made, and showed specimens of the old block-printing which preceded the moveable type and which is still used in China. He pointed out the intimate connection between printing and handwriting – as long as the latter was good the printers had a living model to go by, but when it decayed printing decayed also. He showed in succession on the screen a page from Gutenberg's Bible, (the first printed book – date about 1450–5) and a manuscript of Columella; a printed Livy of 1469, with the abbreviations of handwriting, and a manuscript of the History of Pompeius by Justin of 1451. The latter he regarded as an example of the beginning of Roman type. The resemblance between the manuscripts and the printed books was most curious and suggestive. He then showed a page out of John of Spier's edition of Cicero's Letters, the first book printed at Venice, an edition of the same book by Nicholas Jansen [Jenson] 1470, and a wonderful manuscript Petrarch of the 16th century. He told the audience about Aldus who was the first publisher to start cheap books, who dropped abbreviations, and had his type cut by Francia *pictor et aurifex*, who was said to have taken it from Petrarch's handwriting. He exhibited a page of the copy-book of Vicentino the great Venetian writing master which was greeted with a spontaneous round of applause, and made some excellent suggestions about improving modern copy-books and avoiding slanting writing. A superb Plautus printed at Florence in 1514 for Lorenzo di Medici, Polydore Virgil's History with the fine Holbein designs, printed at Basle in 1556, and other interesting books were also exhibited on the screen, the size of course being very much enlarged. He spoke of Elzevir in the seventeenth century, when handwriting began to fall off, and of the English printer Caslam [Caslon], and of Baskerville, whose type was possibly designed by Hogarth, but is not very good. Latin, he remarked, was a better language to print than English, as the tails of the letters did not so often fall below the line. The wide-spacing between lines occasioned by the use of a lead, he pointed out, left the page in stripes, and made the blanks as important as the lines. Margins should of course be made, except the inner margins, and head lines often robbed the page of its beauty of design. The type used by the *Pall Mall* [Old Style] was, we are glad to say, rightly approved of.

With regard to illustration, the essential thing, Mr. Walker said, is to have harmony between the type and the decoration. He pleaded for true book-ornament as opposed to the silly habit of putting pictures where they are not wanted, and pointed out that mechanical harmony and artistic harmony went hand in hand. No ornament or illustration should be used in a book that cannot be printed in the same way as the type. For his warnings he produced Roger's [*sic*] 'Italy' with a steel-plate engraving, and a page from an American magazine which being florid, pictorial, and bad, was greeted with some laughter. For examples we had a lovely Boccaccio printed at Ulm, and a page out of 'La Mer des Histoires' printed in 1488. Blake and Bewick were also shown, and a page of music designed by Mr. [Herbert P.] Horne.

The lecture was listened to with great attention by a large audience, and was certainly most attractive. Mr. Walker has the keen artistic instinct that comes out of actually working in the art of which he spoke. His remarks about the pictorial character of modern illustration were most well-timed, and we hope that some of the publishers in the audience will take them to heart. Next Thursday Mr. Cobden Saunderson [Sanderson] lectures on book-binding, a subject on which few men in England have higher qualifications for speaking. We are glad to see these lectures are so well attended.

APPENDIX C ❧ KELMSCOTT PRESS EXPENSES

THE document transcribed below came to light after this book was completed. At the Sotheby sale of Sydney Cockerell's books and manuscripts, 10 December 1956, lot 99 was described as 'KELMSCOTT PRESS EXPENSES, 1891. MS. ACCOUNT BOOK in the handwriting of William Morris, his daughter May, and his pressman William Bowden, *bound in at the end are some leaves from Morris's diary for* 1892 *with a few entries in his hand, red morocco, the original limp boards of the account book preserved, in slip-case sm. 8vo* 1891–2'. The buyer was the firm of Bernard Quaritch, which in turn sold the document to Arnold Yates, an English collector. In 1971, near the end of his life, Yates allowed an acquaintance to make a photocopy of it for Dr Joseph R. Dunlap of New York, who has recently given me a copy.

A note in Cockerell's hand appears on the flyleaf: 'This was the original cash book of the Kelmscott Press. It was begun on Jan 17 1891 and the entries as far as Mar 7 are in the hand of William Morris, who was then disabled with a gouty illness. The entries for Mar 14 are by his daughter May, and those for Mar 21 & 28 by the overseer Mr Bowden. The remaining entries to July 7 are by Morris. At the end are bound some leaves from WM's diary of 1892, which was but little used.' On the facing page are three notations that frequently appear in Cockerell's volumes of Morris-related documents: Cockerell's signature; the private shelf-mark 'Morrisiana 28'; and an indication that the expense book was no. 308 in the Morris centenary exhibition at the Victoria and Albert Museum in 1934.

On the cover of the book is a label in Morris's hand that reads 'Kelmscott Press | – Expenses – | January 1891.' Cockerell, following his usual practice, has pasted a signature of Morris (probably clipped from a cheque) over a calendar on the cover verso. At the end of the expenses, on a separate page, is a list, not in Morris's hand, which consists of the names of Morris's printers, though it cannot necessarily be assumed that all of them were employed by Morris during the first half of 1891. The following names appear in the list: Tibbett [probably J. Tippet], [L.] Tasker, [Thomas] Binning, [R.] Eatley and [G.] Heath, Collins jr [two Collinses worked at the Kelmscott Press], Carpenter,

Barks, Eatley and [*illegible*: Freestone?], H. Gage Cole. When the staff of the press was photographed in 1895, all but three were still employed by Morris. The names of three other printers – Stephen Mowlem, H. Howes, and Mrs Pine – appear in the list of expenses.

Figure 56

The three diary entries by Morris alluded to in the Sotheby catalogue are for 20 January, 3 March, and 4 March. The latter records 'in stock paper | big 5 | reams | small 30.- | & retree'. Pasted into the last pages of the document is Quaritch's receipt to Yates for £60 (dated 15 January 1957) for the expense book.

In the following transcription I have recorded all the information in the book but have made no attempt to reproduce its actual form. The expense book serves as a useful supplement to the later ledger in the Morgan Library, because precise documentary evidence of the Press's activities is relatively sparse before Cockerell, an unusually assiduous record-keeper, became the second Secretary of the Kelmscott Press in the summer of 1894. Perhaps the most illuminating entry is that of 20 May, which tells us that Morris bought two manuscripts of the eleventh and twelfth centuries 'for study for type': it is characteristic of Morris as a collector that he purchased early books and manuscripts primarily as an aid in designing his typefaces.

1 January: [William] Bowden, a proving press and various oddments (£13 18*s* 9*d*.)
1 January: [Edward] Prince for punches (£35)
12 February: mould for paper, [Joseph] Batchelor (£7 15*s*. 0*d*.)
12 February: Batchelor, 70 reams of paper (£12)
16 February: Cutley[?], rent of premises (£1 14*s*. 0*d*.)
21 February: a clock for office (6*s*.)
24 February: sundries (10*s*. 4*d*.)
27 February: frames &c (£11 2*s*. 2*d*.)
27 February: matrices &c (£19 14*s*. 3*d*.)
27 February: type (£25 4*s*. 5*d*.)
27 February: ink &c (£2 3*s*. 8*d*.)
27 February: sundries (8*s*. 3*d*.)
28 February: wages W. H. Bowden 2nd (£5)
28 February: W. Bowden (£2)
5 March: Coals (7*s*. 6*d*.)
7 March: wages W. Bowden (£2 10*s*. 0*d*.)
7 March: W. H. Bowden (£2)

7 March: pressman (£1 10*s.* 0*d.*)
14 March: Bowden (£2 10*s.* 0*d.*)
14 March: W. H. Bowden (£2)
14 March: extra for presswork (£1 10*s.* 9*d.*)
21 March: W. Bowden (£2 10*s.* 0*d.*)
21 March: W. H. Bowden (£2)
21 March: pressman (Giles) (£1 10*s.* 0*d.*)
28 March: W. Bowden (£2 10*s.* 0*d.*)
28 March: W. H. Bowden (£2)
28 March: pressman (£1 10*s.* 0*d.*)
28 March: W. H. Bowden (plant material) (£35 1*s.* 4*d.*)
4 April: rent of cottage to 4 April (£2 19*s.* 6*d.*)
4 April: wages (£6)
11 April: plant (£6)
14 April: sundries (15*s.* 3*d.*)
18 April: wages (£4 10*s.* 0*d.*)
23 April: to [W. H.] Hooper for cutting blocks (£13 9*s.* 6*d.*)
25 April: wages (£5 2*s.* 9*d.*)
2 May: wages (£6 6*s.* 0*d.*) [*added in margin*: 'Mrs Pyne' – i.e. Pine]
9 May: wages (£6 6*s.* 0*d.*)
16 May: wages (£8 11*s.* 0*d.*)
16 May: Batchelor for paper for G.L. [*The Golden Legend*] (£68 9*s.* 0*d.*)
16 May: mould for same (£7 15*s.* 0*d.*)
20 May: two books 11th & 12 centuries for study for type (£55)
23 May: wages (£10 13*s.* 0*d.*) [*added in margin*: 'Howes']
29 May: Hooper cutting blocks (£22 8*s.* 0*d.*)
30 May: wages (£12 16*s.* 0*d.*)
1[?] June: wages (£12 18*s.* 0*d.*) [*added in margin*: 'Mowlem for work with Collins']
6 June: wages (£12 18*s.* 0*d.*)
13 June: ink (£3 3*s.* 0*d.*)
13 June: rent of cottage till this date (£2 16*s.* 0*d.*)
20 June: wages (£12 19*s.* 0*d.*)
20 June: sundries (£4 15*s.* 11*d.*)
27 June: sundries (£12 19*s.* 0*d.*)
27 June: tympan parchment (10*s.* 6*d.*)
4 July: wages (£12 19*s.* 0*d.*)
4 July: Hooper block cutting (£6 12*s.* 0*d.*)
7 July: rent (£2 7*s.* 11*d.*)
7 July: ink (red) (15*s.* 0*d.*)

NOTES

THE place of publication for books cited in the notes is London unless otherwise indicated. The following abbreviations are used in the notes and (in a few instances) the text:

Ashbee Memoirs — An abridged version of C. R. Ashbee's unpublished memoirs; typescript copies in LC and elsewhere.
AWS — *William Morris: Artist, Writer, Socialist*, ed. May Morris (2 vols.; Oxford, 1936).
BC — Private collection of Sanford and Helen Berger, Carmel, Calif.
Bibliography — William S. Peterson, *A Bibliography of the Kelmscott Press* (Oxford, 1984).
BL — British Library.
Blunt diary — The diaries of Wilfrid Scawen Blunt (FM).
Bod. — Bodleian Library, Oxford.
BQ — Bernard Quaritch.
BQL — Archives of Bernard Quaritch Ltd., London.
Burne-Jones — *Burne-Jones Talking: His Conversations 1895–1898 Preserved by his Studio Assistant Thomas Rooke*, ed. Mary Lago (1981).
CL — William Andrews Clark Memorial Library, Los Angeles.
Cobden-Sanderson — *The Journals of Thomas James Cobden-Sanderson 1879–1922* (2 vols.; 1928).
Cockerell — Wilfrid Blunt, *Cockerell* (New York, 1965).
Cockerell diary — Diaries of Sydney C. Cockerell, 1886–1900 (BL Add. MSS 52623–37).
Collected Letters — *The Collected Letters of William Morris*, ed. Norman Kelvin (3 vols.; Princeton, 1984– (in progress)).
CUL — Cambridge University Library.
CW — William Morris, *Collected Works*, ed. May Morris (24 vols., 1910–15).
EBJ — Edward Burne-Jones.
EW — Emery Walker.
Fitzgerald — Penelope Fitzgerald, *Edward Burne-Jones: A Biography* (1975).
FM — Department of Manuscripts and Printed Books, Fitzwilliam Museum, Cambridge.
Forman — Harry Buxton Forman, *The Books of William Morris Described with Some Account of his Doings in Literature and the Allied Crafts* (1897).
Franklin — Colin Franklin, *Emery Walker: Some Light on his Theories of Printing and on his Relations with William Morris and Cobden-Sanderson* (Cambridge, 1973).
FSE — Frederick S. Ellis.
HEH — Henry E. Huntington Library, San Marino, Calif.
HPL — Hammersmith Public Libraries.
HRC — Harry Ransom Humanities Research Center, University of Texas at Austin.
Ideal Book — *The Ideal Book: Essays and Lectures on the Arts of the Book by William Morris*, ed. William S. Peterson (Berkeley and Los Angeles, 1982).
Introductions — May Morris, *The Introductions to the Collected Works of William Morris*, with a Preface by Joseph R. Dunlap (2 vols.; New York, 1973).
JAM — Jane (Jenny) Alice Morris.
JC — John Johnson Collection (Bod.).
JJL — J. and J. Leighton.
JM — Jane Morris.
LB — Letterbook of the Kelmscott Press (WMG J531).
LC — Library of Congress.
Ledger — Ledger of the Kelmscott Press (PML).
Mackail — J. W. Mackail, *The Life of William Morris* (1950; first edn. 1899).
Memorials — G[eorgiana] B[urne]-J[ones], *Memorials of Edward Burne-Jones* (2 vols.; 1904).
MM — May Morris.
Morris diary — Diaries of William Morris, 1893, 1895, 1896 (BL Add. MSS 45409–11).
OUA — Oxford University Archives.
PML — Pierpont Morgan Library, New York.
PW — Philip Webb.

Robinson — Duncan Robinson, *A Companion Volume to the Kelmscott Chaucer* (1975).

Rooke diary — Studio diary of T. M. Rooke (owned by Lance Thirkell, London).

RP — Robert Proctor.

SB — St Bride Printing Library, London.

Sparling — H. Halliday Sparling, *The Kelmscott Press and William Morris Master-Craftsman* (1924).

Thompson — Susan O. Thompson, *American Book Design and William Morris* (New York, 1977).

TLS — *Times Literary Supplement.*

UCLA — University of California at Los Angeles.

V & A — Victoria and Albert Museum Library.

WM — William Morris.

WMAB — Paul Needham, Joseph Dunlap, and John Dreyfus, *William Morris and the Art of the Book* (New York, 1976).

WMG — William Morris Gallery, Walthamstow.

WSB — Wilfrid Scawen Blunt.

Yale — Beinecke Library, Yale University.

Introduction

1. *Fleuron*, 3 (1924), 120.
2. *Ideal Book*, p. 1.
3. Morison, *First Principles of Typography* (2nd edn.; Cambridge, 1967), 21.
4. *CW* xxii. 375–90.
5. WM to William Bowden, 3.1.91 (PML).
6. 'A New Era in English Printing', *Inland Printer*, 10 (Mar. 1893), 518.
7. See *Bibliography*, pp. xvii–xxi.

CHAPTER ONE
The World of Victorian Printing

1. Falconer Madan, *The Daniel Press* (Oxford, 1921), 8.
2. *Ideal Book*, pp. 69, 62.
3. Beilenson, 'The Nineteenth Century', *Dolphin*, no. 3 (1938), 233.
4. McLean, *Victorian Book Design* (2nd edn.; Berkeley, Calif., 1972), p. viii.
5. *Ideal Book*, p. 32.
6. See A. F. Johnson, 'The Modern-face Type in England', in his *Selected Essays on Books and Printing* ed. Percy H. Muir (Amsterdam, 1970) 402–8.
7. Updike, *Printing Types* (2nd edn.; Cambridge, Mass., 1937), ii. 197.
8. De Vinne, 'The *Century*'s Printer on the *Century*'s Type' (*Century Magazine*, 1896), quoted in Carl P. Rollins, *Theodore Low De Vinne* (New York, 1968), 67–8.
9. See *Ideal Book*, pp. 42–3.
10. 'A Very Rare and Fine Work of Art', *Journal of the William Morris Society*, 3 (Winter 1978), 34.
11. See A. F. Johnson, 'Old-face Types in the Victorian Age', in his *Selected Essays on Books and Printing*, pp. 423–35; and G. V. Ovink, 'Nineteenth-century Reactions against the Didone Type Model', *Quaerendo*, 1, 2 (1971), 18–31; 2, 1 (1971), 282–301; 3, 2 (1972), 122–43.
12. Janet Ing, 'Foreword', in J. Buechler, *Charles Whittingham, Printer* ([Burlington, Vt.], 1983), p. 8; cf. Arthur Warren, *The Charles Whittinghams, Printers* (New York, 1896), 149–50.
13. Jacobi, *A Note on the Caslon Old-faced Printing Types* (pamphlet printed for the Design and Industries Association, 1920).
14. Bruce Rogers, *Report on the Typography of the Cambridge University Press* (1917; repr. Cambridge, 1950), 5.
15. Emery Walker, 'Printing', in *Transactions of the National Association for the Advancement of Art and its Application to Industry* (1890), 465; *Ideal Book*, p. 62.
16. Rogers, p. 6.
17. Nicolas Barker, *Stanley Morison* (1972), 156.
18. Ing, p. 7.
19. 'Professor Ruskin's Printers visit Mr. Gladstone', *British Bookmaker*, 4 (July 1890), 4.
20. See 'Our Portrait Gallery. Messrs. Geo. Falkner & Sons, Manchester', *British Bookmaker*, 5 (Nov. 1891), 108–9; 'The History of a Printing Firm', *Monotype Recorder*, 23 (Nov.–Dec. 1924), 11–13.
21. Philip Unwin, *The Printing Unwins* (1970), 33.

22. See Ruari McLean, *Joseph Cundall* (1976).

23. Francis Meynell, *English Printed Books* (1946), 30.

24. For an account of Tuer's career, see Cyril Baxter, 'Andrew Tuer of the Leadenhall Press', *Print in Britain*, 11 (Dec. 1963), 31–2.

25. Ruskin, *Works*, ed. E. T. Cook and A. D. O. Wedderburn (39 vols.; 1903–12), xii. 334–5. Ruari McLean, in his *Victorian Book Design*, p. 210, suggests that the book was probably Henry Noel Humphreys, *The Origin and Progress of Writing*.

26. 'The New Edition of "Modern Painters"', *Printing Times and Lithographer*, NS 15 (15 Feb. 1889), 29–30.

27. The best general treatments of 'artistic' printing are John Southward, *Artistic Printing* (1892); Vivian Ridler, 'Artistic Printing: A Search for Principles', *Alphabet and Image*, no. 6 (Jan. 1948), 4–17.

28. On Hailing, see 'Our Portrait Gallery. No. 7 – Thomas Hailing', *British Printer*, 1 (Jan.–Feb. 1889), 5–6; 'Death of Mr. Thomas Hailing', *Printers' Register*, 38 (6 June 1899), 7–8.

29. Jacobi, *Printing* (1890), p. vi.

30. 'Theodore De Vinne on Artistic Printing', *British Printer*, 8 (May–June 1895), 173.

31. *Ideal Book*, p. 110. The rivalry between Morris and Blaikie is described by James G. Nelson, *The Early Nineties: A View from the Bodley Head* (Cambridge, Mass., 1971), 36–9.

32. Emery Walker, 'Private Presses' (letter), *TLS* 17 May 1917, p. 237.

33. Madan, pp. 46–7.

34. Morison, *John Fell* (Oxford, 1967), 200–2; Jacobi, 'The Printing of Modern Illustrated or Decorated Books', *Journal of the Society of Arts*, 50 (6 June 1902), 621.

35. Morison, *John Fell*, p. 203.

36. See Joseph R. Dunlap, *William Caxton and William Morris: Comparisons and Contrasts* (Cambridge [1976]).

37. See Peter Frost, '*The Century Guild Hobby Horse* and its Founders', *Book Collector*, 27 (Autumn 1978), 348–60.

38. Ricketts, *Self-Portrait, Taken from the Letters and Journals of Charles Ricketts, R.A.*, compiled by T. Sturge Moore and ed. Cecil Lewis (1939), 20.

39. Reed, 'Old and New Fashions in Typography', *Journal of the Society of Arts*, 38 (18 Apr. 1890), 534.

40. See Joseph R. Dunlap, 'Two Victorian Voices Advocating Good Book Design. I. Henry Stevens and the Shoddimites', *Printing History*, 1,1 (1979), 18–25.

41. Cf. *Ideal Book*, p. 67.

Chapter Two
William Morris and the Book Arts

1. In this chapter I am especially indebted to Joseph R. Dunlap's unpublished doctoral thesis, 'The Road to Kelmscott: William Morris and the Book Arts before the Founding of the Kelmscott Press', Columbia University, 1972.

2. *Collected Letters*, ii. 28.

3. Mackail, ii. 335.

4. Ibid. i. 18, 30.

5. *Memorials*, i. 98.

6. Sydney Cockerell took a copy of the Kelmscott Press *The Nature of Gothic* to Ruskin at Brantwood in the spring of 1892. Cockerell recorded in his diary: 'He [Ruskin] was pleased with it and interested in my telling him about the new blackletter fount [the Troy type]. He said that Morris was the ablest man of his generation' (17.4.92). A few days later Cockerell conveyed this remark to Morris: 'He seemed very pleased to hear about J.R. and of what he said of him' (21.4.92).

7. Morris, 'The Exhibition of the Royal Academy, by a Rare Visitor', *To-day*, July 1884, quoted in *Introductions*, ii. 457.

8. WM to Gere, 7 and 9.11.[93] (HEH).

9. *Ideal Book*, p. 14.

10. Blunt diary, 21.5.96.

11. F. S. Ellis, 'The Life-Work of William Morris', *Journal of the Society of Arts*, 46 (27 May 1898), 620. It is difficult to determine precisely what manuscripts Morris saw in the Bodleian. As an undergraduate he would not have been allowed to use the Bodleian under normal circumstances, and his name does not appear in the

Notes

library's records during those years. Yet Morris had vivid memories of studying the Bodleian manuscripts, and possibly he might have used the reading room by special permission, and the documents might have been requested by a companion. Morris's name does figure in the Bodleian's 'L.P. Lending Book' after he received his degree, but here the evidence is complicated by the occasional appearance of an 'R. Morris' (probably the Revd Richard Morris, Winchester Lecturer on English Language and Literature at King's College School, London) in the records. 'Mr Morris' (not always, therefore, demonstrably William Morris) called for the following manuscripts (identified by Bodleian shelf-marks): (1) Bod. 264 – 27 Apr. 1856 (certainly William Morris, because 'of Exeter' is added after the name); (2) Auct. F. IV. 32 – 26 Sept. 1856; (3) Ashmole 789, 1114 – 2 Jan. 1863; (4) Dugdale A1. – 4 May 1865; (5) Auct. F. 38 – 18 Nov. 1865; (6) Auct. Q – 25 Nov. 1865; (7) Fairfax 14; Ashmole 43, 50, 53, 781 – 2 Jan. 1866; (8) Auct. L. IV. 30 – 1 June 1866; (9) Auct. π. R. IV. 2, 3 – 9 Nov. 1866; (10) Auct. π. Q. inf. π. 5 – 9 Mar. 1867; (11) Douce 381, 302 – 17 May 1867. In a later 'L.P. Entry Book' it is recorded that Morris (accompanied by Emery Walker) came on 28 Nov. 1894 to inspect the 'Douce Apocalypse' (Douce 180) and several other manuscripts: Auct. D. 4. 17; Douce 366; Bod. 264. On this occasion Morris was investigating the possibility of publishing a facsimile of the Apocalypse: see *Bibliography*, p. 151.

12. *Introductions*, i. 250.

13. *Ideal Book*, p. 10.

14. Hyndman, *The Record of an Adventurous Life* (1911), 355.

15. Mackail, i. 141.

16. Paul Needham, 'William Morris: Book Collector', in *WMAB*, p. 25.

17. Rooke diary, 20.1.97.

18. *Ideal Book*, p. 45; see also p. 25.

19. Printed in *Ideal Book*, pp. 25–44.

20. Mackail, i. 90.

21. *Memorials*, i. 254–5. Cf. this remark by Burne-Jones in 1897: 'Mr. Morris[,] who was always right in matters of art, never liked it [modern wood-engraving]. He loved simplicity and disliked ingenuity in art very much' (Rooke diary, 25.6.97).

22. Bewick himself argued that cross-hatching was unsatisfactory: see *A Memoir of Thomas Bewick, Written by Himself* (Newcastle-on-Tyne and London, 1862), p. 241. Ruskin also made a characteristically furious attack upon cross-hatching in a lecture delivered at Oxford in 1872 (*Works*, xxii. 359–63).

23. *Collected Letters*, i. 410.

24. *Memorials*, i. 157.

25. Ellis (see note 11), p. 627.

26. *William Allingham's Diary*, ed. H. Allingham and D. Radford (Carbondale, Ill., 1967), 139–40. The story of the unfinished edition of *The Earthly Paradise* has been told at length by Joseph R. Dunlap, *The Book that never was* (New York, 1971), and A. R. Dufty, Introduction to *The Story of Cupid and Psyche* by Morris (1974).

27. Arthur Burgess, an engraver employed by Ruskin, was interviewed by Ellis on Morris's behalf, but he did not in the end work on the *Earthly Paradise* illustrations: see *Stray Letters from Professor Ruskin to a London Bibliopole* (1892), 18.

28. Mackail, i. 196.

29. Notes of an interview with Charles Fairfax Murray, in a notebook compiled by J. W. Mackail (WMG J163).

30. *Collected Letters*, i. 61.

31. *Ideal Book*, p. 40.

32. Dufty, pp. 18–20.

33. See 'The Cupid and Psyche Frieze by Sir Edward Burne-Jones, at No. 1 Palace Green', *Studio*, 15 (15 Oct. 1898), 3–13.

34. See Sebastian Carter, *The Book becomes: The Making of a Fine Edition* (Cambridge, 1984).

35. *Memorials*, ii. 23.

36. Trial pages and ornaments for *Love is Enough* are in PML (see *WMAB*, Pls. LXII–XLIV), HEH, BL C. 102. h. 18, and WMG.

37. PML 76910.

38. Rejected designs by Morris for the *Love is Enough* binding, now in the Ashmolean Museum, Oxford, are reproduced in Elizabeth Masterman, 'May Morris: Some Notes for Book Collectors', *Book*

Collector, 33 (Summer 1984), 176–7.

39. The binding of the latter (BL Add. MS 37832) is described and reproduced in H. M. Nixon, 'English Bookbindings LXXIII: A Binding Designed by William Morris in 1872', *Book Collector*, 19 (Summer 1970), 214–15.

40. Mackail, ii. 270.

41. *Ideal Book*, p. 13; *Burne-Jones*, p. 109.

42. *Collected Letters*, i. 219–20.

43. *Letters of Dante Gabriel Rossetti to William Allingham*, ed. George Birkbeck Hill (New York, 1898), 193; a facsimile of Ruskin's letter appears in the *Journal of the William Morris Society*, 2 (Spring 1966), 2. The two most detailed analyses of Morris's manuscripts, upon which I have drawn heavily, are Joseph R. Dunlap, 'William Morris: Calligrapher', in *WMAB*, pp. 48–70 (see also pp. 110–14 and Pls. XVL–LIVe), and Alfred Fairbank, 'A Note on the Manuscript Work of William Morris', in *The Story of Kormak the Son of Ogmund*, trans. William Morris and Eiríkr Magnússon (1970), 53–64. The latter includes Fairbank's 'An Annotated List of the Manuscript Work of William Morris', pp. 65–9. See also Graily Hewitt, 'The Illuminated Manuscripts of William Morris' in an untitled volume of three papers on Morris read to the Double Crown Club, 2 May 1934, pp. 1–12. Full facsimiles of two of Morris's manuscripts have been published: *A Book of Verse*, with Introductions by Roy Strong and Joyce Irene Whalley (1981), and *The Rubáiyát of Omar Khayyam* (Oxford, 1981).

44. Fairbank, p. 55. See also A. S. Osley, 'The Kelmscott Manor Volume of Italian Writing Books', *Antiquaries Journal*, 64, pt. 2 (1984), 351–60.

45. Priscilla Johnston, *Edward Johnston* (1959), 87.

46. Hewitt, pp. 6–7.

47. See Ulla Mickelsson, 'Virgil's Aeneid – the Culminating Achievement of William Morris's Illumination Work', *Libri*, 29 (Oct. 1979), 260–70, and Anna Cox Brinton, *A Pre-Raphaelite Aeneid of Virgil . . . An Essay in Honor of the William Morris Centenary* (Los Angeles, 1934). Two pages of the *Aeneid* are reproduced in *WMAB*, Pls. LIV and LIVe. The manuscript was sold at Christie's, New York, 19 May 1989, lot 2370.

48. *Memorials*, ii. 56.

49. *Ideal Book*, p. 67.

50. Ibid., p. 96. Morris also briefly edited the *Oxford and Cambridge Magazine* during the 1850s, but he seems to have had little to do with the practical details of its production.

51. Rooke diary, 24.10.96.

52. *AWS* ii. 12.

53. Jacobi was the author of *Printing* (1890), *On the Making and Issuing of Books* (1891; revised in 1892 as *Some Notes on Books and Printing*), and *The Printers' Handbook* (1891). See also John Bassett, 'Eminent Living Printers . . . Charles Thomas Jacobi', *Inland Printer*, 8 (Dec. 1890), 265–6.

54. *Stanley Morison and D. B. Updike: Selected Correspondence*, ed. David McKitterick (New York, 1979), 65, 70–1, 101.

55. Jacobi, 'The Work of the Private Presses, I. The Kelmscott Press, 1891–1898', *Penrose's Annual*, 24 (1922), 17–18.

56. BL C. 43. g. 19.

57. John Buechler, 'Compiler's Note', in *Charles Whittingham, Printer* ([Burlington, Vt.], 1983), 35. The only other known use of the Basle Roman type is in three volumes of religious verse by the Revd Orby Shipley, 1863–5 (A. F. Johnson, *Type Design* (3rd edn.; 1966), 85).

58. WM to JM, 4.12.88 (BL Add. MS 45340, fo. 40).

59. Forman, p. 140.

60. Mackail, ii. 226. Morris intended to reprint *The House of the Wolfings* at the Kelmscott Press but never did so; trial pages and a border for such an edition are in BL C. 43. f. 24 (Δb12), C. 102. h. 18, and C. 43. c. 26 (ΣQQ46).

61. 'The Art of Printing . . . A Lecture by Mr. Emery Walker', *British and Colonial Printer*, 13 Feb. 1890, p. 110.

62. WM to FSE, 17.10.[90] (FM: Ellis autograph album, fo. 28).

63. *Introductions*, ii. 412.

64. Mackail, ii. 238–9.

65. Cockerell, who is usually reliable in

such matters of detail, says that seventy-five copies were printed on (Whatman) paper and three copies on vellum (*Ideal Book*, p. 80); Fairfax Murray, in his copy of the book, recorded the figures as one hundred and five respectively (Sotheby catalogue, 7 July 1919, item 73).

66. There are earlier records of 'Caxton' types being cast for this purpose, and Vincent Figgins produced a replica of Caxton's second type which was initially used in an 1855 reprint of *The Game of Chesse*. Towards the end of the century this and other blackletters based – in some cases very loosely – upon Caxton's founts achieved some popularity in display work. (Bernard C. Middleton, *A History of English Craft Bookbinding Technique* (New York, 1963), 207; G. V. Ovink, 'Revivals of Fifteenth-Century Typefaces', in A. R. A. Croiset van Uchelen (ed.), *Hellinga: Festschrift/Feestbundel/Mélanges. Forty-three Studies in Bibliography Presented to Prof. Dr. Wytze Hellinga* (Amsterdam, 1980), 282–3.) A history of the Chiswick Press 'Caxton' face is supplied by an anonymous memorandum in the Press's archives (BL Add. MS 43985, fos. 209–10).

Chapter Three
Founding the Press

1. Mackail, ii. 212.

2. Crane, 'Preface', *Catalogue of the First Exhibition* [of the Arts and Crafts Exhibition Society] (1888), 7–8.

3. For a comprehensive treatment of Morris's relationship with the various arts-and-crafts groups, see Peter Stansky, *Redesigning the World: William Morris, the 1880s, and the Arts and Crafts* (Princeton, 1985).

4. Cockerell repeated this sentence in his life of Walker contributed to the *DNB*. Other important biographical treatments of Walker include R. C. H. Briggs, 'Sir Emery Walker: A Memoir' (mimeographed, Kew, 1959); Colin Franklin, *Emery Walker: Some Light on his Theories of Printing, and on his Relations with William Morris and Cobden-Sanderson* (Cambridge, 1973); Bernard H. Newdigate, 'Contemporary Printers, II. Emery Walker', *Fleuron*, no. 4 (1925), 63–9; Noel Rooke, 'Sir Emery Walker 1851–1933', *Penrose's Annual*, 48 (1954), 39–43.

5. Franklin, p. 23.

6. PW to MM, 6.4.92 (BL Add. MS 45342, fo. 74).

7. WM to JAM, 17.11.88 (BL Add. MS 45340, fo. 37).

8. Franklin, p. 29.

9. Walker's slides continued to be in circulation for a number of years. Cockerell borrowed and studied them in 1889 (Cockerell diary, 10 and 24.11.89), and Cobden-Sanderson used them on an American lecture tour during the winter of 1906–7.

10. *Introductions*, ii. 415.

11. [Cockerell] 'Sir Emery Walker. A Master of Typography', *The Times*, 24 July 1933, p. 7; Franklin, p. 30.

12. *Introductions*, ii. 428.

13. Cockerell, 'Sir Emery Walker'.

14. Needham, 'William Morris: Book Collector', in *WMAB*, pp. 21–2.

15. Needham, 'Catalogue', in *WMAB*, p. 99.

16. *Ideal Book*, p. 92.

17. The Sotheby catalogue (5–10 Dec. 1898) of Morris's library – which does not record some titles sold separately the previous year – includes the following 'books about books' (as booksellers call them): Falconer Madan, *Books in Manuscript: A Short Introduction to Their Study and Use* (1893); Edward Gordon Duff, *Early Printed Books* (1893); A. W. Pollard, *Early Illustrated Books: A History of the Decoration and Illustration of Books in the Fifteenth and Sixteenth Centuries* (1893); H. P. Horne, *The Binding of Books: An Essay in the History of Gold-tooled Bindings* (1894); the Bibliographical Society's Illustrated Monographs, nos. i–iii (G. R. Redgrave, *Erhard Ratdolt and his Work at Venice* [1894]; Robert Proctor, *Jan van Doesborgh* [1894]; and H. S. Ashbee, *The Iconography of 'Don Quixote'* [1895]); *Bookbinding: A Collection of Facsimiles from Examples of Historic or Artistic Bookbinding* (Quaritch catalogue, 1889); W. S. Brassington, *A History of the Art of*

Bookbinding, with Some Account of the Books of the Ancients (1894); Horatio F. Brown, *The Venetian Printing Press* (1891); L. Bickell, *Bucheinbände des XV. bis XVIII. Jahrhunderts aus hessichen Bibliotheken* (Leipzig, 1891); William Bell Scott, *William Blake: Etchings from his Works* (1878); F. A. G. Campbell, *Annales de la typographie Néerlandaise au xve siècle* (2 vols.; La Laye, 1874–90); T. F. Dibdin, *The Bibliographical Decameron*, ii–iii (1817); William Blades, *The Life and Typography of William Caxton* (2 vols.; 1861–3); Edward Gordon Duff, *Hand-lists of English Printers*, Part 1, *1501–6* (Bibliographical Society, Miscellaneous Publications, 1895); *Bookbindings Exhibited at the Burlington Fine Arts Club* (1891); Edward Gordon Duff, *Early English Printing: A Series of Facsimiles of All the Types Used in England during the xvth Century* (1896); Léon Gruel, *Manuel historique et bibliographique de l'amateur de relieurs* (Paris, 1887); J. W. Holtrop, *Monuments typographiques des Pays-Bas au xve siècle* (La Haye, 1868); J. H. Middleton, *Illuminated Manuscripts in Classical and Mediaeval Times* (Cambridge, 1892); F. Lippmann, *The Art of Wood-engraving in Italy in the Fifteenth Century* (1888); Henry Noel Humphreys, *A History of the Art of Printing* (1867); S. T. Prideaux, *An Historical Sketch of Bookbinding* (1893); E. F. Strange, *Alphabets: A Handbook of Lettering* (1895); M. D. Wyatt, *The Art of Illuminating as Practised in Europe from the Earliest Times*, illustrated by W. R. Tymms (1860); C. W. Woolnough, *The Whole Art of Marbling as Applied to Paper Book-edges* (1881); Henry Shaw, *The Art of Illumination in the Middle Ages* (1870); *The British Bookmaker*, iv (1891); *Tregaskis's International Bookbinding Exhibition at the 'Caxton Head'* (1894); O. Thierry-Poux, *Premiers monumens de l'imprimerie en France au xve siècle* (Paris, 1890); Alfred Wallis, *Examples of the Bookbinder's Art of the Sixteenth and Seventeenth Centuries, chiefly from the Royal Continental Libraries* (1890). Morris's copy of Caleb Stower, *The Printer's Grammar; or, Introduction to the Art of Printing* (1808) is in the Humanities Research Center, University of Texas. Morris also owned many facsimiles of early books and manuscripts, which I have not attempted to list here.

18. Mackail, ii. 239–40.

19. Crane, *An Artist's Reminiscences* (1907), 326–7; Crane, *William Morris to Whistler* (1911), p. 34.

20. Franklin, pp. 29–30.

21. *Ideal Book*, pp. 75–6. I have examined Morris's work as a type-designer at greater length in a forthcoming issue of the *Journal of the Printing Historical Society*.

22. Mackail, ii. 262.

23. *Introductions*, ii. 430.

24. *Ideal Book*, p. 108.

25. See G. W. Jones, 'Edward Prince: Punch Cutter', *Caxton Magazine*, 26 (Jan. 1924), 5–7, and Frederick C. Avis, *Edward Philip Prince, Type Punchcutter* (1967).

26. Denys Sutton (ed.), 'Letters from Herbert Horne to Roger Fry', *Apollo*, 72 (Aug. 1985), 141. I am indebted to John Dreyfus for this reference.

27. Sparling, p. 42. For a sketch of Reed's career, see Stanley Morison, *Talbot Baines Reed* (Cambridge, 1960).

28. *Ideal Book*, pp. 63, 69.

29. Ibid., p. 83.

30. De Vinne, *Historic Printing Types* (New York, 1866), 23.

31. Reed, 'Old and New Fashions in Typography', *Journal of the Society of Arts*, 38 (18 Apr. 1890), 531.

32. *Ideal Book*, p. 69.

33. Ibid., p. 79; Franklin, p. 31.

34. Mackail, ii. 268.

35. Ibid. ii. 364.

36. *Stanley Morison and D. B. Updike: Selected Correspondence*, ed. David McKitterick (New York, 1979), 72; Stanley Morrison, *Selected Essays on the History of Letter-forms in Manuscript and Print*, ed. McKitterick (Cambridge, 1981), i. 27.

37. *Ideal Book*, p. 70.

38. 'The Printing of a Book', *Printers' Register – Supplement*, 6 Dec. 1893, p. viii.

39. WM to BQ, 1.6.92 (BQL).

40. *Ideal Book*, p. 76.

41. *Introductions*, ii. 430.

42. WM to EW, 20.12.91 (HRC); the photographs are in BC and BL C. 102. h. 18.

43. *Ideal Book*, p. 76.

44. Ibid., p. 60; Cockerell diary, 5.11.92.

45. BQ to FSE, 11.9.90 (BQ).

46. Mackail, ii. 263–4.

47. Morris described this volume as a Bolognese book of 1473 (*Ideal Book*, p. 75), but in the *Summa. Para 3* (Venice, 1475) by Alexander of Hales, now owned by the Society of Antiquaries, there is a note by Cockerell indicating that it was the book taken by Morris to Little Chart. The Batchelor paper is thicker and more coarsely textured than that used in the *Summa*.

48. Morris's letters to Batchelor have been published in Sydney Ward, 'William Morris and his Papermaker, Joseph Batchelor', *Philobiblon*, 7 (1934), 177–80, and Henry Morris, *A Visit to Hayle Mill* (North Hills, Pa., 1970), pp. 22–45.

49. Sparling, p. 63; *Bibliography*, p. 5.

50. *Ideal Book*, p. 112.

51. Sparling, p. 63.

52. *Bibliography*, p. 191; Sparling, pp. 8–9, 73.

53. Invoice in HEH.

54. Invoice in BL C. 102. h. 18.

55. The present locations of three of Morris's four presses are known. (*a*) The small proofing press, unlike the other three, was not acquired by C. R. Ashbee; it was advertised for sale at 30 guineas by Tregaskis in 1898, apparently without success, for it remained in the basement of the Tregaskis Bookshop in London for a number of years. In the 1920s Dr John W. Bartlett of New York bought it. Theodore M. Lilienthal purchased it from Bartlett in 1939; and it is now in the Huntington Library. (See Lilienthal, *A William Morris Press goes West* (Berkeley, Calif., 1961).) (*b*) Either the first or second of Morris's presses passed from Ashbee in 1913 and through several intermediate owners to the Shakespeare Head Press; Sir Basil Blackwell gave it to the William Morris Society, and it is at present, appropriately, in the basement of Kelmscott House. (See the Morris Society pamphlet *William Morris's Printing Press* (1983).) (*c*) The large *Chaucer* press, after it left Ashbee's hands, wandered about for some years before it came into the possession of Frederic Goudy, the American type-designer, who used it at his Press of the Woolly Whale. It is now in a private collection in New York. (See J. Ben Lieberman, 'The Liberty Bell on the Kelmscott Goudy Press', *Black Art*, 1 (Summer, 1962), 41–9.)

56. Jane Morris's letters to Blunt are in FM.

57. Sparling, p. 73.

58. HEH HM 36678.

59. For a history of the building, see Philip D. Whitting (ed.), *A History of Hammersmith Based Upon that of Thomas Faulkner in 1839* (1965), pp. 204–5.

60. WM to William Bowden, 3.1.91 (PML).

61. Sparling, p. 74.

62. Mackail, ii. 268.

63. EW to T. B. Read, 19.11.91 (SB).

Chapter Four
Morris as Book-designer

1. *Ideal Book*, p. 67.

2. Henry W. and Albert A. Berg Collection, New York Public Library (Astor, Lenox, and Tilden Foundations).

3. See *Ideal Book*, pp. 62–3.

4. *AWS* i. 671–2.

5. SC to T. Edwin Ashworth, 25.9.94 (LB).

6. *Ideal Book*, p. 67.

7. Ibid.

8. See Nikolaus Pevsner, *Pioneers of Modern Design: From William Morris to Walter Gropius* (rev. edn.; Harmondsworth, Middlesex, 1974).

9. An alternative technique on the platen press, used by some twentieth-century printers such as Joseph Blumenthal and Will Carter, is to double-ink the formes – i.e. cause the press to throw off an impression every other time.

10. WM to BQ, 13.8.[90] (BQL).

11. *Ideal Book*, p. 72.

12. WM to Gilbert Redgrave, 13.11.[92], quoted in Colin Franklin, 'On the Binding of Kelmscott Press Books', *Journal of the William Morris Society*, 2 (Summer 1970), 29.

13. SC to Forman, 14.9.97 (Shepherds

Bush Public Library, Archives, DD/341/9).

14. *Ideal Book*, p. 71.

15. Sparling, p. 64. See also *Ideal Book*, p. 104.

16. Morris met W. J. Turney on 1 June 1891 and received two sample skins from him on 18 June; the first regular delivery of vellum was on 8 July (W. J. Turney and Co. to SC, 19.11.97 (HEH HM 36660)).

17. See *Bibliography*, pp. xxiv–xxv. When a reviewer in the *Athenaeum* (15 Jan. 1898) suggested that Morris's vellum was not the genuine article, it evoked thunderous responses from Cockerell and Ellis (22 Jan.). Cockerell sent his letter to the editor at the urging of Burne-Jones (EBJ to SC [Jan. 1898] (BL Add. MS 52708, fo. 86)), though Cockerell rarely needed much encouragement to do battle on behalf of Morris's reputation.

18. *Ideal Book*, p. 87.

19. Peter Guy, 'Reproducing the Kelmscott *Chaucer*', *Penrose Annual*, 69 (1976), 94. For a more lengthy account of Morris's ink problems than I can offer here, see my article 'William Morris and the "Damned Chemists": The Search for an Ideal Ink at the Kelmscott Press', *Printing History*, 3, 2 (1981), 4–11.

20. *Ideal Book*, p. 87.

21. Sparling, pp. 64–6.

22. EW to John Johnson, 20.4.31, quoted in *WMAB*, p. 82; the letter is in JC.

23. The American branch of the Jänecke firm also, more subtly, invoked Morris's name in its advertising, as in this announcement in the July 1925 issue of *Ars typographica*: 'William Morris quite properly believed that fine black ink was an essential to good printing. We can supply black of the highest character for antique papers and clear incisive colors for typographic decoration – as well as a wide selection of inks for halftone and process printing.'

24. Unless otherwise indicated, letters and diaries cited in this discussion of Morris's inks are quoted in 'William Morris and the "Damned Chemists"' (see note 19 above).

25. C. T. Jacobi, 'Printing Inks', *Library*, 2nd ser., 7 (Jan. 1906), 71.

26. B. H. Newdigate, 'Mr. C. H. St. John Hornby's Ashendene Press', *Fleuron*, no. 2 (1924), 78.

27. The *Reynard* proofs are in BC; the proofs in gold ink are (or were) owned by a Chicago collector, as described in *News from Anywhere* (the newsletter of the North American branch of the William Morris Society), no. 11 (Nov. 1969), 16.

28. HEH 'Morrisiana 12'; BL C. 43. c. 26 (ΣQQ46).

29. Proofs in JC.

30. Cobden-Sanderson, i. 94.

31. Ibid. i. 95.

32. WM to Cobden-Sanderson, undated letter quoted in Philip Duschnes catalogue 142, item 250.

33. Cobden-Sanderson, i. 211–12.

34. Ibid. ii. 69. Douglas Cockerell, who had been much more closely associated with Cobden-Sanderson than his brother Sydney was, wrote some years later: 'Cobden-Sanderson's egotism was almost pathological. He lacked the power of co-operation almost entirely and was almost insanely jealous of any reputation, even Morris's, that might rival his own. He lived in a world of his own creation, swayed by emotional storms of great intensity, and I doubt if he was capable of true friendship' (*Cockerell*, p. 85).

35. Cobden-Sanderson, i. 311. Cobden-Sanderson was at this time setting up the Doves Bindery in Upper Mall, Hammersmith; thereafter he produced no more bindings himself but supervised the work of others.

36. WM to BQ, 1.6.92 (BQL).

37. See Hugh William Davies, 'Some Famous English Bookshops, I. Notes on the Firm of J. & J. Leighton and the Old House at Brewer Street', *Library World*, 34 (Jan. 1932), 149–53.

38. WM to BQ, 25.7.[91] (BQL).

39. Cobden-Sanderson, i. 304.

40. 'Bookbinding: Lecture by T. J. Cobden-Sanderson', in *Transactions of the National Association for the Advancement of Art and its Application to Industry* (1890), 467.

41. It came from Francis Jenkinson, University Librarian at Cambridge. For F. S.

Ellis's spirited defence of Morris's practice, see FSE to Jenkinson, 17.11.93 (CUL Add. 6463, no. 2481).

42. SC to JJL, 25.8.94 (LB).

43. WM to Herbert Ellis, 9.6.92 (PML 76893).

44. See SC to Fred J. Hall, 22.10.94 (LB).

45. SC to Brooke, 14.12.94 (LB). See also SC to Dodd, Mead and Co., 29.3.95 (LB).

46. These are reprinted in *Ideal Book*. An excellent brief summary of Morris's typographical principles, based on an interview with him, can be found in T. L. De Vinne, 'The Printing of William Morris', *Book Buyer*, NS 13 (Jan. 1897), 920–1.

47. Catterson-Smith to P. A. Wells, 12.2.06 (WMG J590).

48. EBJ to Frances Horner [?Nov. 1893] (FW).

49. *Ideal Book*, p. 108.

50. Reprinted in T. J. Cobden-Sanderson, *Cosmic Vision*, ed. Richard Cobden-Sanderson (1922), 100–10. It may be observed in passing that Cobden-Sanderson borrowed the phrase 'The Ideal Book' from Morris, who had used it as the title of an 1893 lecture.

51. See A. Charles Sewter, *The Stained Glass of William Morris and His Circle* (New Haven, Conn., 1974), i. 39.

52. For examples of advice to this effect in printers' manuals, see William Savage, *A Dictionary of the Art of Printing* (1841), 179; John Southward, *Practical Printing* (1882), 96; Joseph Gould, *The Letter-press Printer* (4th edn.; 1888), 21.

53. WM to FSE [late 1889], in Mackail, ii. 226.

54. BL C. 102. h. 18.

55. See W. H. Slater, 'Close-spacing versus Wide-spacing: A Fundamental Problem', *Monotype Recorder*, 27 (Mar.–Apr. 1928), 21–6.

56. T. L. De Vinne, *A Treatise on Title-pages* (New York, 1902), p. 331.

57. *Ideal Book*, p. 78. Nevertheless, there is a trial page of *The Order of Chivalry* (1893) in HPL that is distinctly leaded.

58. *Ideal Book*, pp. 70–1.

59. Ibid., p. 78.

60. Pollard, 'Margins', *Dolphin*, no. 1 (1933), 67–80.

61. Tschichold, 'Non-Arbitrary Proportions of Page and Type Area', in A. S. Osley (ed.), *Calligraphy and Paleography: Essays Presented to Alfred Fairbank on his 70th Birthday* (New York, 1966), 179–91.

62. Sotheby catalogue, 19 Dec. 1932, item 34.

63. For a contemporary comment on the treatment of *Maud* as prose, see 'The Printed Page', *Printers' Register – Supplement*, 6 June 1895, p. iv.

64. Emery Walker, 'Typography: Various Printing Processes', in *Report of his Majesty's Commissioners for the Paris International Exhibition* (1901), ii. 29.

Chapter Five
The Decorated Book

1. *Ideal Book*, p. 78.

2. Jackson, *The Eighteen-Nineties* (1913), 319.

3. See Warde, 'The Crystal Goblet, or Printing should be Invisible', in her *The Crystal Goblet: Sixteen Essays on Typography*, ed. Henry Jacob (New York, 1955) 11–17.

4. See Reginald T. Blomfield, 'Of Book Illustration and Book Decoration', in *Catalogue of the Second [Arts and Crafts] Exhibition* (1889), 84–93

5. *Ideal Book*, p. 37.

6. Ibid., p. 40.

7. See SC to MM, 10.2.36 (BL Add. MS 52740, fo. 177); Sparling, p. 67. Lethaby's description (as quoted by Sparling) is the less accurate of the two, since he gives the misleading impression that Morris worked on a ground. Cockerell is correct in his assertion that Morris's normal procedure was to apply the ink directly to the paper. Only two Kelmscott-period initials are known to have been engraved by Morris himself; they are reproduced in a William Morris Society keepsake dated 23 July 1976.

8. See *WMAB*, Pl. LXXXI.

9. See *AWS* ii. 321–3. Two examples are in HEH HM 36915 and BL Add. MS 45891, fos. 68–9.

10. In a private collection in England.

11. PML 76897; CL 'Morrisiana 18'.

12. The only biographical information about Reuter that I have found is in the following sources: Margaret Armour, 'Beautiful Modern Manuscripts', *Studio*, 9 (1897), 52; R.M., 'Studio-talk', *Studio*, 14 (1898), 63–4; W. Irving Way, 'A Visit to William Morris', *Modern Art*, 4 (Summer 1896), 80; SC to MM, 23.3.30 (BL Add. MS 52740, fo. 148); MM to Marion E. Dodd, 13.4.30 (Smith College Library).

13. WM to Gilbert Redgrave (the author of a monograph on Ratdolt published two years later), 13.11.[92], in Colin Franklin, 'On the Binding of the Kelmscott Press Books', *Journal of the William Morris Society*, 2 (Summer 1970), 29. The letter probably refers to *The Golden Legend*.

14. See Oscar Jennings, *Early Woodcut Initials* (1908), 113.

15. WM to FSE, 17.10.[90] (FM: Ellis autograph album, fo. 28).

16. W. H. Hooper to SC, 28.10.99 (BL C. 102. h. 18).

17. WM to EW, 8.4.91 and 2.5.91 (HRC).

18. PML 'Morrisiana 11'.

19. McKerrow, 'Border Pieces Used by English Printers before 1641', *Library*, 4th ser., 5 (1924), 4.

20. The preliminary design for the engraved title-page of *The Golden Legend* is in WMG, and, though the border is identical to the published version, the vine ornament behind the lettering is much lighter than in the book.

21. WM to EW, n.d. (HRC). See *Bibliography*, p. 18.

22. Crane, *William Morris to Whistler* (1911), 35.

23. University of Minnesota Libraries.

24. FM.

25. Dawson, 'The Most Artistic Book', *Library*, 1st ser., 3 (1894), 433–4.

26. *Ideal Book*, p. 40. Morris's friend C. T. Jacobi, in *Some Notes on Books and Printing* (1892), 22, also made a characteristically milder plea for unity between typefaces and illustrations.

27. Crane, 'William Morris', *Scribner's Magazine*, 22 (July 1897), 96.

28. This anecdote appears in Emery Walker's 1924 Sandars Lectures (typescript, HRC; photocopy, CUL).

29. WM to EW, 4.2.92 (HRC).

30. WM to EW, 2.5.[92] (HRC).

31. Aymer Vallance, 'The Revival of Tapestry-weaving: An Interview with Mr. William Morris', *Studio*, 3 (1894), 101. See also Linda Parry, *William Morris Textiles* (1983), 104.

32. EBJ to Frances Horner [22.9.94] (FM).

33. See Carole Cable, 'Charles Fairfax Murray, Assistant to Dante Gabriel Rossetti', *Library Chronicle of the University of Texas*, NS 10 (1978), 81–9.

34. Mackail, ii. 293. Murray also persuaded Morris to print a letter by Savonarola at the Kelmscott Press (*Epistola de contemptu mundi*, 1894) for which he designed a wood-engraving (see *Bibliography*, pp. 80–2). However, I have not included Murray in my discussion of Kelmscott illustrators later in this chapter, since his design is little more than a clever pastiche.

35. EW to MM, 6.6.22 (BL Add. MS 45347, fo. 176).

36. CL.

37. An entry in Morris's diary (28.1.93) implies that the illustration redrawn by Murray is the one which appears on page 22 of the *Chaucer*.

38. Note by MM (BL Add. MS 45347, fo. 171).

39. The retouched photograph and a proof are in the Victoria and Albert Museum, Department of Prints and Drawings, E. 582–1921, E. 583–1921.

40. EW to MM, 6.6.22 (BL Add. MS 45347, fo. 176); Robert Catterson-Smith, undated memorandum (PML). Walker's letter is summarized and Catterson-Smith's memorandum quoted in *Bibliography*, pp. xxix–xxxi.

41. Robert Catterson-Smith to SC, 1.3.02 (BL Add. MS 52709, fo. 135). The illustrations were published in *The Beginning of the World: Twenty-five Pictures by Edward Burne-Jones* (1902).

Notes

42. The marginalia quoted all appear on Catterson-Smith's redrawings of the *Chaucer* illustrations (PML).

43. Fitzgerald, p. 247.

44. See Robert Catterson-Smith, 'Sir Edward Burne-Jones', in J. H. Muirhead (ed.), *Nine Famous Birmingham Men: Lectures Delivered in the University* (Birmingham, 1909), pp. 225–51.

45. 'Art in South London: A Black and White Exhibit', *Daily Chronicle*, 14 Nov. 1898, p. 3. Pennell repeated the charge in his book *The Graphic Arts* (Chicago, 1921), p. 88, provoking some angry correspondence among Pennell, May Morris, Cockerell, and Walker. Bernard Newdigate also became involved in the controversy: see his 'Book Production Notes', *London Mercury*, 6 (July 1922), 299–300. Pennell's wife was actually the first to suggest in print that Morris had treated Hooper and his other craftsmen unfairly by not mentioning them in the *Chaucer* colophon ([E. R. Pennell], 'The Kelmscott Chaucer', *Nation*, 63 (3 Sept. 1896), 173–4).

46. *Bibliography*, pp. xxix–xxx.

47. SC to Frederick H. Evans, 3.11.94 (LB).

48. Vallance, 'The Invention of Aubrey Beardsley', *Magazine of Art*, May 1898, pp. 363–4.

49. *Bibliography*, p. 193. The letter which I have there conjecturally dated 1892 should, I think, be dated 1890 instead.

50. PML; LC (in copy 2 of *The Story of the Glittering Plain* (1894)); Brotherton Library, University of Leeds.

51. WM to Walter Crane, 17.2.94 (letter privately printed by William P. Barlow, Jr., 1975); 'The Work of Walter Crane', *Art Journal*, Easter Annual, 1898, p. 10. Crane's remark about being insufficiently Gothic in outlook was a commonplace in the Morris circle. Cockerell said it to Crane's face: 'In m[ornin]g called on Walter Crane at Shepherd's Bush. . . . Urged him to go to Beauvais, and to pay more attention to Gothic forms and traditions – as I consider him to be Gothic in temper, and to have been led by accident into the acceptance of classical ornament and ideas' (diary, 2.6.89). And C. R. Ashbee quoted Morris as saying that Crane 'was a Gothic soul with a Renaissance training' (*Ashbee Memoirs*, vii. 266).

52. David Cox, 'Edmund New's "Diary of a Visit to Kelmscott Manor"', *Journal of the William Morris Society*, 3 (Spring 1974), 7; *Bibliography*, p. 61.

53. *Bibliography*, p. 60.

54. See Alan Crawford (ed.), *By Hammer and Hand: The Arts and Crafts Movement in Birmingham* (Birmingham, 1984); 'The Birmingham Municipal School of Art. With Many Illustrations of Its Students' Work – I', *Studio*, 2 (1894), 90–9.

55. A group of letters from Gaskin to Cockerell is in BL Add. MS 52716, fos. 11–15.

56. *Ideal Book*, p. 110.

57. *AWS* i. 296.

58. See Philippe Mairet, 'The Landscape Paintings of C. M. Gere and H. A. Payne', *Studio*, 80 (1890), 23–5; 'Mr. Charles Gere' (obituary), *The Times*, 6 Aug. 1957, p. 8.

59. HEH.

60. See the exhibition catalogue *Arthur and Georgie Gaskin* (Birmingham, 1981).

61. Gordon N. Ray, *The Illustrator and the Book in England from 1790 to 1914* (New York, 1976), 159; Colin Franklin, *The Private Presses* (1969), 43.

62. WM to JAM, 12.7.93 (BL Add. MS 45340, fo. 122).

63. EBJ to Frances Horner, n.d. (FM).

64. Carr, *Some Eminent Victorians* (1908), 72; *Memorials*, ii. 45.

65. Fitzgerald, p. 177.

66. JM to WSB, 12.3.[94] (FM).

67. EBJ to Helen Gaskell [before Apr. 1894] (BL Add. MS 54217, fo. 187).

68. See John Christian, 'Burne-Jones Studies, I. Burne-Jones's Drawings for "The Fairy Family"', *Burlington Magazine*, 115 (Feb. 1973), 93–100.

69. Forrest Reid, *Illustrators of the Sixties* (1928), 99.

70. Fitzgerald, p. 182.

71. Rooke diary, 7.1.96.

72. Martin Harrison and Bill Waters, *Burne-Jones* (1973), 164.

CHAPTER SIX
The Business of the Press

1. WM to JAM, 23.9.91, in Mackail, ii. 291.

2. *Cockerell*, p. 66; Elizabeth R. Pennell, *The Life and Letters of Joseph Pennell* (Boston, 1929), i. 159; Elizabeth R. Pennell, 'Some Memories of William Morris', *American Magazine of Art*, 11 (Feb. 1920), 125; W. B. Yeats, *Memoirs*, ed. Denis Donoghue (New York, 1973), 34.

3. Shaw, 'Morris as I knew him', in *AWS* ii, pp. xxix–xxxi.

4. JM to WSB, 24.5.[94] and 26.5.[94] (FM).

5. EW to N. L. McMinn, 13.2.28 (Kelmscott Manor).

6. Cockerell, 'Private Diaries' (letter), *Spectator*, 168 (16 Jan. 1942), 60.

7. *Cockerell*, p. 54; FSE to SC, 6.5.00 (BL Add. MS 52715, fo. 69).

8. PW to SC, 20.8.94 (V & A 86TT. 14).

9. For accounts of Ellis's career, see the *DNB* and the Introduction to *The Letters of Dante Gabriel Rossetti to his Publisher, F. S. Ellis*, ed. Oswald Doughty (1928).

10. WM to EW, 14.7.[91] (HRC).

11. FSE to SC, 13.4.00 (BL Add. MS 52715, fo. 61).

12. Sparling, p. 85.

13. BL C. 43. f. 24 (Δ.b.9).

14. FSE to WM, 6.1.94 (BL Add. MS 52715, fo. 11).

15. Sotheby catalogue, 18 Dec. 1919, item 402.

16. See *Bibliography*, p. 47.

17. Jacobi's annotated copy of *Gothic Architecture* was sold at the Christie's auction of the John A. Saks collection (New York, 20 May 1983).

18. CUL Sel. 3. 231.

19. HPL.

20. SC to MM, 19.1.13 (BL Add. MS 52740, fo. 14).

21. V & A R. N. 7, 8.

22. FSE to Edward Gordon Duff, 20.3.92 (HEH DF 268).

23. BL C. 43. c. 26 (ΣFF20); PML.

24. FM.

25. Ellis offered an energetic apologia for his decision to adopt 'old-spelling' in a letter to T. R. Lounsbury, 16.12.93 (Yale): 'It would be presumptuous of me to offer any argument as to the adoption of modern spelling for Chaucer, but I am quite convinced an edition so printed would not gain him one additional reader. . . . I have myself gone so far in the other direction with regard to orthography as to have printed Shakespeare's Poems, for the first time in a collected form with the author's spelling. Though the book was published at 25/- the 500 copies were all sold on the day of publication. I did the same good work this year for Cavendish's Life of Wolsey, with a like satisfactory result, holding that an author's orthography is a part of his work. Furnivall I am happy to say is now engaged in editing the whole of Shakespeare in like manner – at last we shall be able to read him with the full 15th–17th century flavour.'

26. Sparling, p. 84.

27. See *Bibliography*, pp. 65–6.

28. *Ideal Book*, p. 112.

29. Mackail, ii. 288.

30. 'Morris as I knew him' (see note 3 above), ii, p. xxviii.

31. Mackail, ii. 325.

32. Ibid. ii. 326.

33. For a brief account of William Bowden's career, see Nicolas Barker and John Collins, *A Sequel to an Enquiry into the Nature of Certain Nineteenth Century Pamphlets* (1983), 107.

34. Sparling, p. 82; Emily Hill, 'Women as Compositors', *British and Colonial Printer*, 41 (14 Oct. 1897), 11.

35. The minutes of the Hammersmith Socialist League are in BL Add. MSS 45891–4; the minutes of the Guild of Handicraft are in V & A 86. DD. 15.

36. *Ashbee Memoirs*, ii. 333.

37. *WMAB*, p. 82; L. T. Owens, *J. H. Mason* (1976), 28.

38. Frank Colebrook, 'William Morris, Master Printer', *British and Colonial Printer*, 39 (10 Dec. 1976), 374.

39. Sparling, p. 87.

40. Programme and bill in CL.

41. *WMAB*, p. 82.

42. Sparling, p. 83.
43. Colebrook, 'William Morris, Master Printer', p. 373.
44. Sparling, p. 14.
45. 'The Poet, William Morris, and the "Kelmscott Press"', *British Bookmaker*, 5 (Nov. 1891), 104.
46. 'Art and its Producers', *CW* xxii. 352.
47. Frank Colebrook, *William Morris: Master Printer* (Tunbridge Wells [?1897]), 16.
48. HEH HM 36678.
49. J. Thorne to SC, 11.11.97 (HEH HM 36679).
50. Emily Preston, 'Cobden-Sanderson and the Doves Bindery', *Craftsman*, 2 (Apr. 1902), 29.
51. Way, 'A Visit to William Morris', *Modern Art*, 4 (Summer 1896), 78–9.
52. See W. Roberts, *The Book-hunter in London* (Chicago, 1895), pp. 226–7.
53. *Burne-Jones*, p. 56; JM to WSB, 20.4.92 (FM).
54. SC to Reeves and Turner, 20.12.95 (LB); Cockerell diary, 23.12.95.
55. PML 'Morrisiana 11'.
56. Mackail, ii. 262.
57. Charlotte Quaritch Wrentmore, 'Foreword', *A Catalogue of Books and Manuscripts to Commemorate the 100th Anniversary of the Firm of Bernard Quaritch* (1947), p. vi.
58. SC to Walter T. Spencer, 14.11.94 (LB).
59. This and other letters cited below from Bernard Quaritch to B. Alfred Quaritch are in Bod. MS Eng. lett. c. 435.
60. WM to BQ, 18.7.93 (BQL).
61. Ellis and Elvey to WM, 25.5.95 (Ellis and Elvey letterbook, UCLA Library).
62. Sparling, p. 86.
63. SC to W. Glaisher, 24.10.95 (LB).
64. The letter appears in an undated *Hatchard's List* and is reprinted in the *Book Buyer*, NS 12 (Apr. 1895), 141.
65. SC to Dodd, Mead and Co., 25.9.94 (LB).
66. John R. Nash, 'Mr. Cobden-Sanderson's Two-handed Engine', *Book Collector*, 25 (Winter 1976), 497.
67. WM to PW, 27.8.94. Webb had this letter printed in the Golden type and inserted in each of his Kelmscott Press books, which are now in the Library of Trinity College, Cambridge.
68. WM to BQ, 25.7.91 (BQL).
69. SC to JM, 2.3.98 (BL Add. MS 52738, fo. 37).
70. SC to JJL, 22.3.95 (LB).
71. SC to P. A. Valladaris, 19.1.95 (LB); SC to JJL, 22.3.95 (LB).
72. EW to T. B. Reed, 19.11.91 (SB).
73. Albert L. Cotton, 'The Kelmscott Press and the New Printing', *Contemporary Review*, 74 (Aug. 1898), 224. A similar phrase appears in Herbert H. Horne, 'William Morris as Printer', *Saturday Review*, 82 (24 Oct. 1896), 438.
74. Frank Rinder, 'The Kelmscott Press', *Connoisseur*, 1 (Dec. 1901), 265–7; 'The Kelmscott Press: A Review and Some Statistics', *Scottish Review*, 36 (July 1900), 30–3. See also W. G. Menzies, 'The Present Value of the Publications of the Kelmscott Press', *Connoisseur*, 12 (May 1905), 42–4; Frederick M. Hopkins, 'Kelmscott Books and the Revival of Fine Printing', *Publishers' Weekly*, 100 (1 Oct. 1921), 1175–7.
75. Slater, 'The Book Sales of 1898 – II', *Athenaeum*, 7 Jan. 1899, p. 19.
76. Muir, *Book-Collecting as a Hobby* (New York, 1947), 34.
77. FSE to SC, 31.3.99 and 7.5.99 (BL Add. MS 52715, fos. 21, 27).
78. *Ideal Book*, p. 104.
79. *AWS* i. 133.
80. SC to PW, 20.12.99 (BL Add. MS 52769, fo. 38).
81. Colebrook, *William Morris*, p. 17.
82. SC to Strange, 18.12.95 (LB). See also SC to Percy Cockran, 19.12.95 (LB).
83. SC to George Allen, 23.11.95 (LB).
84. WM to Reeves and Turner, 16.6.91 (BC).
85. Ellis and Elvey to William Michael Rossetti, 1.2.94 and 9.2.94 (Ellis and Elvey letterbook, UCLA Library).
86. JM to WSB, 24 Apr. 1894 (FM); Blunt diary, 30.4.94.
87. *Burne-Jones*, p. 87.
88. Bullen, 'William Morris, Regen-

erator of the Typographic Art', *Inland Printer*, 69 (June 1922), 372; 'Mr. H. L. Bullen on Type – Mr. George W. Jones on William Morris', *Caxton Magazine*, 26 (Apr. 1924), 233.

89. SC to RP, 13.12.97 (BL Add. MS 52743, fo. 277). My summary of this episode is based on various passages in Cockerell's letters and diary during December 1897.

90. [Gilder], *Critic*, 29 (10 Oct. 1896), 214, quoted in Thompson, p. 35.

91. See Emma Lazarus, 'A Day in Surrey with William Morris', *Century Illustrated Monthly Magazine*, 32 (July 1886), 396–7.

92. Crane to C. R. Ashbee, 12.2.90 (V & A 86. DD. 17a).

93. WM to Crane, 17.2.94 (privately printed by William P. Barlow, Jr., 1975).

94. SC to A. J. R. Gooderson, 6.8.94 (LB).

95. *Ideal Book*, p. 111.

96. WM to Furnivall, 12.7.95 (HEH FU 686).

97. WM to BQ, 1.12.92 (BQL).

98. See *Bibliography*, p. 47.

99. Ibid., pp. xxxi–xxxv.

100. Paul Needham, 'Catalogue', *WMAB*, p. 126; Emery Walker, 'Private Presses' (letter), *TLS* 17 May 1917, p. 237.

Chapter Seven
The Story of Three Books

1. Mackail, ii. 261–2.

2. BQ to B. A. Quaritch, 5.6.90 (Bod. MS Eng. Lett. c. 435).

3. BQ to FSE, 8.9.90 (BC).

4. Ibid.

5. For the full text of the contract, see *Bibliography*, p. 191.

6. F. Colebrook, *William Morris: Master Printer* (Tunbridge Wells [?1897], 5. Two leaves of her manuscript (BL C. 43. c. 26 (ΣQQ44)) survived only because proofs were later pulled on their blank versos.

7. FSE to Duff, 4.9.92 (HEH DF 270).

8. FSE to Duff, 20.3.92 (HEH DF 268).

9. Mackail, ii. 263–4.

10. WM to FSE, 31.9.90, in ibid., pp. 265–6.

11. Ibid., p. 268.

12. JC MS 7, fo. 70.

13. JC; BL C. 102. h. 18; Victoria and Albert Museum, Department of Prints and Drawings, Box I 95. R.

14. Sydney Ward, 'William Morris and his Papermaker, Joseph Batchelor', *Philobiblon*, 7 (1934), 138.

15. WM to FSE, early Apr. 1891, in Mackail, ii. 269.

16. Sparling, p. 80.

17. JM to WSB, 11.12.[91] (FM).

18. WM to BQ, 25.7.[91] (BQL).

19. *Memorials*, ii. 217.

20. FSE to Francis Jenkinson, 17.7.92 (CUL Add. 6463, no. 2107).

21. Receipt in Bod. MS Eng. lett. c. 436, fo. 184.

22. *Library*, 4 (1892), 346.

23. A. C. Swinburne to FSE, 18.10.92 (FM: Ellis autograph album, fo. 42).

24. Mackail, ii. 273.

25. *Introductions*, i. 270.

26. Mackail, ii. 279.

27. SC to JJL, 8.9.97 (JC MS 7, fo. 100).

28. Elizabeth Longford, *A Pilgrimage of Passion: The Life of Wilfrid Scawen Blunt* (1979), 278–80; Peter Faulkner, *Wilfrid Scawen Blunt and the Morrises* (1981).

29. An undated passage in Blunt's 'My Secret Diaries', xix. 82, 84 (FM).

30. The letters are now in FM. Peter Faulkner's edition of them, *Jane Morris to Wilfrid Scawen Blunt* (Exeter, 1986), appeared after this chapter was written.

31. May Morris, in an undated letter [1918?] to Walker (WMG J440), remarked, 'I'm afraid I always thought Mr Blunt, with all his qualities, an Egotist and vain. He is a very interesting man, but unsympathetic.'

32. Rowland Elzea (ed.), *The Correspondence between Samuel Bancroft, Jr. and Charles Fairfax Murray 1892–1916* (Wilmington, Del., 1980), 80.

33. WM to JAM, 22.10.91 (BL Add. MS 45340, fo. 73).

34. Entry of 10.9.91 (BL Add. MS 53985).

35. PML 'Morrisiana 11'.

36. *Cockerell*, p. 177. Cockerell, in old age, insisted that the Morris marriage was

an entirely happy one: 'It pains me to have aspersions cast on my heroes and heroines,' he wrote (Cockerell, 'Rossetti and Mrs Morris' (letter), *TLS* 24 Aug. 1951, p. 533).

37. FSE to Francis Jenkinson, 14.3.92 (CUL Add. 6463, no. 2018).

Chapter Eight
The Kelmscott Chaucer

1. Rooke diary, 3.11.96; F. S. Ellis, 'The Life-work of William Morris', *Journal of the Society of Arts*, 46 (27 May 1898), 627.

2. *One Hundred Books* (Bridwell Library, Southern Methodist University, 1 May 1986), no. 64.

3. Even Burne-Jones was struck by the similarities between Morris and Chaucer (*Memorials*, ii. 265). For a dissenting viewpoint which emphasizes their differences, see 'Geoffrey Chaucer and William Morris', *New Monthly Magazine*, 149 (1871), 280–6.

4. *Collected Letters*, i. 473.

5. See Charles Muscatine, *The Book of Geoffrey Chaucer: An Account of the Publication of Geoffrey Chaucer's Works from the Fifteenth Century to Modern Times* (San Francisco, 1963); Elwin F. Piper, 'The Miniatures of the Ellesmere Chaucer', *Philological Quarterly*, 3 (1924), 241–56; Alice Miskimin, 'The Illustrated Eighteenth-century Chaucer', *Modern Philology*, 77 (1979), 26–55. Morris himself owned Chaucer's *Works*, published by Thomas Godfray in 1532 (the first collected edition), Skeat's six-volume edition (Clarendon Press, 1894), and many of the publications of the Chaucer Society, including Furnivall's six-text edition of *The Canterbury Tales*.

6. FSE to T. R. Lounsbury, 24.12.00 (Yale).

7. *The Cambridge MS . . . of Chaucer's Canterbury Tales*, ed. F. J. Furnivall (1868–9). For Hooper's comments on his part in this undertaking, see M. H. Spielmann, 'The Portraits of Geoffrey Chaucer – III', *Magazine of Art*, 23 (Sept. 1900), 494.

8. Furnivall, 'Recent Work at Chaucer', *Macmillan's Magazine*, 27 (Mar. 1872), 389.

9. See Skeat, *A Student's Pastime* (Oxford, 1896), p. xxv.

10. FSE to T. R. Lounsbury, 16.12.93 (Yale); Rooke diary, 28.7.97. Some even more indignant comments by Burne-Jones on Skeat are recorded in Rooke's diary on 23.9.95.

11. FSE to Francis Jenkinson, 17.4.92 (CUL Add. 6463, no. 2046).

12. Doble to Skeat, 29.11.92 (OUA DUP/PUB/1/56).

13. OUA DUP/PUB/11/23.

14. WM to P. Lyttelton Gell, 13.8.94 (OUA DUP/PUB/11/23).

15. EBJ to Frances Horner, n.d. [June 1896?] (FM).

16. Corrected proof in BL C. 102. h. 18.

17. FSE to Francis Jenkinson, 14.3.92 (CUL Add. 6463, no. 2018).

18. Mackail, ii. 288.

19. *Ideal Book*, p. 76.

20. BC.

21. Yale Folio 286.

22. *WMAB*, Pl. CII.

23. WM to FSE, 25.2.95, quoted in a notebook compiled by Mackail (WMG J164).

24. Sparling, p. 164.

25. WM to JAM, [11.8.94] (BL Add. MS 45340, fo. 145).

26. Rowland Elzea (ed.), *The Correspondence between Samuel Bancroft, Jr. and Charles Fairfax Murray 1892–1916* (Wilmington, Del., 1980), 69.

27. Morris diary, 11–12.5.96; Cockerell diary, 12.5.96. The retouched title is in PML.

28. *WMAB*, p. 140.

29. I am convinced the design is by Morris, because no binder except Cobden-Sanderson or Leighton would have had access to the cancelled signature (which is the same one used in the experimental dummy copy produced at the Doves Bindery, now in HEH).

30. This and other preliminary drawings for the *Chaucer*, as well as the finished designs, are in the Department of Prints and Drawings, Fitzwilliam Museum.

31. Robinson, p. 27.

32. Walker's lecture notes (HRC) are undated.

33. Robinson, p. 27.

34. *Burne-Jones*, pp. 68–9.

35. Edward Hodnett, *Image and Text: Studies in the Illustrations of English Literature* (1982), 218.

36. The miniature is in MS 2617, fo. 53, the Österreichische Nationalbibliothek, Vienna.

37. EBJ to Helen Gaskell [22.7.94] (BL Add. MS 54217, fo. 251).

38. Fitzgerald, p. 282.

39. Chaucer's authorship of *The Romaunt of the Rose* is still a matter of dispute.

40. EBJ to Helen Gaskell [23.7.94] (BL Add. MS 54217, fo. 255).

41. Robinson, pp. 43, 48.

42. EBJ to Frances Horner, 29.8.95 (FM).

43. Rooke diary, 7.1.96.

44. EBJ to Lady Rayleigh [29.12.95] (FM).

45. JM to WSB, 15.12.[94] (FM).

46. *The Letters of William Morris to his Family and Friends*, ed. Philip Henderson (1950), 360.

47. Sanford Berger has succeeded in finding the sources of several of the motifs on the *Chaucer* binding: see Anthony Bliss, 'William Morris and Book Design', in *William Morris: The Sanford and Helen Berger Collection* (Berkeley, Calif., 1948), 44.

48. *Introductions*, ii. 710.

49. EBJ to Frances Horner, 29.8.95 (FM).

50. Viola Meynell (ed.), *Friends of a Lifetime: Letters to Sydney Carlyle Cockerell* (1940), 269.

Chapter Nine
Closing the Press

1. F. S. Ellis, 'The Life-work of William Morris', *Journal of the Society of Arts*, 46 (27 May 1898), 628.

2. JM to WSB, 9.6.[96] (FM).

3. Rooke diary [25 or 26].10.95.

4. EBJ to WSB [6.6.96] (FM).

5. Memorandum by BQ, 17.9.96 (Bod. MS Eng. lett. c. 437).

6. Blunt diary, 28.9.96.

7. *Cockerell*, p. 63.

8. A proof of the colophon with the earlier reading is in BL C. 43. c. 26 (FF20).

9. Sotheby catalogue, 10 Dec. 1956, item 77.

10. BC; BL C. 102. l. 22.

11. *Ideal Book*, pp. 111–12.

12. BL C. 102. I. 22. Morris could have been familiar with this style of initial from Günther Zainer's books or from Fust and Schoeffer's *Mainz Psalter* (1457): see A. F. Johnson, *Decorative Initial Letters* (1931), p. 8.

13. A. C. Swinburne to SC, 20.1.97 (PML).

14. *Memorials*, ii. 276.

15. Rooke diary, 7.1.96.

16. Ibid., 5.5.96.

17. SC to RP, 12.11.96 (BL Add. MS 52743, fo. 239.

18. See A. W. Pollard, 'Memoir', in Proctor, *Bibliographical Essays* (1905), pp. ix–xli; Cockerell, 'Robert Proctor's Diaries' (letter), *Library*, 5th ser., 6 (Dec. 1951), 219. Proctor's Morris bibliography is in BL C. 43. c. 28. Proctor was so mesmerized by Morris, whom he met in October 1894, that he took up the study of Icelandic sagas and designed a Greek fount much influenced by the Golden type.

19. SC to JM, 18.12.97 (BL Add. MS 52738, fo. 28).

20. JM to SC, 17.3.[98] (V & A 86ss, fo. 15).

21. SC to JAM, 24.12.97 (BL Add. MS 52738, fo. 36).

22. SC to JAM, 7.1.98 (BL Add. MS 52739, fo. 40).

23. SC to H. C. Marillier, 17.2.98 (Walsdorf Collection, University of Maryland Libraries).

24. SC to JJL, 31.3.98 (JC MS 7, fo. 134).

25. SC to JJL, 30.11.97 (JC MS 7, fo. 116).

26. Copy of Hollings's notice in PML 76961.

27. *Ashbee Memoirs*, vii. 290.

28. Unidentified newspaper cutting, *c.*

Jan. 1899, in a William Morris scrapbook (SB), pp. 32–3.

29. FSE to SC, 8.6.99 (BL Add. MS 52715, fo. 26).

30. FSE to SC, 11.6.99 (BL Add. MS 52715, fo. 28).

31. SC to MM, 26.5.19 (BL Add. MS 52740, fo. 84).

32. SC to MM, 18.1.29 (BL Add. MS 52740, fo. 118).

33. SC to MM, 31.12.33 (BL Add. MS 45334 fo. 189); T. Balston, *The Cambridge University Press Collection of Private Press Types* (Cambridge, 1951), p. v.

Chapter Ten
The Legacy of Morris and His Press

1. Cobden-Sanderson, ii. 25.

2. *Ashbee Memoirs*, i. 218.

3. Walker, 'Printing Fine Editions: Some Governing Principles', in *Printing in the Twentieth Century: A Survey* (1930), p. 72; *Printers' Register – Supplement*, 6 Nov. 1896, p. vi (reprinted from the *Daily Chronicle*).

4. See Alan Crawford, *C. R. Ashbee: Architect, Designer and Romantic Socialist* (New Haven, Conn., 1985); Fiona MacCarthy, *The Simple Life: C R. Ashbee in the Cotswolds* (1981); Lionel Lambourne, *Utopian Craftsmen: The Arts and Crafts Movement from the Cotswolds to Chicago* (Salt Lake City, Utah, 1980).

5. B. G. Burrough, 'Three Disciples of William Morris, 2. Charles Robert Ashbee', *Connoisseur*, 172 (Oct. 1969), 90. It is not clear what the 'gages' were: Ashbee admitted that he was unfamiliar with printers' terminology.

6. Ashbee, *A Bibliography of the Essex House Press* (Chipping Campden, 1904), 3. For Ashbee's other comments on his Press, see Ashbee, 'The Essex House Press: And the Purpose or Meaning of a Private Press', *Book Collector's Quarterly*, 3 (July–Sept. 1933), 69–88; Ashbee, *The Private Press: A Study in Idealism* (Chipping Campden, 1909).

7. Ashbee, 'The Essex House Press', p. 73.

8. MacCarthy, pp 151–2.

9. See John Dreyfus, 'New Light on the Design of Types for the Kelmscott and Doves Presses', *Library*, 5th ser., 29 (Mar. 1974), 36–41.

10. *J. H. Mason, R.D.I.: A Selection from the Notebooks* (Leicester, 1961), unpaginated.

11. Viola Meynell (ed.), *Friends of a Lifetime: Letters to Sydney Carlyle Cockerell* (1940), 233.

12. Hans Schmoller, Introduction to Mardersteig's *The Officina Bodoni: An Account of the Work of a Hand Press, 1923–1977* (Verona, 1980), pp. xix–xx.

13. Viola Meynell (ed.), *The Best of Friends: Further Letters to Sydney Carlyle Cockerell* (1956), 75.

14. Noel Carrington, *Industrial Design in Britain* (1967) 55–6.

15. Ricketts, *A Defence of the Revival of Printing* (1899), 11–12.

16. Ibid., p. 19.

17. Taylor, *The Art Nouveau Book in England* (1966), 56. Thompson also emphasizes and explains clearly the differences between the two styles.

18. Beardsley, *Letters*, ed. Henry Maas, J. L. Duncan, and W. G. Good (Rutherford, N.J., 1970), 44.

19. Ibid., p. 413.

20. Thompson, pp. 10–13.

21. See William Dana Orcutt, *In Quest of the Perfect Book* (Boston, 1926), 68–71; Shaw, 'The Author's View: A Criticism of Modern Book Printing', *Caxton Magazine*, 2 (Jan. 1902), 119–21; Shaw, 'The Typography of G.B.S.' (letter), *London Mercury*, 12 (Sept. 1925), 524; Joseph R. Dunlap, 'Morrisian and Shavian Typography', *Independent Shavian*, 3 (Autumn 1964), 6–7; James Shand, 'Author and Printer: G.B.S. and R. & R. Clark : 1898–1948', *Alphabet and Image*, no. 8 (Winter 1948), 3–38.

22. *Library*, 10 (1898), 135,

23. Morris's interview is reprinted in *Ideal Book*, pp. 95–8; 'A Master Critic on the Master Printer' is reprinted in the *British Printer*, 6 (Apr.–May 1893), 98–9, and the *British Bookmaker*, 6 (31 Mar. 1893), 194. For Pennell's possible involvement, see Grant

Richards, *Author Hunting* (New York, 1934), 31–2.

24. 'The Late William Morris', *Printers' Register – Supplement*, 6 Nov. 1896, p. vi (reprinted from the *Daily Chronicle*); 'The Late Mr. William Morris, Poet and Printer', *British Printer*, 9 (Sept.–Oct. 1896), 224; 'Culture and Honest Work', *British and Colonial Printer*, 39 (22 Oct. 1896), 250.

25. Frederic W. Goudy, 'William Morris: His Influence on American Printing', *Philobiblon*, 7 (1934), 187.

26. Herbert Simon, *Song and Words, A History of the Curwen Press* (1973), 131.

27. Jackson, *The Printing of Books* (2nd edn., 1947), 177–85.

28. James Moran, *The Double Crown Club: A History of Fifty Years* (1948), 33.

29. Joseph Thorp, *B. H. Newdigate: Scholar–Printer, 1869–1944* (Oxford, 1950), 26.

30. Oliver Simon, *Printer and Playground: An Autobiography* (1956), 24.

31. Huxley, Introduction to Oliver Simon and Julius Rodenberg (eds.), *Printing of To-Day* (1928), 3.

32. Reed to EW, 24.10.88 (HRC).

33. See Peter Faulkner, *William Morris and Eric Gill* (1975).

34. Sparling, p. 120.

35. Quoted in G. W. Ovink and F. Mayer, 'Holland', in Charles Ede (ed.), *The Art of the Book* (1951), 63.

36. Jackson, 'The Nonage of Nineteenth-century Printing in England', *Fleuron*, 2 (1924), 88.

37. Morison, *The Typographic Book* (1963), 11.

38. See L. T. Owens, *J. W. Mason, 1875–1951: Scholar–Printer* (1976).

39. L. W. Wallis, *Leonard Jay* (1963), 38.

40. Meynell's comment was quoted by John C. Tarr in a letter to the *Listener*, 37 (30 Jan. 1947), 203. A letter from Meynell himself was published on 24 Apr. (p. 631) in which he explained, 'My use of the word "calamity" is the measure of my admiration of Morris'.

41. Meynell, 'Some Collectors Read', *Colophon*, 1, pt 4 (1930), unpaginated.

42. De Vinne, *A Treatise on Title-pages* (New York, 1902) 398.

43. *Pi: A Hodge-podge of the Letters and Addresses Written during the Last Sixty Years by Bruce Rogers* (Cleveland, 1953), 18–19.

44. Norman H. Strouse, *The Passionate Printer* (North Hills, Pa., 1964), 65.

45. Felix Shay, *Elbert Hubbard of East Aurora* (New York, 1926), 249.

46. Dard Hunter, *My Life with Paper* (New York, 1958), 35–6.

47. Elbert Hubbard II, 'William Morris: His Influence on Elbert Hubbard', *Roycrofter*, 6, no. 3 (Dec. 1931), 97.

48. Freeman Champney, *Art and Glory: The Story of Elbert Hubbard* (New York, 1968), 120.

49. Ransom, *Private Presses and their Books* (New York, 1929), 126.

50. Rollins, 'Morris's Typographical Adventure', *Printing and Graphic Arts*, 6 (June 1958), 30–1.

51. *Daniel Berkeley Updike to William Morris* (San Marino, Calif., 1968).

52. Updike, *Printing Types* (2nd edn.; 2 vols.; Cambridge, Mass., 1937), ii. 209.

53. 'Paul Beaujon' (pseud. of Beatrice Warde), review of *The Nonesuch Century*, in: *Signature*, no. 3 (July 1936), 48.

54. Moriz Sondheim, 'William Morris', *Zeitschrift für Bücherfreunde*, 2 (Apr. 1898), 12–20. See also Colin Franklin, 'The Kelmscott Press: An Album from the Nineties', *Journal of the William Morris Society*, 2 (Summer 1968), 16.

55. Quoted in Fernand Baudin, 'Books in Belgium', in Kenneth Day (ed.), *Book Typography 1815–1965* (Chicago, 1966), 20. The chapters of this volume, incidentally, are the best source for tracing Morris's international impact on book-design; see also Erik Dal, *Scandinavian Bookmaking in the Twentieth Century* (Urbana, Ill., 1968).

56. Pollard, 'Private Presses and their influence on the Art of Printing', *Ars typographica*, 1 (Autumn 1934), 41. This is a revised version of a similar paragraph which had appeared in a Monotype Corporation pamphlet announcing the Centaur type, 1929, reprinted in Paul A. Bennett (ed.), *Books and Printing* (Cleveland, 1951), 188–9.

57. Moxon, *Mechanick Exercises on the*

Whole Art of Printing, ed. Herbert Davis and Harry Carter (2nd edn.; 1962), 11–12.

58. Floud, 'The Inconsistencies of William Morris', *Listener*, 14 Oct. 1954, p. 616.

59. EBJ to Helen Gaskell [12.11.94] (BL Add. MS 54217, fo. 338). In a report of a lecture by Morris published in the *Printers' Register – Supplement*, 6 Dec. 1893, p. viii, he is quoted as saying, 'Mr. Morris concluded by a prediction that within fifty years printing books would be an extinct art – we should be carrying all our books about in bottles with patent stoppers. While there was still a chance, they should try and produce a few specimens of what was really good printing.'

SELECT BIBLIOGRAPHY

SINCE the quantity of words written about Morris and the Kelmscott Press is beyond measure, the following lists are deliberately brief. The reader who wishes further information will find many additional sources listed in the notes above and in *A Bibliography of the Kelmscott Press*, pp. xvii–xxi.

Published Sources

Avis, Frederick, C., *Edward Philip Prince, Type Punchcutter* (1967).

Balston, Thomas, *The Cambridge University Press Collection of Private Press Types: Kelmscott, Ashendene, Eragny, Cranach* (Cambridge, 1951).

Barker, Nicholas, *Stanley Morison* (1972).

The Beginning of the World: Twenty-five Pictures by Edward Burne-Jones, with an Introductory Note by Georgiana Burne-Jones (1902).

Blunt, Wilfrid, *Cockerell: Sydney Carlyle Cockerell, Friend of Ruskin and William Morris and Director of the Fitzwilliam Museum* (New York, 1965).

Buechler, John, *Charles Whittingham, Printer, 1795–1876: An Exhibit of his Works at Bailey/Howe Library at the University of Vermont*, with a Foreword by Janet Ing ([Burlington, Vt.], 1983).

Burne-Jones, Edward, *Burne-Jones Talking: His Conversations 1895–1898 Preserved by his Studio Assistant Thomas Rooke*, ed. Mary Lago (1981).

B[urne]-J[ones], G[eorgiana], *Memorials of Edward Burne-Jones* (2 vols.; 1904).

Cave, Roderick, *The Private Press* (2nd edn., New York, 1983).

Cobden-Sanderson, Thomas J., *The Journals of Thomas James Cobden-Sanderson 1879–1922* (2 vols.; 1928).

Colebrook, Frank, *William Morris: Master Printer* (Tunbridge Wells [?1897]; repr., with some alterations, from *British and Colonial Printer*, 39 (3 Dec. 1895), 362–3, 365; (10 Dec. 1896), 373–4. Colebrook's lecture has been reprinted more recently by the Yellow Barn Press of Council Bluffs, Iowa.

Crane, Walter, *An Artist's Reminiscences* (1907).

—— *Of the Decorative Illustration of Books Old and New* (2nd edn.; 1902).

Day, Kenneth, ed., *Book Typography 1815–1965 in Europe and the United States of America* (Chicago, 1966).

Dreyfus, John, 'New Light on the Design of Types for the Kelmscott and Doves Presses', *Library*, 5th ser., 29 (1974), 36–41.

Dunlap, Joseph R., *The Book that never was* (New York, 1971).

—— 'The Road to Kelmscott: William Morris and the Book Arts before the Founding of the Kelmscott Press', Doctoral thesis, Colombia University, 1972. For an abridged version, see Dunlap, 'Morris and the Book Arts before the Kelmscott Press', *Victorian Poetry*, 13 (1975) 141–57.

Fitzgerald, Penelope, *Edward Burne-Jones: A Biography* (1975).

Floud, Peter, 'William Morris as an Artist: A New View', *Listener*, 7 Oct. 1954, pp. 562–4; 'The Inconsistencies of William Morris', 14 Oct. 1954, pp. 615–17.

Forman, Harry Buxton, *The Books of William Morris Described with Some Account of his Doings in Literature and the Allied Crafts* (1897).

Franklin, Colin, *Emery Walker: Some Light on his Theories of Printing and on his Relations with William Morris and Cobden-Sanderson* (Cambridge, 1973).

—— *The Private Presses* (1969).

Gill, Eric, *An Essay on Typography* (4th edn., 1960; 1st edn., 1931).

Harrison, Martin, and Bill Waters, *Burne-Jones* (1973).

Select Bibliography

In Fine Print: William Morris as Book Designer. Catalogue of an Exhibition 18 December–19 March 1977 (William Morris Gallery, 1977).

Jackson, Holbrook, *The Eighteen-nineties* (1913); see 'The Revival of Printing'.

Johnson, A. F., *Selected Essays on Books and Printing*, ed. Percy H. Muir (Amsterdam, 1970); see 'The Modern-face Type in England' and 'Old-face Types in the Victorian Age'.

Keynes, Geoffrey, *William Pickering, Publisher: A Memoir and a Check-list of his Publications* (rev. edn., 1969).

Kraus, Joe W., *Messrs. Copeland & Day, 69 Cornhill, Boston, 1893–1899: A History and Bibliography* (Philadelphia, 1979).

Longford, Elizabeth, *A Pilgrimage of Passion: The Life of Wilfrid Scawen Blunt* (1979).

Mackail, J. W., *The Life of William Morris* (1950; 1st edn., 1899).

McLean, Ruari, *Victorian Book Design and Colour Printing* (2nd edn., Berkeley, Calif., 1972).

Morison, Stanley, *Selected Essays on the History of Letter-forms in Manuscript and Print*, ed. David McKitterick (2 vols.; Cambridge, 1981).

—— and D. B. Updike, *Stanley Morison and D. B. Updike: Selected Correspondence*, ed. David McKitterick (New York, 1979).

Morris, May, 'Reminiscences about my Father,' *Philobiblon*, 7 (1934), 169–72.

Morris, William, *The Collected Letters*, ed. Norman Kelvin (3 vols.; Princeton, 1984– (in progress)).

—— *Collected Works*, ed. May Morris (24 vols.; 1910–14); May Morris's introductions are reprinted in *Introductions to the Collected Works of William Morris*, with a Preface by Joseph R. Dunlap (2 vols.; New York, 1973).

—— *The Ideal Book: Essays and Lectures on the Arts of the Book*, ed. William S. Peterson (Berkeley and Los Angeles, 1982).

—— *A Note . . . on his Aims in Founding the Kelmscott Press, together with a Short Description of the Press by S. G. Cockerell and an Annotated List of the Books Printed Thereat* (Hammersmith, 1898).

—— *The Story of Cupid and Psyche*, with an Introduction by A. R. Dufty (3 vols.; London and Cambridge, 1974).

—— *William Morris: Artist, Writer, Socialist*, ed. May Morris (2 vols.; Oxford, 1936).

Needham, Paul, Joseph Dunlap, and John Dreyfus, *William Morris and the Art of the Book* (New York, 1976).

Peterson, William S., *A Bibliography of the Kelmscott Press* (Soho Bibliographies; Oxford, 1984).

—— 'William Morris and the "Damned Chemists": The Search for an Ideal Ink at the Kelmscott Press', *Printing History*, 3, 2 (1981), 4–11.

Pevsner, Nikolaus, *Pioneers of Modern Design: From William Morris to Walter Gropius* (rev. edn.; Harmondsworth, Middlesex, 1974).

Ray, Gorsdon N., *The Illustrator and the Book in England from 1790 to 1914* (New York, 1976).

Ricketts, Charles, *A Defence of the Revival of Printing* (1899).

Robinson, Duncan, *A Companion Volume to the Kelmscott Chaucer* (1975).

Sparling, H. Halliday, *The Kelmscott Press and William Morris Master-Craftsman* (1924).

Stansky, Peter, *Redesigning the World: William Morris, the 1880s, and the Arts and Crafts* (Princeton, 1985).

Stevens, Henry, *Who spoils our New English Books?* (1884).

Taylor, John Russell, *The Art Nouveau Book in England* (1966).

Thompson, Susan O., *American Book Design and William Morris* (New York, 1977).

The Typographical Adventure of William Morris: An Exhibition Arranged by the William Morris Society (1957).

Updike, Daniel Berkeley, *Printing Types: Their History, Forms, and Use* (2nd edn., 2 vols.; Cambridge, Mass., 1937).

Ward, Sydney, 'William Morris and his Papermaker, Joseph Batchelor', *Philobiblon*, 7 (1934), 177–80.
Watkinson, Ray, *William Morris as Designer* (1967).

Unpublished Sources

Detailed lists of Kelmscott Press proofs and trial pages are supplied in *A Bibliography of the Kelmscott Press*. Only a few of the most important repositories are included below.

BERNARD QUARITCH LTD., LONDON Morris's letters to Bernard Quaritch.

BODLEIAN LIBRARY, OXFORD Morris's letters to J. and J. Leighton; Bernard Quaritch's letters to his son; miscellaneous correspondence; several folders of KP material in the John Johnson Collection.

BRITISH LIBRARY in the *Department of Printed Books* many volumes of KP proofs and trial pages assembled by Sydney Cockerell and Robert Proctor; in the *Department of Manuscripts* diaries and correspondence of Cockerell and Morris; letters of Burne-Jones.

FITZWILLIAM MUSEUM, CAMBRIDGE in the *Department of Manuscripts and Books* letters of Burne-Jones and correspondence and diaries of Wilfrid Scawen Blunt; in the *Department of Prints and Drawings* Burne-Jones's designs for the *Chaucer* illustrations.

HAMMERSMITH PUBLIC LIBRARIES proofs and trial pages.

HARRY RANSOM HUMANITIES RESEARCH CENTER, UNIVERSITY OF TEXAS AT AUSTIN papers of Emery Walker.

HENRY E. HUNTINGTON LIBRARY, SAN MARINO, CALIF. miscellaneous trial pages and correspondence.

OXFORD UNIVERSITY ARCHIVES Morris's correspondence with the Oxford University Press.

PIERPONT MORGAN LIBRARY, NEW YORK the largest collection of KP material in the USA, including proofs, trial pages, designs by Burne-Jones, correspondence, and the KP ledger.

PRIVATE COLLECTION OF SANFORD AND HELEN BERGER, CARMEL, CALIF. proofs, trial pages, correspondence.

UCLA LIBRARY archives of Ellis and Elvey.

VICTORIA AND ALBERT MUSEUM LIBRARY miscellaneous correspondence and proofs.

WILLIAM ANDREWS CLARK MEMORIAL LIBRARY, LOS ANGELES trial pages, proofs, and ephemera.

WILLIAM MORRIS GALLERY, WALTHAMSTOW correspondence, designs.

INDEX

The only abbreviations used in this index are 'KP' for Kelmscott Press and 'WM' for William Morris. The Notes section is indexed selectively.

Index

Index

Index

Index

Index

Index

CREDITS

BRISTOL: National Westminster Bank – 91.
CAMBRIDGE: Fitzwilliam Museum, Department of Printed Books and Manuscripts – 68; Department of Prints and Drawings – 76, 78.
CARMEL, CALIF.: Private collection of Sanford and Helen Berger – 31.
CHICAGO: Newberry Library – 92, 94.
COLLEGE PARK, MD.: University of Maryland Libraries – 61, 62; Special Collections – 20, 23, 72, 82.
LONDON: Alfred Dunhill Ltd. – 4.
LONDON: Bernard Quaritch Ltd. – 59.
LONDON: British Library, Department of Manuscripts – 15, 87; Department of Printed Books – 17, 28, 30, 33, 34, 39, 44, 54, 56, 58, 60, 63, 64, 84, 86.
LONDON: Hammersmith and Fulham Archives, Shepherds Bush Library – 67.
LONDON: National Portrait Gallery – 52, 53.
LONDON: Private collection of John Dreyfus – 25.
LONDON: Royal Commission on Historical Monuments – 24.
LONDON: St Bride Printing Library – 27, 74.
LONDON: Victoria and Albert Museum – 19.
LOS ANGELES: William Andrews Clark Memorial Library – 26, 51, 57, 77.
MERRITT ISLAND, FLA.: Private collection of Freeman Champney – 95.
NEW YORK: Pierpont Morgan Library – 16, 22, 42, 75.
OXFORD: Bodleian Library – 7, 13, 41.
PORTLAND, ORE.: Private collection of John J. Walsdorf – 85.
SAN MARINO, CALIF.: Henry E. Huntington Library – 32, 35, 43, 65, 66, 73, 80.
WASHINGTON, D.C.: Library of Congress – 5, 9, 10, 11, 12, 21, 29, 36, 37, 38, 40, 45, 55, 69, 83, 88, 89, 93, 97.
WASHINGTON, D.C.: Private collection of Mark Samuels Lasner – 18.

The manuscript in Figure 20 was in the Estelle Doheny Collection, St John's Seminary, Camarillo, Calif., but was sold while this book was in the press and is now in a private collection. The photograph is the property of Christie, Manson & Woods International Inc. (copyright) to whom we are grateful for permission to reproduce it. The location of the drawing in Figure 81 is unknown; it is reproduced here by permission of Barrie & Jenkins Ltd. from Martin Harrison and Bill Waters, *Burne-Jones* (1973). The Cambridge University Library supplied reproduction proofs of titles set in Morris's Golden type.